HUMAN DEVELOPMENT REPORT 1990

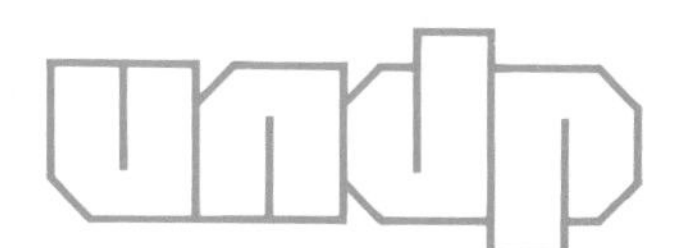

Published
for the United Nations
Development Programme
(UNDP)

New York Oxford
Oxford University Press
1990

Oxford University Press
Oxford New York Toronto
Delhi Bombay Calcutta Madras Karachi
Petaling Jaya Singapore Hong Kong Tokyo
Nairobi Dar es Salaam Cape Town
Melbourne Auckland

and associated companies in
Berlin Ibadan

ISBN 0-19-506481-X (paper)
ISBN 0-19-506480-1 (cloth)

Printing (last digit): 9 8 7 6 5 4 3 2
Printed in the United States of America on acid-free paper

Editing, desktopping and production management: Bruce Ross-Larson and Eileen Hanlon, American Writing Corporation, Washington, D.C.
Design: Gerald Quinn, Quinn Information Design, Cabin John, Maryland

Foreword

We live in stirring times. An irresistible wave of human freedom is sweeping across many lands. Not only political systems but economic structures are beginning to change in countries where democratic forces had been long suppressed. People are beginning to take charge of their own destiny in these countries. Unnecessary state interventions are on the wane. These are all reminders of the triumph of the human spirit.

In the midst of these events, we are rediscovering the essential truth that people must be at the centre of all development. The purpose of development is to offer people more options. One of their options is access to income — not as an end in itself but as a means to acquiring human well-being. But there are other options as well, including long life, knowledge, political freedom, personal security, community participation and guaranteed human rights. People cannot be reduced to a single dimension as economic creatures. What makes them and the study of the development process fascinating is the entire spectrum through which human capabilities are expanded and utilised.

UNDP has undertaken to produce an annual report on the human dimension of development. This *Human Development Report 1990* is the first such effort.

The central message of this *Human Development Report* is that while growth in national production (GDP) is absolutely necessary to meet all essential human objectives, what is important is to study how this growth translates — or fails to translate — into human development in various societies. Some societies have achieved high levels of human development at modest levels of per capita income. Other societies have failed to translate their comparatively high income levels and rapid economic growth into commensurate levels of human development. What were the policies that led to such results? In this line of enquiry lie promising seeds of a much better link between economic growth and human development, which is by no means automatic.

The orientation of this Report is practical and pragmatic. It aims to analyse country experience to distill practical insights. Its purpose is neither to preach nor to recommend any particular model of development. Its purpose is to make relevant experience available to all policymakers.

The Report is of a seminal nature. It makes a contribution to the definition, measurement and policy analysis of human development. It is the first in a series of annual reports. It opens the debate. Subsequent reports will go into further detail regarding the planning, management and financing of human development.

The Report is accompanied by the human development indicators, which assemble all available social and human data for each country in a comparable form. UNDP will undertake, along with other agencies, a programme of action to compile the missing country data and to improve the existing statistics so that these human development indicators come to be used over time as a standard reference for country and global analysis.

The preparation of this Report has been a United Nations systemwide initiative. I am personally grateful to all the specialised agencies and other organisations in the UN system, including the World Bank and the IMF, for their wholehearted support of the preparation of this Report. One of the incidental benefits of such collaboration has been the emergence of a close intellectual network within the UN system which will also be helpful for future reports.

The *Human Development Report 1990* has been prepared by a team of UNDP staff and eminent outside consultants under the overall guidance of Mahbub ul Haq, former Finance and Planning Minister of Pakistan, in his capacity as Special Adviser to me. The views expressed in this Report are those of the team and not necessarily shared by UNDP or its Governing Council or the member governments of UNDP. The essence of any such report must be its independence and its intellectual integrity.

I hope that this Report — and its annual sequels — will make a significant contribution to the development dialogue in the 1990s and lead to a serious exploration of human development programming at the country level. UNDP stands ready to assist this process both at the intellectual and operational levels.

New York
May 1, 1990

William H. Draper III
Administrator UNDP

Team for the preparation of the *Human Development Report 1990*

Project director
Mahbub ul Haq

UNDP team
Inge Kaul, Leo Goldstone, Bernard Hausner, Saraswathi Menon and Jin Wei, assisted by Shabbir Cheema, Beth Ebel, Akhtar Mahmood, Ragnar Gudmundsson, Martin Krause and Roman Schremser

Panel of consultants
Gustav Ranis, Amartya K. Sen, Frances Stewart, Keith Griffin, Meghnad Desai, Aziz Khan, Paul Streeten, Shlomo Angel, Pietro Garau and Mahesh Patel

Acknowledgements

The preparation of the Report would have been impossible without the valuable contributions that the authors received from a large number of organisations and individuals.

Particular thanks are due to the agencies and offices of the United Nations system, which provided generous assistance, sharing their accumulated experience, studies and statistical data with the Report team. Their assistance made it possible for the Report to be a genuine UN systemwide initiative. Special mention must be made of the collaboration of Habitat (United Nations Centre for Human Settlements), in the preparation of chapter 5. The other contributing UN system and affiliated organisations were FAO, IFAD, ILO, UN Statistical Office and Population Division, UNESCO, UNFPA, UNHCR, UNICEF, UNIDO, UNOV, UNRISD, UNSO, WFP, WHO, and the World Bank. Further inputs were received from various UNDP offices, in particular UNDP's country offices, the Regional Bureaux, the Division for Women in Development, the Division for Nongovernmental Organisations and the Office of Project Services. Ian Steele assisted in editing the first draft of the Report.

The Report draws on the statistical data bases established by the UN Statistical Office and Population Division, the World Bank, the IMF, and the OECD. These have been complemented, and in part updated, by selected statistical data collected from government sources by UNDP country offices.

Many colleagues in the UNDP contributed to the evolution of the Report through comments and observations on earlier drafts. Thanks are due to G. Arthur Brown, Denis Benn, Pierre-Claver Damiba, Gary Davis, Luis Gomez-Echeverri, Trevor Gordon-Somers, Michael Gucovsky, Arthur Holcombe, Andrew J. Joseph, Uner Kirdar, Sarah Papineau, Jehan Raheem, Augusto Ramirez-Ocampo, Elizabeth Reid, Sarah Timpson and Gustavo Toro.

Secretarial and administrative support for the Report's preparation was provided by Linda Grahek, Gwen Halsey, Ida Simons, Odette Tin-Aung, Carol Joseph and Karin Svadlenak-Castro.

Abbreviations

ECE	Economic Commission for Europe
ECLAC	Economic Commission for Latin America and the Caribbean
ESCAP	Economic and Social Commission for Asia and the Pacific
EUROSTAT	Statistical Office of the European Communities
FAO	Food and Agriculture Organization of the United Nations
GATT	General Agreement on Tariffs and Trade
IFAD	International Fund for Agricultural Development
ILO	International Labour Organisation
IMF	International Monetary Fund
OECD	Organisation for Economic Co-operation and Development
UNDP	United Nations Development Programme
UNESCO	United Nations Educational, Scientific, and Cultural Organization
UNFPA	United Nations Fund for Population Activities
UNHCR	Office of the United Nations High Commissioner for Refugees
UNICEF	United Nations Children's Fund
UNOV	United Nations Office at Vienna
UNRISD	United Nations Research Institute for Social Development
UNSO	United Nations Sudano-Sahelian Office
USAID	United States Agency for International Development
WFC	World Food Council
WFP	World Food Programme
WHO	World Health Organization
IBRD	International Bank for Reconstruction and Development (World Bank)

Contents

CHAPTER FIVE: A SPECIAL FOCUS

BOXES

TABLES

Text tables

Annex tables

FIGURES

Overview

This Report is about people — and about how development enlarges their choices. It is about more than GNP growth, more than income and wealth and more than producing commodities and accumulating capital. A person's access to income may be one of the choices, but it is not the sum total of human endeavour.

Human development is a process of enlarging people's choices. The most critical of these wide-ranging choices are to live a long and healthy life, to be educated and to have access to resources needed for a decent standard of living. Additional choices include political freedom, guaranteed human rights and personal self-respect.

Development enables people to have these choices. No one can guarantee human happiness, and the choices people make are their own concern. But the process of development should at least create a conducive environment for people, individually and collectively, to develop their full potential and to have a reasonable chance of leading productive and creative lives in accord with their needs and interests.

Human development thus concerns more than the formation of human capabilities, such as improved health or knowledge. It also concerns the use of these capabilities, be it for work, leisure or political and cultural activities. And if the scales of human development fail to balance the formation and use of human capabilities, much human potential will be frustrated.

Human freedom is vital for human development. People must be free to exercise their choices in properly functioning markets, and they must have a decisive voice in shaping their political frameworks.

Starting with this perspective, human development is measured in this Report not by the yardstick of income alone but by a more comprehensive index — called the human development index — reflecting life expectancy, literacy and command over the resources to enjoy a decent standard of living. At this stage, the index is an approximation for capturing the many dimensions of human choices. It also carries some of the same shortcomings as income measures. Its national averages conceal regional and local distribution. And a quantitative measure of human freedom has yet to be designed.

The index does, however, have the virtue of incorporating human choices other than income, and consequently is a move in the right direction. It also has the potential for refinement as more aspects of human choice and development are quantified. This Report lays out a concrete priority agenda for better data collection that will enable the human development index to be used increasingly as a more genuine measure of socioeconomic progress.

The Report analyses the record of human development for the last three decades and the experience of 14 countries in managing economic growth and human development. Several policy conclusions from this experience underpin a detailed analysis of human development strategies during the 1990s. The Report ends with a special focus on the problems of human development in an increasingly urban setting. The orientation of the Report is practical, looking not just at what is to be done — but also at how.

The Report's central conclusions and policy messages are clear, and some of their salient features are summarised here.

1. The developing countries have made significant progress towards human development in the last three decades.

Life expectancy in the South rose from 46 years in 1960 to 62 years in 1987. The adult literacy rate increased from 43% to 60%. The under-five mortality rate was halved. Primary health care was extended to 61% of the population, and safe drinking water to 55%. And despite the addition of 2 billion people in developing countries, the rise in food production exceeded the rise in population by about 20%.

Never before have so many people seen such significant improvement in their lives. But this progress should not generate complacency. Removing the immense backlog of human deprivation remains the challenge for the 1990s. There still are more than a billion people in absolute poverty, nearly 900 million adults unable to read and write, 1.75 billion without safe drinking water, around 100 million completely homeless, some 800 million who go hungry every day, 150 million children under five (one in three) who are malnourished and 14 million chil dren who die each year before their fifth birthday. In many countries in Africa and Latin America, the 1980s have witnessed stagnation or even reversal in human achievements.

2. North-South gaps in basic human development have narrowed considerably in the last three decades, even while income gaps have widened.

In 1987 the average per capita income in the South was still only 6% of that in the North. But its average life expectancy was 80% of the northern average and its average literacy rate 66%.

Developing countries reduced their average infant mortality from nearly 200 deaths per 1,000 live births to about 80 in about four decades (1950-88), a feat that took the industrial countries nearly a century to accomplish. This is clearly a message of hope. The essential task of taking the developing world to an acceptable threshold of human development can be accomplished in a fairly manageable period and at a modest cost — if national development efforts and international assistance are properly directed.

But this promising trend must be seen in its proper perspective. While North-South gaps have narrowed in basic human survival, they continue to widen in advanced knowledge and high technology.

3. Averages of progress in human development conceal large disparities within developing countries — between urban and rural areas, between men and women, between rich and poor.

Rural areas in the developing countries have on the average half the access to health services and safe drinking water that urban areas have, and only a quarter of the access to sanitation services.

Literacy rates for women are still only two-thirds of those for men. And the maternal mortality rate in the South is 12 times that in the North — the largest gap in any social indicator and a sad symbol of the deprived status of women in the Third World.

High-income groups often preempt many of the benefits of social services. Levels of health, education and nutrition among higher income groups far exceed those of the poor in many countries. There is thus considerable room for improvement to ensure that the benefits of social expenditures are more evenly distributed, flowing to the very poor. The rationale for government intervention greatly weakens if social spending, rather than improving the distribution of income, makes it worse.

4. Fairly respectable levels of human development are possible even at fairly modest levels of income.

Life does not begin at $11,000, the average per capita income in the industrial world. Sri Lanka managed a life expectancy of 71 years and an adult literacy rate of 87% with a per capita income of $400.

By contrast, Brazil has a life expectancy of only 65 years, and its adult literacy rate is 78% at a per capita income of $2,020. In

Saudi Arabia, where the per capita income is $6,200, life expectancy is only 64 years and the adult literacy rate is an estimated 55%.

What matters is how economic growth is managed and distributed for the benefit of the people. The contrast is most vivid in rankings of developing countries by their human development index and by their GNP per capita. Sri Lanka, Chile, Costa Rica, Jamaica, Tanzania and Thailand, among others, do far better in human development than in income, showing that they have directed more of their economic resources towards human progress. Oman, Gabon, Saudi Arabia, Algeria, Mauritania, Senegal, Cameroon and the United Arab Emirates, among others, do considerably worse, showing that they have not yet translated their income into human progress.

The valuation given to similar human development achievements is quite different depending on whether they were accomplished in a democratic or an authoritarian framework. A simple quantitative measure to reflect the many aspects of human freedom — such as free elections, multiparty political systems, an uncensored press, adherence to the rule of law, guarantees of free speech, personal security and so on — will be designed over time and incorporated in the human development index. Meanwhile, the Report lists the top 15 countries that have achieved relatively high levels of human development within a reasonably democratic political and social framework: Costa Rica, Uruguay, Trinidad and Tobago, Mexico, Venezuela, Jamaica, Colombia, Malaysia, Sri Lanka, Thailand, Turkey, Tunisia, Mauritius, Botswana and Zimbabwe.

5. The link between economic growth and human progress is not automatic.

GNP growth accompanied by reasonably equitable distribution of income is generally the most effective path to sustained human development. The Republic of Korea shows what is possible. But if the distribution of income is unequal and if social expenditures are low (Pakistan and Nigeria) or distributed unevenly (Brazil), human development may not improve much, despite rapid GNP growth.

Even in the absence of satisfactory economic growth or a relatively even income distribution, countries can achieve significant improvements in human development through well-structured public expenditures. For example, during the last three decades, Sri Lanka experienced relatively slow growth, rather equally distributed, and Botswana and Malaysia had adequate growth, unequally distributed. Yet all these countries have made impressive achievements in their human development levels because they have had well-structured social policies and expenditures.

Costa Rica and Chile, too, have demonstrated that dramatic human progress can be achieved — in a short time and even without rapid GNP growth.

But distributive policies can compensate for the effects of low GNP growth or unequal income distribution only in the short and medium run. These policy interventions do not work indefinitely without the nourishment that well-distributed growth provides. In the long run, economic growth is crucial for determining whether countries can sustain progress in human development or whether initial progress is disrupted or reversed (as in Chile, Colombia, Jamaica, Kenya and Zimbabwe).

6. Social subsidies are absolutely necessary for poorer income groups.

The distribution of income is fairly uneven in most of the Third World. Simply stated, economic growth seldom trickles down to the masses. Free market mechanisms may be vital for allocative efficiency, but they do not ensure distributive justice. That is why added policy actions are often necessary to transfer income and other economic opportunities to the very poor.

Food and health subsidies serve that purpose — as long as they are properly targetted to low-income beneficiaries and efficiently administered. They establish an essential safety net in poor societies that generally do not have the social security schemes that are familiar in the industrial nations. Generally amounting to less than

3% of GNP, these subsidies have not been too costly. And when they are removed without an alternative safety net, the ensuing political and social disturbance has cost far more than the subsidies themselves.

Social subsidies will serve the interest of developing countries much better if more effort is devoted to designing them as efficient tools of income redistribution, without hurting the efficiency of resource allocation. Such effort is far preferable to the usual acrimonious debate supporting or rejecting all subsidies arbitrarily and across the board.

7. Developing countries are not too poor to pay for human development *and* take care of economic growth.

The view that human development can be promoted only at the expense of economic growth poses a false tradeoff. It misstates the purpose of development and underestimates the returns on investment in health and education. These returns can be high, indeed. Private returns to primary education are as high as 43% in Africa, 31% in Asia and 32% in Latin America. Social returns from female literacy are even higher — in terms of reduced fertility, reduced infant mortality, lower school dropout rates, improved family nutrition and lower population growth.

Most budgets can, moreover, accommodate additional spending on human development by reorienting national priorities. In many instances, more than half the spending is swallowed by the military, debt repayments, inefficient parastatals, unnecessary government controls and mistargetted social subsidies. Since other resource possibilities remain limited, restructuring budget priorities to balance economic and social spending should move to the top of the policy agenda for development in the 1990s.

Special attention should go to reducing military spending in the Third World — it has risen three times as fast as that in the industrial nations in the last 30 years, and is now approaching $200 billion a year. Developing countries as a group spend more on the military (5.5% of their combined GNP) than on education and health (5.3%). In many developing countries, current military spending is sometimes two or three times greater than spending on education and health. There are eight times more soldiers than physicians in the Third World.

Governments can also do much to improve the efficiency of social spending by creating a policy and budgetary framework that would achieve a more desirable mix between various social expenditures, particularly by reallocating resources:

- from curative medical facilities to primary health care programmes,
- from highly trained doctors to paramedical personnel,
- from urban to rural services,
- from general to vocational education,
- from subsidising tertiary education to subsidising primary and secondary education,
- from expensive housing for the privileged groups to sites and services projects for the poor,
- from subsidies for vocal and powerful groups to subsidies for inarticulate and weaker groups and
- from the formal sector to the informal sector and the programmes for the unemployed and the underemployed.

Such a restructuring of budget priorities will require tremendous political courage. But the alternatives are limited, and the payoffs can be enormous.

8. The human costs of adjustment are often a matter of choice, not of compulsion.

Since there is considerable room for reallocating expenditures within existing budgets, the human costs of adjustment are often a matter of choice, not compulsion. When there is a sudden squeeze on resources, it is for policymakers to decide whether budgetary cuts will fall on military spending, parastatals and social subsidies for the privileged groups — or on essential health, education and well-targetted food subsidies. The evidence of the 1980s shows that some countries (such as Indonesia and Zimbabwe) protected their human development programmes during the process of adjustment by reorienting their budgets. Yet in some countries where education and health expenditures were cut, military expenditures actually rose. Obviously, the poverty of their economies was no barrier to the affluence of their armies.

External donors can help protect human development by providing additional resources to ease the pain of adjustment and by agreeing with developing countries on new and benign conditions for adjustment assistance — conditions that would make it clear that external assistance will be reduced if a country insists on spending more on its army than on its people. They could stress the right of the recipient country — indeed its obligation — not to cut social expenditures and subsidies that benefit poorer income groups and other vulnerable segments of the population. And they could specify that human development programmes should be the last, not the first, to be reduced in an adjustment period after all other options have been explored and exhausted.

9. A favourable external environment is vital to support human development strategies in the 1990s.

The outlook is not good. The net transfer of resources to the developing countries has been reversed — from a positive flow of $42.6 billion in 1981 to a negative flow of $32.5 billion in 1988. Primary commodity prices have reached their lowest level since the Great Depression of the 1930s. The foreign debts of developing countries, more than $1.3 trillion, now require nearly $200 billion a year in debt servicing alone.

In the 1990s the rich nations must start transferring resources to the poor nations once again. For this to happen, there must be a satisfactory solution to the lingering debt crisis — with debts written down drastically, and a debt refinancing facility created, within the existing structures of the IMF and the World Bank, to foster an orderly resolution of the debt problem.

10. Some developing countries, especially in Africa, need external assistance a lot more than others.

The least developed countries, particularly those south of the Sahara, suffer the greatest human deprivation. Africa has the lowest life expectancy of all the developing regions, the highest infant mortality rates and the lowest literacy rates. Its average per capita income fell by a quarter in the 1980s.

There is thus a growing trend towards a concentration of poverty in Africa. Between 1979 and 1985 the number of African people below the poverty line increased by almost two-thirds, compared with an average increase of one-fifth in the entire developing world. That number is projected to rise rapidly in the next few years — from around 250 million in 1985 to more than 400 million by the end of the century.

In any concerted international effort to improve human development in the Third World, priority must go to Africa. The concept of short-term adjustment is inappropriate there. Required, instead, is long-term development restructuring. Also required is a perspective of at least 25 years for Africa to strengthen its human potential, its national institutions and the momentum of its growth. The international community should earmark an overwhelming share of its concessional resources for Africa and display the understanding and patience needed to rebuild African economies and societies in an orderly and graduated way.

11. Technical cooperation must be restructured if it is to help build human capabilities and national capacities in the developing countries.

The record is not reassuring. In many developing countries the amount of technical assistance flowing each year into the salaries and travel of foreign experts exceeds by far the national civil service budget. Unemployment of trained personnel and a national civil service demoralised by low salary levels often exist side by side with large numbers of foreign, high-priced experts and consultants. In some countries, there continues to be an acute lack of trained national personnel. Technical assistance to Africa amounts to $4 billion a year — as much as $7 a person. But institution-building and the expansion of human capabilities has been grossly inadequate in most of the region.

More successful technical cooperation in the 1990s requires that programmes focus more on human development issues. This will broaden the basis for more effective national capacity-building — through the exchange of experience, the transfer of competence and expertise and the fuller mobilisation and use of national development capacities. Emphasis must be placed on improving the availability of relevant social indicators and on assisting developing countries in formulating their own human development plans. The yardstick for measuring the success and impact of technical assistance programmes must be the speed with which they phase themselves out.

12. A participatory approach — including the involvement of NGOs — is crucial to any strategy for successful human development.

Many overplanned, overregulated economies are now embracing greater market competition. Increasingly, the role of the state is being redefined: it should provide an enabling policy environment for efficient production and equitable distribution, but it should not intervene unnecessarily in the workings of the market mechanism.

The movement of nongovernmental organisations (NGOs) and other self-help organisations has gained considerable momentum and proven its effectiveness in enabling people to help themselves. NGOs are generally small, flexible and cost-effective, and most of them aim at building self-reliant development. They recognise that when people set their own goals, develop their own approaches and take their own decisions, human creativity and local problem-solving skills are released, and the resulting development is more likely to be self-sustaining. A comprehensive policy for the participation of NGOs is essential for any viable strategy of human development.

13. A significant reduction in population growth rates is absolutely essential for visible improvements in human development levels.

The number of people in developing countries — having increased from 2 billion in 1960 to an estimated 4 billion in 1990 — will probably reach 5 billion in 2000. The decline in the population growth rate — from 2.3% a year during 1960-88 to an estimated 2.0% during 1988-2000 — is insufficient to make a dent in the overall demographic picture. More vigorous efforts are required to reduce population growth in the developing world, above all in Africa and South Asia. There is an urgent need to strengthen programmes of family planning, female literacy, fertility reduction and maternal and child health care.

The world's demographic balance is shifting fast. The share of the developing countries in world population is expected to grow from 69% in 1960 to 84% by 2025, and that of the industrial nations to shrink from 31% to 16%. Even more telling, 87% of all new births are in the Third World, and only 13% in the industrial nations.

If the developing world's new generations cannot improve their conditions through liberal access to international assistance, capital markets and the opportunities for trade, the compulsion to migrate in search of better economic opportunities will be overwhelming — a sobering thought for the 1990s, one that spotlights the urgent need for a better global distribution of development opportunities.

14. The very rapid population growth in the developing world is becoming concentrated in cities.

Between 1950 and 1987 the number of urban dwellers in developing countries more than quadrupled, from 285 million to one and a quarter billion. Their number is likely to increase to nearly 2 billion by 2000, when eight of the 10 largest mega-cities (each with 13 million people or more) will be in the Third World. This process of urbanisation seems to be inevitable, as various attempts to discourage urban migration have for the most part failed.

The urban challenge for planners and policymakers in developing countries during the 1990s will be to identify and implement innovative programmes to deal with four critical issues.

- Decentralising power and resources from the central government to municipalities.
- Mobilising municipal revenue from local sources with the active participation of private and community organisations.
- Emphasising "enabling" strategies for shelter and infrastructure, including assistance targeted to weaker groups.
- Improving the urban environment, especially for the vast majority of urban poor in slums and squatter settlements.

The effectiveness of government responses to these issues will largely determine human development in the urban setting.

15. Sustainable development strategies should meet the needs of the present generation without compromising the ability of future generations to meet their needs.

On this, the consensus is growing. But the concept of sustainable development is much broader than the protection of natural resources and the physical environment. After all, it is people, not trees, whose future choices have to be protected. Sustainable development therefore must also include the protection of future economic growth and future human development. Any form of debt — financial debt, the debt of human neglect or the debt of environmental degradation — is like borrowing from the next generations. Sustainable development should aim at limiting all these debts.

Poverty is one of the greatest threats to the environment. In poor countries, poverty often causes deforestation, desertification, salination, poor sanitation and polluted and unsafe water. And this environmental damage reinforces poverty. Many choices that degrade the environment are made in the developing countries because of the imperative of immediate survival, not because of a lack of concern for the future. Any plans of action for environmental improvement must therefore include programmes to reduce poverty in the developing world.

If environmental problems are seen in the above perspective, it will help ensure that global ecological security is viewed as a unifying link, not a divisive issue, between the North and the South. Further, the additional costs of environmental protection must come largely from the rich nations since they are responsible for a major part of environmental degradation. With 20% of the world's population, they emit more than half the greenhouse gases that warm our planet. It is mainly the willingness of the rich nations to change their environmental policies, to transfer environmentally sound technologies and to provide additional resources that can ensure the protection of our global commons.

• • •

These, then, are the main policy conclusions and policy messages of this first *Human Development Report*. Far from answering all questions in this first effort, the findings and conclusions often point to issues requiring deeper analysis and more meticulous research: What are the essential elements of strategies for planning, managing, and financing human development? What are the requirements of a practical framework for participatory development? What is a conducive external environment for human development? These and related questions will set the agenda for future Human Development Reports.

CHAPTER 1

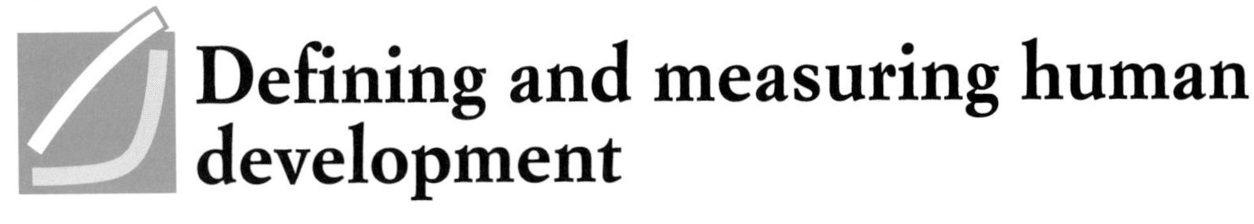

Defining and measuring human development

People are the real wealth of a nation. The basic objective of development is to create an enabling environment for people to enjoy long, healthy and creative lives. This may appear to be a simple truth. But it is often forgotten in the immediate concern with the accumulation of commodities and financial wealth.

Technical considerations of the means to achieve human development — and the use of statistical aggregates to measure national income and its growth — have at times obscured the fact that the primary objective of development is to benefit people. There are two reasons for this. First, national income figures, useful though they are for many purposes, do not reveal the composition of income or the real beneficiaries. Second, people often value achievements that do not show up at all, or not immediately, in higher measured income or growth figures: better nutrition and health services, greater access to knowledge, more secure livelihoods, better working conditions, security against crime and physical violence, satisfying leisure hours, and a sense of participating in the economic, cultural and political activities of their communities. Of course, people also want higher incomes as one of their options. But income is not the sum total of human life.

This way of looking at human development is not really new. The idea that social arrangements must be judged by the extent to which they promote "human good" goes back at least to Aristotle. He also warned against judging societies merely by such things as income and wealth that are sought not for themselves but desired as means to other objectives. "Wealth is evidently not the good we are seeking, for it is merely useful and for the sake of something else."

Aristotle argued for seeing "the difference between a good political arrangement and a bad one" in terms of its successes and failures in facilitating people's ability to lead "flourishing lives". Human beings as the real end of all activities was a recurring theme in the writings of most of the early philosophers. Emmanuel Kant observed: "So act as to treat humanity, whether in their own person or in that of any other, in every case as an end withal, never as means only."

The same motivating concern can be found in the writings of the early leaders of quantification in economics — William Petty, Gregory King, François Quesnay, Antoine Lavoisier and Joseph Lagrange, the grandparents of GNP and GDP. It is also clear in the writings of the leading political economists—Adam Smith, David Ricardo, Robert Malthus, Karl Marx and John Stuart Mill.

But excessive preoccupation with GNP growth and national income accounts has obscured that powerful perspective, supplanting a focus on ends by an obsession with merely the means.

Recent development experience has once again underlined the need for paying

TABLE 1.1
GNP per capita and selected social indicators

Country	GNP per capita (US$)	Life expectancy (years)	Adult literacy (%)	Infant mortality (per 1,000 live births)
Modest GNP per capita with high human development				
Sri Lanka	400	71	87	32
Jamaica	940	74	*82*	18
Costa Rica	1,610	75	93	18
High GNP per capita with modest human development				
Brazil	2,020	65	78	62
Oman	5,810	57	30	40
Saudi Arabia	6,200	64	*55*	70

close attention to the link between economic growth and human development — for a variety of reasons.

- Many fast-growing developing countries are discovering that their high GNP growth rates have failed to reduce the socioeconomic deprivation of substantial sections of their population.
- Even industrial nations are realizing that high income is no protection against the rapid spread of such problems as drugs, alcoholism, AIDS, homelessness, violence and the breakdown of family relations.
- At the same time, some low-income countries have demonstrated that it is possible to achieve high levels of human development if they skilfully use the available means to expand basic human capabilities.
- Human development efforts in many developing countries have been severely squeezed by the economic crisis of the 1980s and the ensuing adjustment programmes.

Human development is the process of enlarging people's choices

Recent development experience is thus a powerful reminder that the expansion of output and wealth is only a means. The end of development must be human well-being. How to relate the means to the ultimate end should once again become the central focus of development analysis and planning.

How can economic growth be managed in the interest of the people? What alternative policies and strategies need to be pursued if people, not commodities, are the principal focus of national attention? This Report addresses these issues.

BOX 1.1

Human development defined

Human development is a process of enlarging people's choices. In principle, these choice can be infinite and change over time. But at all levels of development, the three essential ones are for people to lead a long and healthy life, to acquire knowledge and to have access to resources needed for a decent standard of living. If these essential choices are not available, many other opportunities remain inaccessible.

But human development does not end there. Additional choices, highly valued by many people, range from political, economic and social freedom to opportunities for being creative and productive, and enjoying personal self-respect and guaranteed human rights.

Human development has two sides: the formation of human capabilities — such as improved health, knowledge and skills — and the use people make of their acquired capabilities — for leisure, productive purposes or being active in cultural, social and political affairs. If the scales of human development do not finely balance the two sides, considerable human frustration may result.

According to this concept of human development, income is clearly only one option that people would like to have, albeit an important one. But it is not the sum total of their lives. Development must, therefore, be more than just the expansion of income and wealth. Its focus must be people.

Defining human development

Human development is a process of enlarging people's choices. The most critical ones are to lead a long and healthy life, to be educated and to enjoy a decent standard of living. Additional choices include political freedom, guaranteed human rights and self-respect — what Adam Smith called the ability to mix with others without being "ashamed to appear in publick" (box 1.1).

It is sometimes suggested that income is a good proxy for all other human choices since access to income permits exercise of every other option. This is only partly true for a variety of reasons:

- Income is a means, not an end. It may be used for essential medicines or narcotic drugs. Well-being of a society depends on the uses to which income is put, not on the level of income itself.
- Country experience demonstrates several cases of high levels of human development at modest income levels and poor levels of human development at fairly high income levels.
- Present income of a country may offer little guidance to its future growth prospects. If it has already invested in its people, its potential income may be much higher than what its current income level shows, and vice versa.
- Multiplying human problems in many industrial, rich nations show that high income levels, by themselves, are no guarantee for human progress.

The simple truth is that there is no automatic link between income growth and human progress. The main preoccupation of development analysis should be how such a link can be created and reinforced.

The term *human development* here denotes both the *process* of widening people's choices and the *level* of their achieved well-being. It also helps to distinguish clearly between two sides of human development. One is the formation of human capabilities, such as improved health or knowledge. The other is the use that people make of their acquired capabilities, for work or leisure.

This way of looking at development differs from the conventional approaches to economic growth, human capital formation, human resource development, human welfare or basic human needs. It is necessary to delineate these differences clearly to avoid any confusion:

- GNP growth is treated here as being necessary but not sufficient for human development. Human progress may be lacking in some societies despite rapid GNP growth or high per capita income levels unless some additional steps are taken.
- Theories of human capital formation and human resource development view human beings primarily as means rather than as ends. They are concerned only with the supply side — with human beings as instruments for furthering commodity production. True, there is a connection, for human beings *are* the active agents of all production. But human beings are more than capital goods for commodity production. They are also the ultimate ends and beneficiaries of this process. Thus, the concept of human capital formation (or human resource development) captures only one side of human development, not its whole.
- Human welfare approaches look at human beings more as the beneficiaries of the development process than as participants in it. They emphasise distributive policies rather than production structures.
- The basic needs approach usually concentrates on the bundle of goods and services that deprived population groups need: food, shelter, clothing, health care and water. It focuses on the provision of these goods and services rather than on the issue of human choices.

Human development, by contrast, brings together the production and distribution of commodities and the expansion and use of human capabilities. It also focusses on choices — on what people should have, be and do to be able to ensure their own livelihood. Human development is, moreover, concerned not only with basic needs satisfaction but also with human development as a participatory and dynamic process. It applies equally to less developed and highly developed countries.

Human development as defined in this Report thus embraces many of the earlier approaches to human development. This broad definition makes it possible to capture better the complexity of human life — the many concerns people have and the many cultural, economic, social and political differences in people's lives throughout the world.

The broad definition also raises some questions: Does human development lend itself to measurement and quantification? Is it operational? Can it be planned and monitored?

Measuring human development

In any system for measuring and monitoring human development, the ideal would be to include many variables, to obtain as comprehensive a picture as possible. But the current lack of relevant comparable statistics precludes that. Nor is such comprehensiveness entirely desirable. Too many indicators could produce a perplexing picture — perhaps distracting policymakers from the main overall trends. The crucial issue therefore is of emphasis.

The key indicators

This Report suggests that the measurement of human development should for the time

BOX 1.2

What price human life?

The use of life expectancy as one of the principal indicators of human development rests on three considerations: the intrinsic value of longevity, its value in helping people pursue various goals and its association with other characteristics, such as good health and nutrition.

The importance of life expectancy relates primarily to the value people attach to living long and well. That value is easy for theorists to underestimate in countries where longevity is already high. Indeed, when life expectancy is very high, the challenge of making the lives of the old and infirm happy and worthwhile may be regarded by some as an exacting task. For the less fortunate people of the world, however, life is battered by distress, deprivation and the fear of premature death. They certainly attach a higher value to longer life expectancy.

Longevity also helps in the pursuit of some of life's other most valued goals. Living long may not be people's only objective, but their other plans and ambitions clearly depend on having a reasonable life span to develop their abilities, use their talents and carry out their plans.

A long life correlates closely with adequate nutrition, good health and education and other valued achievements. Life expectancy is thus a proxy measure for several other important variables in human development.

being focus on the three essential elements of human life — longevity, knowledge and decent living standards.

For the first component — longevity — life expectancy at birth is the indicator. The importance of life expectancy lies in the common belief that a long life is valuable in itself and in the fact that various indirect benefits (such as adequate nutrition and good health) are closely associated with higher life expectancy. This association makes life expectancy an important indicator of human development, especially in view of the present lack of comprehensive information about people's health and nutritional status (box 1.2).

For the second key component — knowledge — literacy figures are only a crude reflection of access to education, particularly to the good quality education so necessary for productive life in modern society. But literacy is a person's first step in learning and knowledge-building, so literacy figures are essential in any measurement of human development. In a more varied set of indicators, importance would also have to be attached to the outputs of higher levels of education. But for basic human development, literacy deserves the clearest emphasis.

BOX 1.3

What national averages conceal

Averages of per capita income often conceal widespread human deprivation. Look at Panama, Brazil, Malaysia and Costa Rica in the table below. That is the order of their ranking by GNP per capita.

If the GNP figures are corrected for variations in purchasing power in different countries, the ranking shifts somewhat — to Brazil, Panama, Malaysia and Costa Rica.

But if distributional adjustments are made using each country's Gini coefficient, the original ranking reverses to Costa Rica, Malaysia, Brazil, Panama.

The average value of literacy, life expectancy and other indicators can be similarly adjusted. There is a great deal of technical literature on the subject, but the basic approach is simple. If inequality is seen as reducing the value of average achievement as given by an unweighted mean, that average value can be adjusted by the use of inequality measures. Such distributional corrections can make a significant difference to evaluations of country performance.

Country	GNP per capita (US$) 1987	Real GDP per capita (PPP$) 1987	Gini coefficient of inequality	Distribution-adjusted GDP per capita (PPP$)
Panama	2,240	4,010	.57	1,724
Brazil	2,020	4,310	.57	1,852
Malaysia	1,810	3,850	.48	2,001
Costa Rica	1,610	3,760	.42	2,180

The third key component of human development — command over resources needed for a decent living — is perhaps the most difficult to measure simply. It requires data on access to land, credit, income and other resources. But given the scarce data on many of these variables, we must for the time being make the best use of an income indicator. The most readily available income indicator — per capita income — has wide national coverage. But the presence of nontradable goods and services and the distortions from exchange rate anomalies, tariffs and taxes make per capita income data in nominal prices not very useful for international comparisons. Such data can, however, be improved by using purchasing-power-adjusted real GDP per capita figures, which provide better approximations of the relative power to buy commodities and to gain command over resources for a decent living standard.

A further consideration is that the indicator should reflect the diminishing returns to transforming income into human capabilities. In other words, people do not need excessive financial resources to ensure a decent living. This aspect was taken into account by using the logarithm of real GDP per capita for the income indicator.

All three measures of human development suffer from a common failing: they are averages that conceal wide disparities in the overall population. Different social groups have different life expectancies. There often are wide disparities in male and female literacy. And income is distributed unevenly.

The case is thus strong for making distributional corrections in one form or another (box 1.3). Such corrections are especially important for income, which can grow to enormous heights. The inequality possible in respect of life expectancy and literacy is much more limited: a person can be literate only once, and human life is finite.

Reliable and comparable estimates of inequality of income are hard to come by, however. Even the Gini coefficient, probably the most widely used measure of income inequality, is currently available for fewer than a quarter of the 130 countries in the Human Development Indicators at the end of this Report — and many of those estimates are far from dependable. Distributional data for life expectancy and literacy by income group are not being collected, and those available on rural-urban and male-female disparities are still too scant for international comparisons.

The conceptual and methodological problems of quantifying and measuring human development become even more complex for political freedom, personal security, interpersonal relations and the physical environment. But even if these aspects largely escape measurement now, analyses of human development must not ignore them. The correct interpretation of the data on quantifiable variables depends on also keeping in mind the more qualitative dimensions of human life. Special effort must go into developing a simple quantitative measure to capture the many aspects of human freedom.

Attainments and shortfalls

Progress in human development has two perspectives. One is attainment: what has been achieved, with greater achievements meaning better progress. The second is the continuing shortfall from a desired value or target.

In many ways the two perspectives are equivalent — the greater the attainments, the smaller the shortfalls. But they also have some substantive differences. Disappointment and dismay at low performance often originate in the belief that things could be much better, an appraisal that makes the concept of a shortfall from some acceptable level quite central. Indeed, human deprivation and poverty inevitably invoke shortfalls from some designated value, representing adequacy, acceptability or achievability.

The difference between assessing attainments and shortfalls shows up more clearly in a numerical example. Performances often are compared in percentage changes: a 10-year rise in life expectancy from 60 years to 70 years is a 17% increase, but a 10-year rise in life expectancy from 40 years to 50 years is a 25% increase. The less the attainment already achieved, the higher the percentage value of the same absolute increase in life expectancy.

Raising a person's life expectancy from 40 years to 50 years would thus appear to be a larger achievement than going from 60 years to 70 years. In fact, raising life expectancy from the terribly low level of 40 years to 50 years is achievable through such relatively easy measures as epidemic control. But improving life expectancy from 60 years to 70 years may often be a much more difficult and more creditable accomplishment. The shortfall measure of human

BOX 1.4

Constructing a human development index

Human deprivation and development have many facets, so any index of human progress should incorporate a range of indicators to capture this complexity. But having too many indicators in the index would blur its focus and make it difficult to interpret and use. Hence the need for compromise — to balance the virtues of broad scope with those of retaining sensitivity to critical aspects of deprivation.

This Report has chosen three types of deprivation as the focus of attention: people's deprivation in life expectancy, literacy and income for a decent living standard. Each measure could have been further refined (especially by making distributional adjustments) if there had been adequate comparable data. But in the absence of such data, the focus here represents a move in the right direction — away from the narrow and misleading attention to only one dimension of human life, whether economic or social.

The first two indicators — life expectancy and adult literacy — are commonly used concepts. But the third — the purchasing power to buy commodities for satisfying basic needs — is not as well understood. The GNP figures typically used for international comparisons do not adequately account for national differences in purchasing power or the distorting effect of official exchange rates. To overcome these inadequacies, we use here the purchasing-power-adjusted GDP estimates developed in the International Price Comparison Project, a collaborative effort of the UN Statistical Office, the World Bank, EUROSTAT, OECD, ECE and ESCAP, now being expanded by USAID. And since there are diminishing returns in the conversion of income into the fulfilment of human needs, the adjusted GDP per capita figures have been transformed into their logarithms.

To construct a composite index, a *minimum* value (the maximum deprivation set equal to one) and a *desirable* or *adequate* value (no deprivation set equal to zero) had to be specified for each of the three indicators.

The minimum values were chosen by taking the lowest 1987 national value for each indicator. For life expectancy at birth, the minimum value was 42 years, in Afghanistan, Ethiopia and Sierra Leone. For adult literacy, it was 12%, in Somalia. For the purchasing-power-adjusted GDP per capita, the value was $220 (log value 2.34), in Zaire.

The values of desirable or adequate achievement were Japan's 1987 life expectancy at birth of 78 years, an adult literacy rate of 100%, and the average official "poverty line" income in nine industrial countries, adjusted by purchasing power parities, of $4,861. The nine countries are Australia, Canada, the Federal Republic of Germany, the Netherlands, Norway, Sweden, Switzerland, the United Kingdom and the United States.

The minimum and desirable or adequate values are the end-points of a scale indexed from one to zero for each measure of deprivation. Placing a country at the appropriate point on each scale and averaging the three scales gives its average human deprivation index, which when subtracted from 1 gives the human development index (HDI). A mathematical formulation of the HDI is given in technical note 3.

progress captures this better than the attainment measure does.

Taking once again the example of life expectancy, if 80 years is the target for calculating shortfalls, a rise of life expectancy from 60 years to 70 years is a 50% reduction in shortfall — halving it from 20 years to 10 years. That is seen as a bigger achievement than the 25% reduction in shortfall (from 40 years to 30 years) when raising life expectancy from 40 years to 50 years.

The shortfall thus has two advantages over the attainment in assessing human progress. It brings out more clearly the difficulty of the tasks accomplished, and it emphasises the magnitude of the tasks that still lie ahead.

The human development index

People do not isolate the different aspects of their lives. Instead, they have an overall sense of well-being. There thus is merit in trying to construct a composite index of human development.

FIGURE 1.1
GNP per capita and the HDI

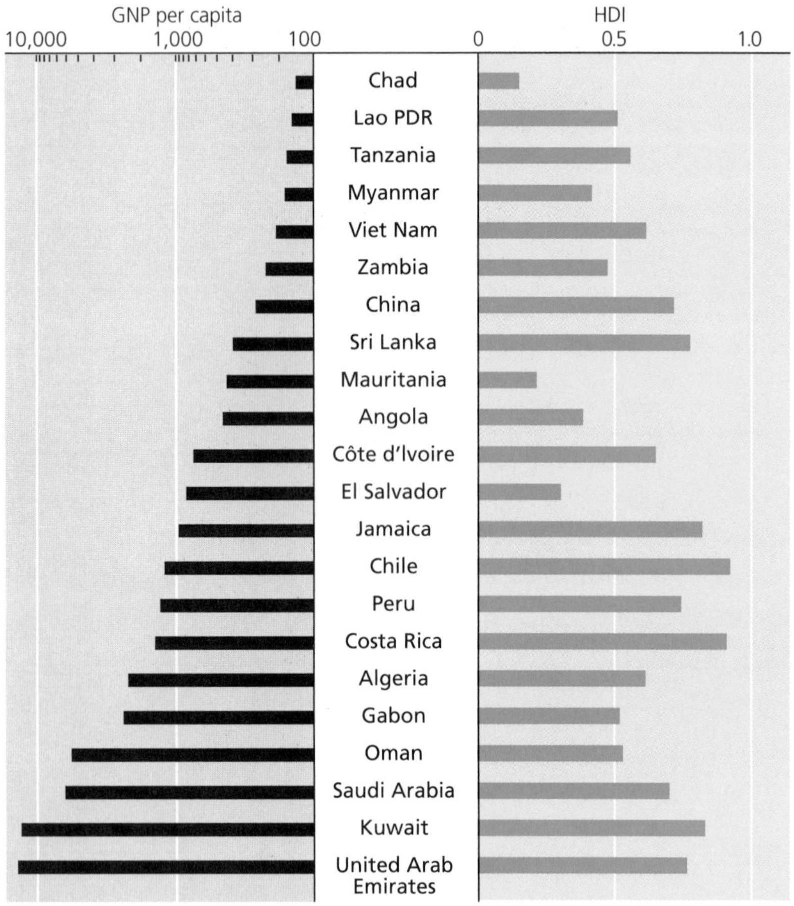

Past efforts to devise such an index have not come up with a fully satisfactory measure (see technical note 1). They have focussed either on income or on social indicators, without bringing them together in a composite index. Since human beings are both the means and the end of development, a composite index must capture both these aspects. This Report carries forward the search for a more appropriate index by suggesting an index that captures the three essential components of human life — longevity, knowledge and basic income for a decent living standard. Longevity and knowledge refer to the formation of human capabilities, and income is a proxy measure for the choices people have in putting their capabilities to use.

The construction of the human development index (HDI) starts with a deprivation measure (box 1.4). For life expectancy, the target is 78 years, the highest average life expectancy attained by any country. The literacy target is 100%. The income target is the logarithm of the average poverty line income of the richer countries, expressed in purchasing-power-adjusted international dollars. Human development indexes for 130 countries with more than a million people are presented in the Human Development Indicators, table 1. Those for another 32 countries with fewer than a million people are in the Human Development Indicators, table 25.

Country ranking by HDI and GNP

The human development index ranks countries very differently from the way GNP per capita ranks them. The reason is that GNP per capita is only one of life's many dimensions, while the human development index captures other dimensions as well.

Sri Lanka, Chile, Costa Rica, Jamaica, Tanzania and Thailand, among others, do far better on their human development ranking than on their income ranking, showing that they have directed their economic resources more towards some aspects of human progress. But Oman, Gabon, Saudi

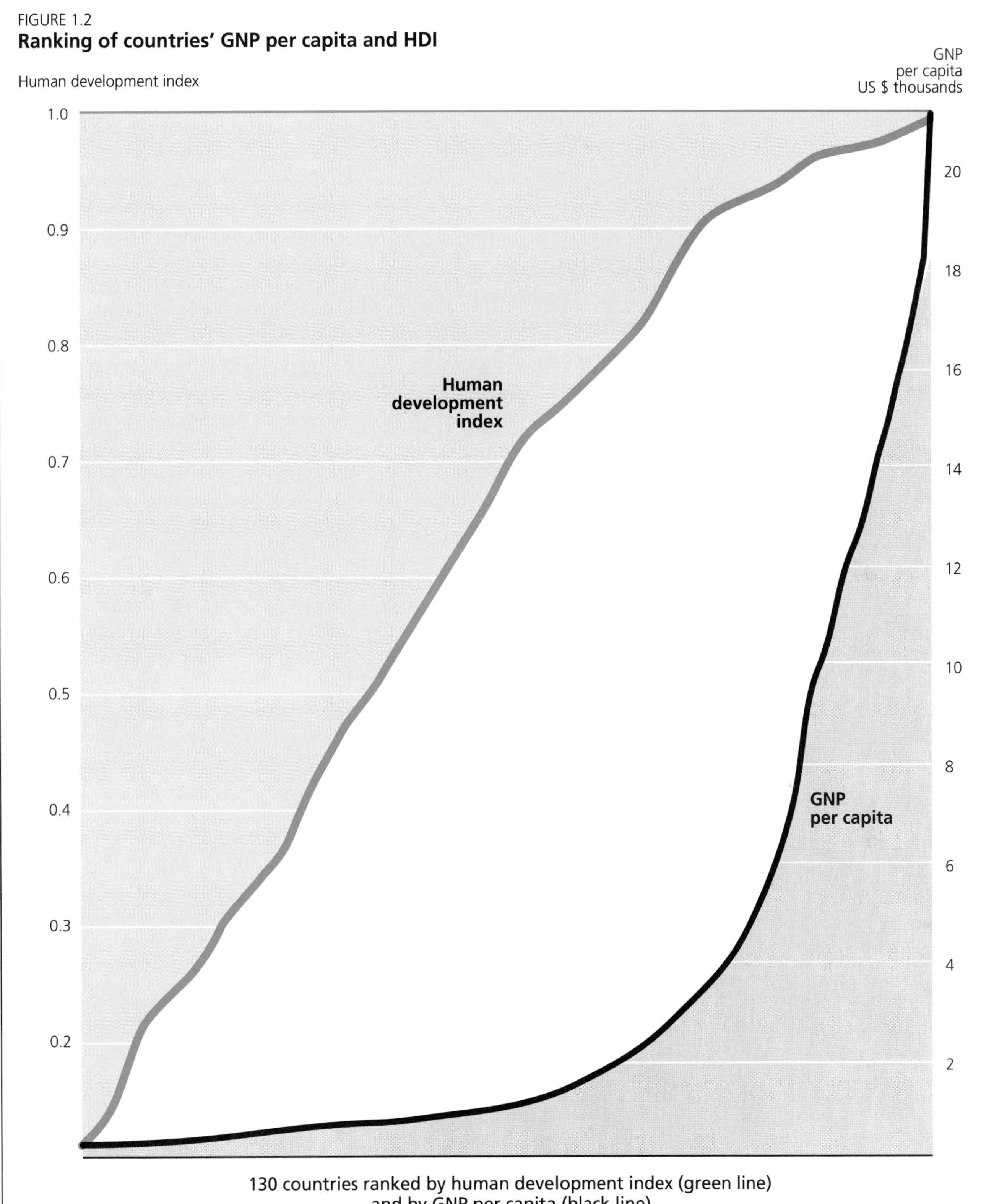

The chart shows two separate distributions of countries. The upper curve represents their ranking according to the human development index while the lower curve shows their ranking according to GNP per capita. The two curves reveal that the disparity among countries is much greater in income than in human development. There is no automatic link between the level of per capita income in a country and the level of its human development.

BOX 1.5

Freedom and human development

Human development is incomplete without human freedom. Throughout history, people have been willing to sacrifice their lives to gain national and personal liberty. We have witnessed only recently an irresistible wave of human freedom sweep across Eastern Europe, South Africa and many other parts of the world. Any index of human development should therefore give adequate weight to a society's human freedom in pursuit of material and social goals. The valuation we put on similar human development achievements in different countries will be quite different depending on whether they were accomplished in a democratic or an authoritarian framework.

While the need for qualitative judgement is clear, there is no simple quantitative measure available yet to capture the many aspects of human freedom — free elections, multiparty political systems, uncensored press, adherence to the rule of law, guarantees of free speech and so on. To some extent, however, the human development index (HDI) captures some aspects of human freedom. For example, if the suppression of people suppresses their creativity and productivity, that would show up in income estimates or literacy levels. In addition, the human development concept, adopted in this Report, focusses on people's capabilities or, in other words, people's strength to manage their affairs — which, after all, is the essence of freedom.

For illustrative purposes, the table below shows a selection of countries (within each region) that have achieved a high level of human development (relative to other countries in the region) within a reasonably democratic political and social framework. And a cursory glance at the ranking of countries in table 1 of the Human Development Indicators, given at the end of this report, shows that countries ranking high in their HDI also have a more democratic framework — and vice versa — with some notable exceptions.

What is needed is considerable empirical work to quantify various indicators of human freedom and to explore further the link between human freedom and human development.

Top 15 countries in democratic human development

Country	HDI	Country	HDI
Latin America and the Caribbean		*Middle East and North Africa*	
Costa Rica	0.916	Turkey	0.751
Uruguay	0.916	Tunisia	0.657
Trinidad and Tobago	0.885		
Mexico	0.876	*Sub-Saharan Africa*	
Venezuela	0.861	Mauritius	0.788
Jamaica	0.824	Botswana	0.646
Colombia	0.801	Zimbabwe	0.576
Asia			
Malaysia	0.800		
Sri Lanka	0.789		
Thailand	0.783		

Arabia, Algeria, Mauritania, Senegal and Cameroon, among others, do considerably worse on their human development ranking than on their income ranking, showing that they have yet to translate their income into corresponding levels of human development.

To stress again an earlier point, the human development index captures a few of people's choices and leaves out many that people may value highly — economic, social and political freedom (box 1.5), and protection against violence, insecurity and discrimination, to name but a few. The HDI thus has limitations. But the virtue of broader coverage must be weighed against the inconvenience of complicating the basic picture it allows policymakers to draw. These tradeoffs pose a difficult issue that future editions of the *Human Development Report* will continue to discuss.

CHAPTER 2

Human development since 1960

The developing countries have made significant progress towards human development in the last three decades. They increased life expectancy at birth from 46 years in 1960 to 62 years in 1987. They halved the mortality rates for children under five and immunised two-thirds of all one-year-olds against major childhood diseases. The developing countries also made primary health care accessible to 61% of their people and safe water to 55% (80% in urban areas). In addition, they increased the per capita calorie supply by about 20% between 1965 and 1985.

Their progress in education was equally impressive. Adult literacy rates rose from 43% in 1970 to 60% in 1985 — male literacy from 53% to 71% and female literacy from 33% to 50%. The South's primary educational output in 1985 was almost six times greater than that in 1950, its secondary educational output more than 18 times greater. The results were 1.4 billion literate people in the South in 1985, compared with nearly a billion in the North.

North-South gaps in human development narrowed considerably during this period even while income gaps tended to widen. The South's average per capita income in 1987 was still only 6% of the North's, but its average life expectancy was 80% and its average literacy rate 66% of the North's. The North-South gap in life expectancy narrowed from 23 years in 1960 to 12 years in 1987, and the literacy gap from 54 percentage points in 1970 to less than 40 percentage points in 1985. The developing countries also reduced their average infant mortality from 200 deaths per 1,000 live births to 79 between 1950 and 1985, a feat that took nearly a century in the industrial countries.

This progress must be put in perspective, however.

First, tremendous human deprivation remains. There still are nearly 900 million adults in the developing world who cannot read or write, 1.5 billion people without access to primary health care, 1.75 billion people without safe water, around 100 million completely homeless, some 800 million people who still go hungry every day and more than a billion who survive in absolute poverty.

Children and women suffer the most. Some 40 million newborns still are not properly immunised. Fourteen million children under the age of five die each year and 150 million are malnourished. The mater-

FIGURE 2.1
Disparities between developing and industrial countries

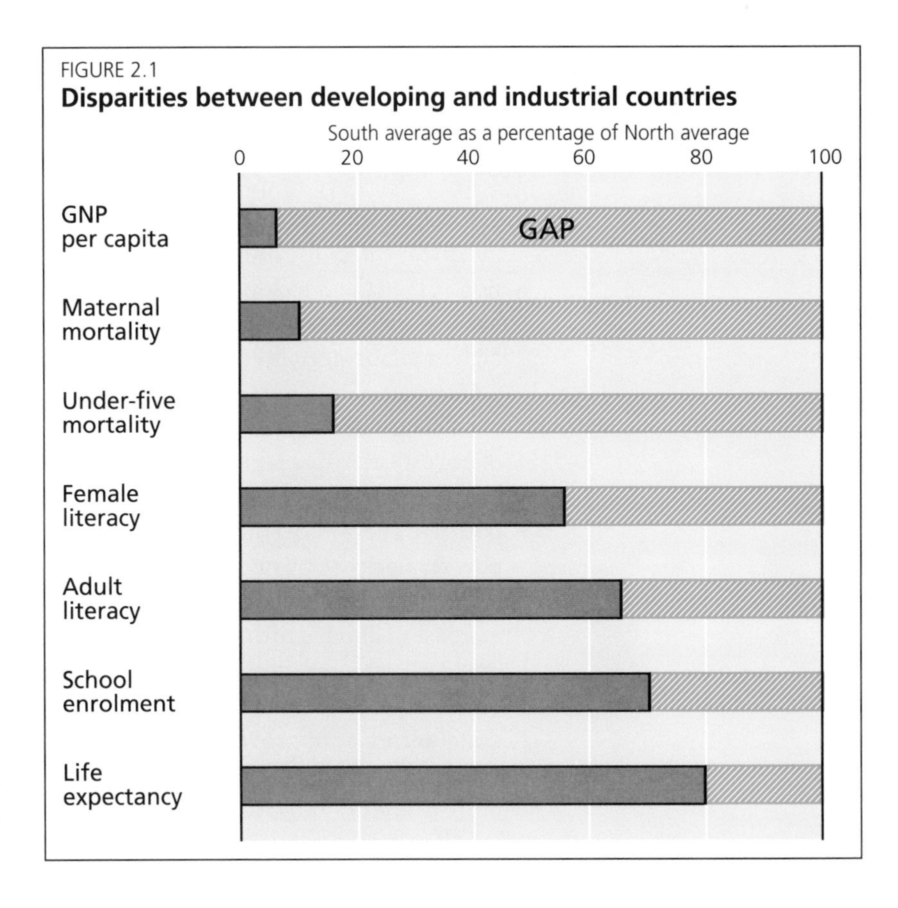

nal mortality rate in the South is 12 times higher than that in the North, and the female illiteracy rate is at least 15 times higher. Obviously, the backlog of human deprivation presents a challenging agenda for the next decade.

Second, the recent progress in narrowing human development gaps between North and South raises hope — and a question mark. The hope is that the developing world can be taken to a basic level of human development in a fairly short period — if national development efforts and international assistance are properly directed. The question mark relates to the fact that four-fifths of the people in the Third World are leading longer, better educated lives, but they lack opportunities to tap their full potential. Unless economic opportunities are created in the South, more human talent will be wasted, and pressures for international migration are likely to increase dramatically. Moreover, while gaps in basic survival have narrowed, the widening gaps in science and technology threaten the South's future development.

Significant human progress co-exists with tremendous human deprivation

Third, the average figures for human development hide considerable disparities among countries in the South. Life expectancy exceeds 70 years in 13 developing countries but is still less than 50 years in another 20 countries. Similarly, seven countries have literacy rates over 90%, but another seven have rates less than 25%. In general, the least developed countries, many in Africa, suffer the most human deprivation. Of all the developing regions, Africa has the lowest life expectancy figures, the highest infant mortality rates and the lowest literacy rates.

This trend towards the concentration of poverty in Africa is growing: more than half the people in Africa live in absolute poverty. The number of Africans below the poverty line rose by two-thirds in the first half of the 1980s — compared with an increase of about a fifth for the developing world as a whole — and is projected to rise rapidly in the next decade. Any international effort to improve human development in the Third World must thus give priority attention to Africa and the other least developed countries.

Fourth, the gaps in human development within countries are also great — between urban and rural areas, between men and women and between rich and poor. For developing countries as a whole, urban areas have twice the access to health services and safe water as rural areas and four times the access to sanitation services. Female literacy rates are only two-thirds those of men. And the rich often appropriate a major share of social subsidies. These wide disparities show the considerable room for improvement in distributing social expenditures.

Fifth, human progress over the last three decades has been neither uniform nor smooth. Many countries recorded major reverses in the 1980s — with rising rates of child malnutrition and infant mortality, particularly in Sub-Saharan Africa and Latin America. Budget cuts greatly squeezed social spending. Some countries avoided reductions in social programmes through better economic management, but most countries in Africa and Latin America paid a heavy social price during the adjustment period of the 1980s.

The 1990s present the challenge of rectifying the damage to human development in many developing countries and then building up momentum to achieve essential human goals by the year 2000. The responses to this challenge will require more resources, mobilised both domestically and internationally, and in many instances they will also require major shifts in budget priorities. Needed most are cuts in spending on the military, on inefficient public enterprises and on mistargeted social subsidies. To create the enabling framework for more broadly based development, macroeconomic policy formulation and management must improve, and popular participation and private initiatives must increase.

The remainder of this chapter documents the record of human development in the developing world since 1960. The last section also takes up some of the human problems now confronting both developed and developing countries. The discussion throughout reinforces this Report's basic thesis: income alone is not the answer to human development.

Expanding human capabilities

The key components of the human development index — life expectancy, literacy and basic income — are the starting point for this review of the formation of human capabilities. Basic income is used here as a proxy for access to resources for a decent living standard. The review also examines some major contributing factors, especially people's access to food and such social services as water, education and primary health care.

Life expectancy

Life expectancy in the developing countries has risen on the average by nearly a third since 1960, from 46 years to 62 years. But this average masks important interregional and intercountry differences. Africa's average life expectancy is only 51 years, ranging from 42 years in Ethiopia and Sierra Leone to 69 years in Mauritius. Asia's average life expectancy is 64 years, reflecting the rise in China's life expectancy from 47 years to 70 years in three decades. Latin America's average life expectancy is 67 years, fairly close to the industrial nations' average of 69 years in 1960. Nine Latin American and Caribbean countries fall into the group of 18 developing countries that already have a life expectancy of 70 years.

Life expectancy generally is well correlated with a country's income, but important exceptions show that significant gains in life expectancy can be made even at modest incomes. Sri Lanka ($400 per capita) enjoys a life expectancy of 70 years, as high as that in the Republic of Korea ($2,690), Venezuela ($3,230) and the United Arab Emirates ($15,830). Rapid advances in health and nutrition made these exceptional gains possible.

Until the mid-1970s the average life expectancy in low-income countries was increasing three times faster than that in the middle- and high-income countries, but since then the increase has been only slightly faster. As a result, the life expectancy gap between least developed countries and the developing countries as a whole has widened from seven years to 12 years.

FIGURE 2.2
Life expectancy trends

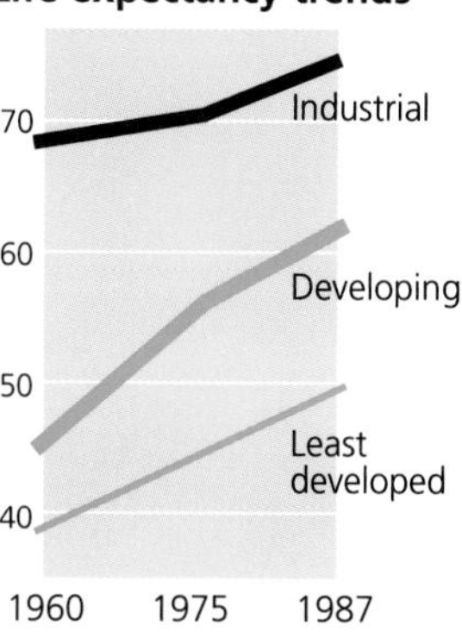

TABLE 2.1
Life expectancy, 1960-87

	Annual rate of reduction in shortfall (%) 1960-87		Life expectancy (years) 1987
Fastest progress		*Highest life expectancy*	
Hong Kong	4.99	Hong Kong	76
Costa Rica	4.55	Costa Rica	75
China	4.33	Jamaica	74
United Arab Emirates	4.06	Singapore	73
Jamaica	4.00	Kuwait	73
Kuwait	3.93	Panama	72
Chile	3.70	Chile	72
Malaysia	3.48	Uruguay	71
Korea, Rep.	3.43	United Arab Emirates	71
Panama	3.38	Sri Lanka	71
Slowest progress, among countries with a life expectancy of less than 60 years		*Lowest life expectancy*	
Ethiopia	0.52	Sierra Leone	42
Paraguay	0.78	Ethiopia	42
Rwanda	0.79	Afghanistan	42
Kampuchea, Dem.	0.80	Guinea	43
Afghanistan	0.81	Mali	45
Sierra Leone	0.84	Angola	45
Burundi	0.85	Niger	45
Guinea	0.88	Somalia	46
Central African Rep.	0.90	Central African Rep.	46
Mali	0.91	Chad	46
South	2.33		
North	2.22		

	1960	1987
South	46	62
North	69	74
South as % of North	67	84

FIGURE 2.3
Infant mortality trends

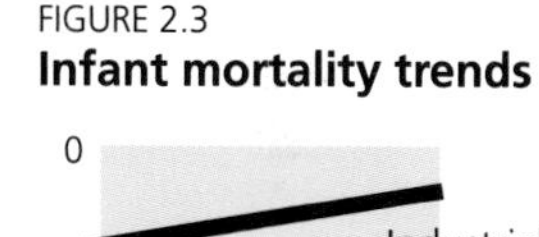
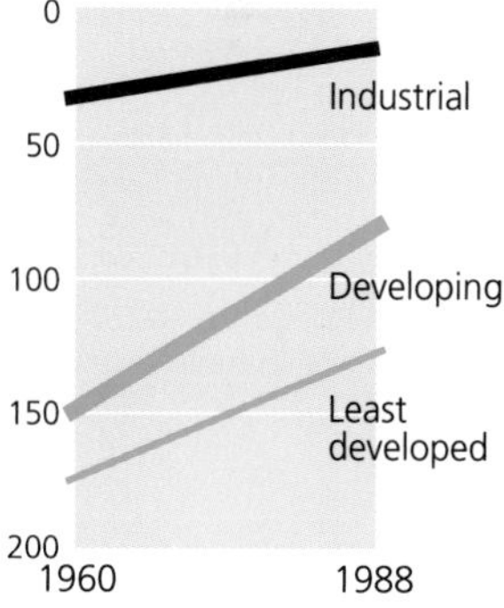

TABLE 2.2

Infant mortality rate, 1960-88

	Annual rate of reduction in shortfall (%) 1960-88		Infant mortality rate (per 1,000 live births) 1988
Fastest progress		*Lowest infant mortality rate*	
Chile	6.20	Hong Kong	8
United Arab Emirates	6.09	Singapore	9
Hong Kong	5.91	Costa Rica	18
Oman	5.81	Jamaica	18
China	5.48	Kuwait	19
Kuwait	5.37	Chile	19
Costa Rica	5.35	Trinidad and Tobago	20
Singapore	4.83	Mauritius	22
Korea, Rep.	4.42	Panama	23
Jamaica	4.32	Malaysia	24
Slowest progress		*Highest infant mortality rate*	
Mozambique	0.35	Mozambique	172
Ethiopia	0.48	Angola	172
Kampuchea, Dem.	0.50	Afghanistan	171
Rwanda	0.67	Mali	168
Angola	0.68	Sierra Leone	153
Mali	0.79	Ethiopia	153
Afghanistan	0.81	Malawi	149
Uganda	0.94	Guinea	146
Bangladesh	0.99	Burkina Faso	137
Somalia	1.03	Niger	134
South	2.18		
North	3.08		

	1960	1988
South	150	81
North	36	15
South as % of North (Survival)	88	93

TABLE 2.3

Adult literacy rate, 1970-85

	Annual rate of reduction in shortfall (%) 1970-85		Adult literacy rate (%) 1985
Fastest progress		*Highest literacy rate*	
Iraq	11.26	Chile	98
Chile	10.74	Trinidad and Tobago	96
Mexico	6.29	Argentina	96
Thailand	5.48	Uruguay	95
Jordan	4.86	Costa Rica	93
Botswana	4.70	Korea, Rep.	*93*
Trinidad and Tobago	4.52	Thailand	91
Zambia	4.48	Mexico	90
Peru	4.41	Panama	89
Venezuela	4.27	Iraq	89
Slowest progress, among countries with an adult literacy rate of less than 50%		*Lowest literacy rate*	
		Somalia	12
Burkina Faso	0.42	Burkina Faso	14
Sudan	0.54	Niger	14
Somalia	0.67	Mali	17
Mali	0.73	Mauritania	17
Niger	0.73	Sudan	23
Bangladesh	0.84	Afghanistan	24
Pakistan	0.84	Yemen Arab Rep.	25
Benin	0.94	Bhutan	*25*
India	0.97	Nepal	26
Nepal	1.07		
South	2.33		
North	..		

	1970	1985
South	43	60
North	..	..
South as % of North	..	..

Progress in reducing the deaths of children under five, especially infants, has contributed greatly to higher life expectancy. Developing countries reduced their infant (under age one) mortality rate from nearly 200 deaths per 1,000 births in 1960 to 79 in 1988 — and their child (under five) mortality rate from 243 deaths per thousand to 121.

Some countries have done particularly well, often despite modest incomes. Jamaica's child mortality rate was 22 in 1988, compared with 85 in Brazil, a country with more than twice the per capita income of Jamaica. Similarly, Mauritius has the lowest infant and child mortality rates in Africa — having reduced the deaths of children under five from 104 per thousand to 29 since 1960, a performance much better than that of countries at considerably higher per capita incomes, such as Gabon and South Africa. Some developing countries with the lowest infant mortality rates in 1988 — Hong Kong, Singapore, Costa Rica, Kuwait and Chile — are also among the countries that reduced their infant mortality rates fastest between 1960 and 1988.

Literacy

Rapid improvements in education have sharply increased the ability of people in developing countries to read and write. The literacy rate for men rose from 53% in 1970 to 71% in the first half of the 1980s. Although the female literacy rate was still only 50% in 1985, enrolment rates for girls have been increasing far more rapidly than those for boys, an encouraging sign.

Several developing countries already have adult literacy rates above 90%, comparable to the rates in many industrial nations. Despite such successes, some of the most populous countries, such as India, Bangladesh, and Pakistan, have been extremely slow in reducing widespread illiteracy.

Sub-Saharan Africa has witnessed especially fast progress in adult literacy, but since it started from a very low point, its average literacy rate of 48% in 1985 was still far below the average of 60% for the developing world. Low-income Kenya made spectacular progress in extending universal primary education and raised its literacy rate from 32% in 1970 to 60% in 1985.

Literacy rates in Latin America continue to be well ahead of those for all other developing countries, having risen from 72% in 1970 to 83% in 1985. Asia's literacy rates closely follow the developing country average. They have moved from 41% to 59%. Holding down the region's average are four South Asian countries: Bangladesh (33%), Pakistan (30%), Nepal (26%) and Afghanistan (24%). South Asia's literacy rate was only 41% in 1985 — the lowest of all the regional rates.

The least developed countries have an average literacy rate of only 37%. As with other human development indicators, the disparity is growing between their performance and that of the developing countries as a whole. Their literacy gap widened from 18 percentage points in 1970 to 23 percentage points in 1985.

The number of illiterate people in the developing world, just under 900 million in 1985, may well reach a billion by the end of the century. Three-quarters of them live in the five most populous Asian countries: India, China, Pakistan, Bangladesh and Indonesia. Any attack on global illiteracy will thus need to concentrate on these countries.

FIGURE 2.4
Adult literacy trends

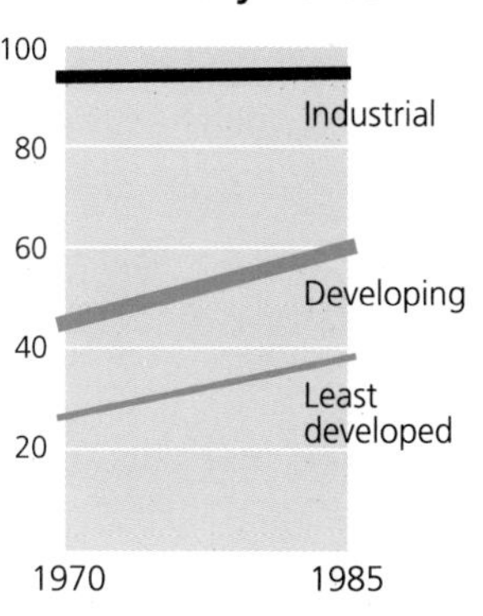

Income

The growth of per capita income, one of the critical elements in improving human development, was 2.9% a year on average for all developing regions between 1965 and 1980. This trend broke sharply in the 1980s. Sub-Saharan Africa's per capita income grew by only 1.6% a year between 1965 and 1980, but it has since been declining by 2.4% a year. Latin America, because of persistent debt problems, moved sharply from 3.8% annual growth in per capita income in 1965-80 to an annual decline of 0.7% in the 1980s.

For human development, the distribution of GNP is as important as the growth of GNP. One measure of the distribution of income is the Gini coefficient, which cap-

FIGURE 2.5
GNP per capita trends

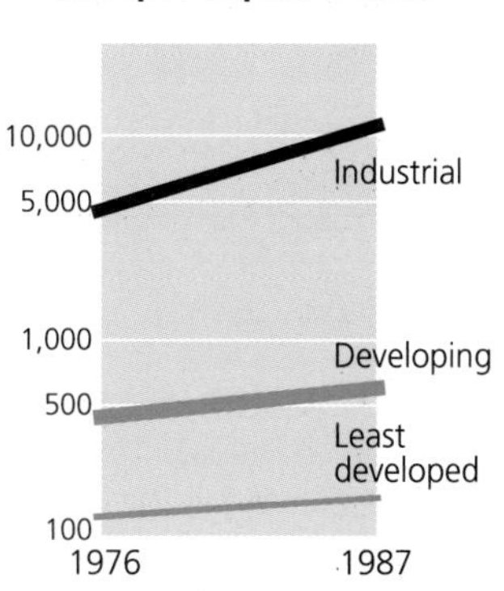

FIGURE 2.6
Absolute poverty by region

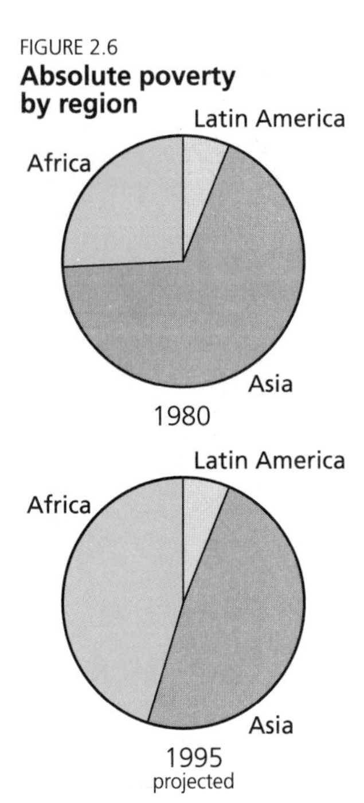

tures disparities in the percentages of income that each 1% (percentile) of the population receives. If each percentile receives 1% of the income, there is no disparity, and the Gini coefficient is zero. If one percentile receives all the income, there is maximum disparity, and the Gini coefficient is 1. In nine of the 28 developing countries for which the Gini coefficient is available, it is 0.50 or higher, showing that a small part of the population in these countries is getting a very large part of the income.

Another indicator of inequality is the per capita income of the richest 20% of the population compared with that of the poorest 20%. In 12 of the 23 developing countries where such a comparison is available, the income of the richest group was 15 times or more that of the poorest group.

Yet another distributional indicator in predominantly agricultural economies is the concentration of land, which is highly skewed in Latin America. Of 17 countries surveyed, 10 show land concentration indices (Gini coefficients) above 0.8, and five between 0.7 and 0.8. The FAO estimates that about 30 million agricultural households have no land and about 138 million are almost landless, two-thirds of them in Asia.

Most poverty estimates for developing countries use the income needed to meet minimum food needs and thus measure absolute poverty (see technical note 2). Country data are sparse, however, and not always comparable. The available data reveal an overall reduction in the *percentage* of people living in absolute poverty between 1970 and 1985. But owing to population growth, the *absolute number* of poor increased by about a fifth. In 1985 more than a billion people in the Third World were trapped in absolute poverty (box 2.1).

In Latin America more than 110 million people, about 40% of the population, lived in poverty in 1970, a quarter of them in extreme poverty. Fifteen years later, nearly 150 million people, more than a third of the population, were still poor, largely as a result of the economic stagnation in the 1980s. Poverty is so widespread in Latin America, despite its high average income, because of inadequate distribution of income in many countries. Brazil's GNP per capita was $2,020 in 1987, but the poorest 40% of Brazilians received only 7% of the household income. The top 2% of landowners control 60% of the arable land, while the bottom 70% of rural households are landless or nearly landless.

For Africa the ILO estimates that the number of absolute poor rose in the five years between 1980 and 1985 to more than 270 million, about half the total population. If nothing is done to reverse this ominous trend, nearly 400 million people will be living in extreme poverty in Africa by 1995.

In Asia the percentage of poor people is decreasing, but the greatest number of the world's poor, three-quarters of a billion people, still live there. Poverty is extensive in Bangladesh (where more than 80% of the people are poor), Nepal, India and the Lao People's Democratic Republic. The 1980s have been especially harsh for some countries: in Sri Lanka and Bangladesh, the poorest income groups had their shares of household income fall. Some East and Southeast Asian economies have nevertheless made tremendous progress in alleviating poverty.

BOX 2.1

Who the poor are

The renewed concern about human deprivation in recent years has generated a growing body of research on poverty. Here is a summary of some of the salient facts.

First, the poor are not a homogeneous group. The *chronic poor* are at the margin of society and constantly suffering from extreme deprivation. The *borderline poor* are occasionally poor, such as the seasonally unemployed. The *newly poor* are direct victims of structural adjustment of the 1980s, such as retrenched civil servants and industrial workers.

Second, over 1 billion people live in absolute poverty in the Third World. Asia has 64% of the developing countries' people in absolute poverty, Africa 24% and Latin America and the Caribbean 12%. Poverty is growing fastest in Africa, with the number of absolute poor having increased by about two-thirds between 1970 and 1985.

Third, three-quarters of the developing countries' poor people live in rural areas. There is, however, a recent trend towards the urbanisation of poverty, owing to the rapid increase in urban slums and squatter settlements, expanding by about 7% a year.

Fourth, there is a close link between poverty and the environment. About three-quarters of the developing countries' poor people are clustered in ecologically fragile areas, with low agricultural potential. Owing to a lack of employment and income-earning opportunities outside agriculture, environmental degradation and poverty continuously reinforce each other.

Fifth, poverty has a decided gender bias. A large proportion of poor households are headed by women, especially in rural Africa and in the urban slums of Latin America. Female members of a poor household are often worse off than male members because of gender-based differences in the distribution of food and other entitlements within the family. In Africa women produce 75% of the food — yet they suffer greater deprivation than men.

Poverty is by no means a problem of the developing countries alone, nor can consistent rates of economic growth guarantee its alleviation. In the United States, after 200 years of economic progress, nearly 32 million people, about 13% of the population, are still below the official poverty line.

Access to basic goods and services

The extent to which people can improve their capabilities depends largely on the access that they have to basic goods and services.

Food. There has been a general global improvement in food production and calorie supplies. The daily supply of calories in the developing world improved from 90% of total requirements in 1965 to 107% in 1985. Confirming this evidence, food production data show a roughly 20% increase in average calorie supplies per person between 1965 and 1985.

Countries having the most urgent need for food show the slowest progress. For the poorest countries, the daily per capita calorie supply increased only from 87% of the total requirements to 89% between 1965 and 1985.

Regional disparities in daily calorie supplies are stark. Sixteen African countries, of the 34 having data, recorded declines in their supply of calories per capita, while Gabon, Niger and Mauritius had theirs increase by 15% or more. In Latin America, the disparities are similar. The best progress was in the Middle East and in Asia where the per capita calorie supplies went up by 30% and 23%, respectively.

Estimates of world hunger vary. According to the World Food Council, more than half a billion people were hungry in the mid-1980s. The World Bank, in a study of 87 developing countries with 2.1 billion people, put the number of undernourished people—whose diet does not provide them with enough calories for an active working life—at 730 million in 1980. The figure is growing constantly, with as many as eight million people said to have joined the ranks of the hungry each year during the first half of the 1980s. Hunger today may be stunting the lives of as many as 800 million people in the Third World.

FIGURE 2.7
Nutrition trends

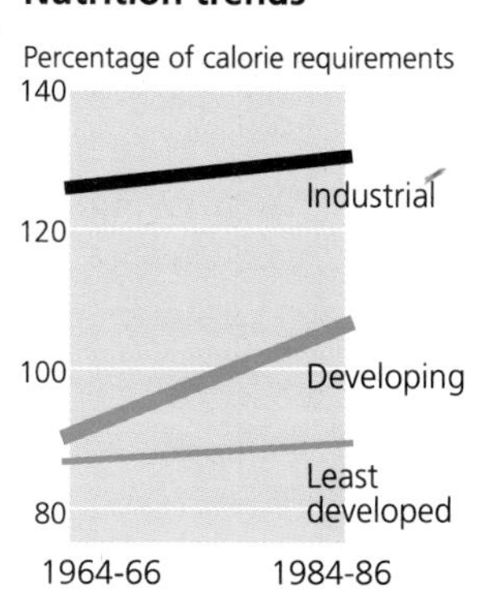

TABLE 2.4
Access to safe water, 1975-86

	Annual rate of reduction in shortfall (%) 1975-86		Percentage with access to safe water 1986
Fastest progress		*Most access*	
Saudi Arabia	20.22	Mauritius	100
Chile	13.61	Singapore	100
Colombia	12.78	Trinidad and Tobago	98
Malaysia	12.09	Saudi Arabia	97
Jamaica	10.76	Jamaica	96
Trinidad and Tobago	10.76	Jordan	96
Costa Rica	9.80	Chile	94
Iraq	8.37	Lebanon	93
Burkina Faso	7.19	Colombia	92
Thailand	6.45	Costa Rica	91
Slowest progress		*Least access*	
Rwanda	-4.14	Kampuchea, Dem.	3
Algeria	-3.05	Ethiopia	16
Argentina	-2.37	Mozambique	16
Congo	-2.23	Mali	17
Uganda	-1.91	Guinea	19
Bangladesh	-1.88	Côte d'Ivoire	19
Somalia	-0.57	Uganda	20
El Salvador	-0.19	Afghanistan	21
Guatemala	-0.15	Sudan	21
Nicaragua	0.52	Congo	21

			1975	1986
South	3.29	South	35	55
North	..	North	..	..
		South as % of North	..	..

FIGURE 2.8
Access to health services, 1986

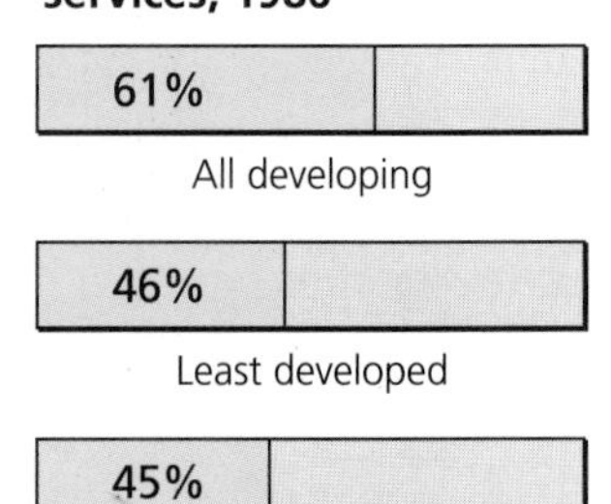

FIGURE 2.9
Access to safe water trends

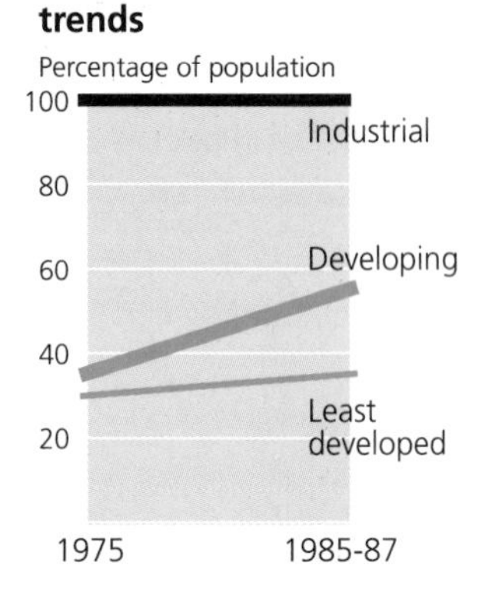

FIGURE 2.10
North-South distribution of school enrolment

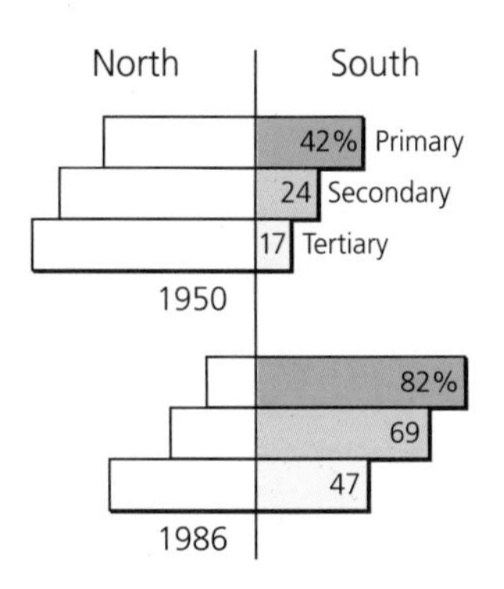

Two-thirds of those hungry in the developing countries live in Asia, and a quarter in Africa. Mirroring this distribution is the number of low-birth-weight infants in different regions, with Asian countries having some of the highest figures.

A major challenge for the 1990s is thus to ensure that food production increases rapidly, particularly in Africa, and that food is well distributed — supplemented where necessary by targetted nutritional programmes for the poorest and most vulnerable groups.

Health services. Ready and affordable access to health services is vital for human development. Most countries collect data on the percentage of people with easy access to health services and on the number of doctors and nurses. But these data do not mean that health services are actually available to people. Doctors may be concentrated in urban areas, possibly specialising in expensive tertiary medicine. People may be close to health services but unable to afford them. Despite the current limitations of available data, some broad conclusions are possible.

Several developing countries came close to the objective of primary health care for all during the 1980s. Many of them also stand out in life expectancy — for example, the Republic of Korea, Costa Rica, Jamaica, Tunisia and Jordan. On the average, however, only 61% of the people in developing countries have access to primary health care services today. For the least developed countries and Sub-Saharan Africa, the corresponding figures are 46% and 45%, respectively.

Access to health care, according to every available measure, is worst in Africa. In Latin America, which has the most doctors and nurses per person in the developing world, only 61% of the people have access to health services, well below the averages for Asia, North Africa and the Middle East.

Exemplifying the considerable progress in the Middle East and North Africa, Kuwait now has more doctors per person than Switzerland. But Kuwait's infant mortality rate is still four times that in Switzerland, reinforcing the argument that the availability of doctors is no guarantee of good health.

Water and sanitation. Progress in water and sanitation has generally been much slower than that in health, and it has been slower in sanitation than in water. More than half the people in developing countries had access to safe water in 1986, up from 35% in 1975. In the best-performing countries, practically every person has access to safe water. For the least developed countries, however, the rise was a mere four percentage points: only a third of their people have a source of potable water within reach.

Latin America has made good general progress, with nearly three-quarters of the people there having access to safe water in 1980-87. Chile and Trinidad have reached developed country standards.

Progress in access to safe water has also been impressive in the Middle East and North Africa. Several countries there report that more than 90% of their people have access to safe water, and only in Sudan and the Yemen Arab Republic do fewer than half the people have access.

Asia made good progress between 1975 and 1985, increasing the access to safe water to more than half from less than a third of the population. But in Bangladesh the access has declined by 10 percentage points since 1975.

Africa shows the least progress. In a third of the countries having current data, the access to safe water declined, and in eight African countries fewer than a fifth of the people have access to safe water.

For sanitation, about a third of the South's population had access to proper facilities in the second half of the 1980s.

Education. The enrolment gains have been impressive in most developing countries, despite their rapid population growth. Well over 80% of the children of primary school age were enrolled in primary schools in 1987, and several developing countries are close to the goal of universal primary enrolment.

The progress has been considerable in every region. Despite stagnant economies and rapid population growth, half the children of primary and secondary school age in Africa now attend school. Asia, the Middle East and North Africa also show steadily rising trends, with net primary school enrolment ratios of well over 80% for males. Further progress has been held back by low enrolments of females, an imbalance that future education programmes must redress. In Latin America and the Caribbean, the net primary school enrolment ratio reached 75% in 1985, with equal participation by boys and girls.

The experience of developing countries with secondary and tertiary education has been varied. In East and Southeast Asia, secondary enrolment ratios in the newly industrialising countries rose to 90% for both females and males. Tertiary enrolments also increased considerably. Some Latin American countries surpass even the Asian newly industrialising countries — and even some of the old industrial countries — in tertiary enrolment. By contrast, tertiary enrolment in the least developed countries is 1% for females and 4% for males, showing how much they have to catch up during the next few decades.

The global distribution of basic education has changed radically since 1960. The South now has more than four times as many students in primary education as the North (480 million compared with 105 million) and about twice as many secondary-level students (190 million compared with 87 million). But the South still has to catch up in tertiary education — and in science and technology. It also has to improve the quality and relevance of students' knowledge, for which part of the groundwork has been laid in the past three decades.

More people sharing scarce resources

Life has become more liveable for most of the world's people, with millions finding access to improved goods and services. Disappointing, however, is the equally staggering number of people suffering severe deprivation (box 2.2 on p. 27). This does not mean, however, that development has failed. It means that population growth has outpaced part of development's success.

Two billion people have joined the world's population since 1960, bringing the total to more than 5 billion people today. Most of the population growth has occurred in developing countries, where the population has doubled, and this trend is likely to continue for decades.

The developing countries' overall population growth is expected to decline from 2.3% a year between 1960 and 1988 to 2.0% a year between 1988 and 2000. But some parts of the world will not achieve even this modest slowdown in growth — Africa's population is projected to continue growth by 3.1% a year between now and 2000, and the least developed countries' population, by 2.8% a year. The developing countries' share of world population, now 77%, is projected to rise to 80% by 2000 and 84% by 2025.

For most developing countries, human development thus poses a triple challenge. They have *to expand the development opportunities for a growing number of people.* They have *to upgrade living standards*. And they often have *to achieve more with less* — to meet the first two challenges with stagnating or even declining resources.

Between 1980 and 1987 the developing countries' share in world GDP fell almost two percentage points (from 18.6% to 16.8%), while their share in world population moved up one percentage point (from 74.5% to 75.6%). The combined impact of these changes proved difficult for them to accommodate.

FIGURE 2.11
World population trend and North-South distribution

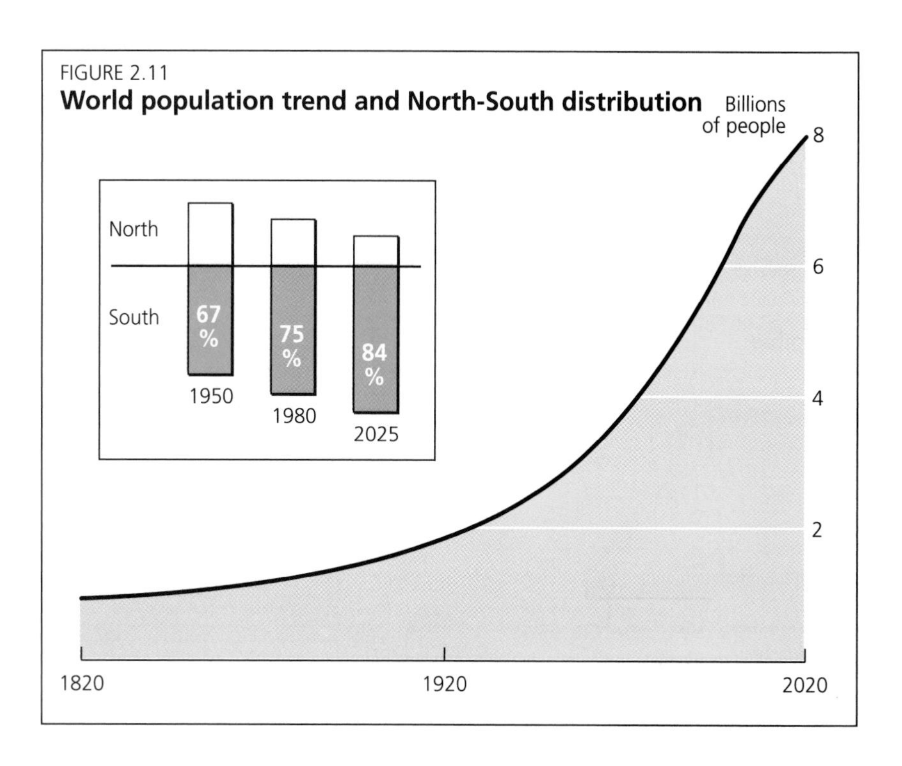

The developing countries' decline in income must be halted to avoid the growing risk of sharp reversals in human development. Early solution of the debt crisis and better opportunities for trade will be as necessary as stronger efforts by the developing countries to improve their economic performance with scarce resources.

Using human capabilities

Skilled, healthy and well-educated people are in a better position than others to take their lives into their own hands. They are generally more likely to find employment and earn better wages. They have better access to information, such as that gained through agricultural or business training, and are thus more likely to succeed as farmers or entrepreneurs. The educated can also contribute more to the advancement of culture, politics, science and technology. They are more valuable to society and better equipped to help themselves.

The use of human capabilities, as conceptualized here, encompasses the use people decide to make of their abilities as well as their usefulness to society.

FIGURE 2.12
Wage employment and labour force outside agriculture

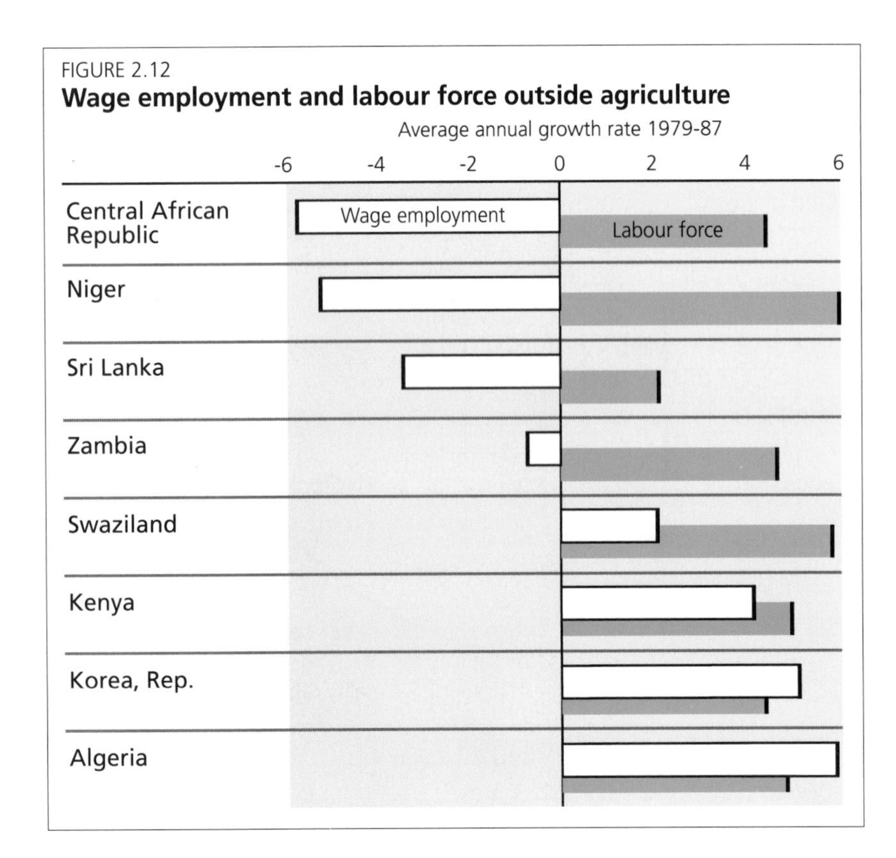

Employment

More than 900 million people have joined the developing countries' labour force in the past three decades. High population growth was not the only reason. The ranks of the labour force were increased by women seeking jobs and by poorer families trying to increase the number of income earners in the family. During the 1990s another 400 million are likely to join the labour force.

Economic growth has failed to provide enough employment opportunities for the job-seekers of the last three decades. Reliable data on open unemployment do not exist, but it is common knowledge that unemployment and underemployment are extensive in many developing countries.

The 1980s saw rapid rises in informal sector employment. In Africa the informal sector accommodated about 75% of the new entrants into the labour force between 1980 and 1985, and the formal sector only 6%. In Latin America between 1980 and 1987, the informal sector absorbed 56% of the new workers.

Governments have long ignored the informal sector, but that is beginning to change. It is increasingly being realised that the informal sector needs active political and economic support. It is, after all, absorbing the bulk of new workers, particularly women, youth and the poor.

The fuller use of human capabilities requires sustained economic growth and considerable investment in human beings. The returns from such investment are extremely high. A World Bank study showed private returns to primary education as high as 43% in Africa, 31% in Asia and 32% in Latin America. For developing countries as a whole, average social returns for every level of education exceed 10% to 15%. Differences in technical and general education accounted for about a third of the differences in agricultural productivity in the 1960s in the United States and a sample of developing countries. The special returns to female education are even higher, in terms of reduced fertility, lower population growth, reduced child mortality, reduced school dropout rates and improved family nutrition.

BOX 2.2

Balance sheet of human development

HUMAN PROGRESS	HUMAN DEPRIVATION
Life expectancy	
• Average life expectancy in the South increased by a third during 1960-87 and is now 80% of the North's average.	• Average life expectancy in the South is still 12 years shorter than that in the North.
Education	
• The South now has more than five times as many students in primary education as the North, 480 million compared with 105 million.	• There still are about 100 million children of primary-school age in the South not attending school.
• The South has 1.4 billion literate people, compared with nearly one billion in the North.	• Nearly 900 million adults in the South are illiterate.
• Literacy rates in the South increased from 43% in 1970 to 60% in 1985.	• Literacy rates are still only 41% in South Asia and 48% in Sub-Saharan Africa.
Income	
• Average per capita income in developing countries increased by nearly 3% a year between 1965 and 1980.	• More than a billion people still live in absolute poverty. • Per capita income in the 1980s declined by 2.4% a year in Sub-Saharan Africa and 0.7% a year in Latin America.
Health	
• More than 60% of the population of the developing countries has access to health services today.	• 1.5 billion people are still deprived of primary health care.
• More than 2 billion people now have access to safe, potable water.	• 1.75 billion people still have no access to a safe source of water.
Children's health	
• Child (under five) mortality rates were halved between 1960 and 1988.	• 14 million children still die each year before reaching their fifth birthday.
• The coverage of child immunisation increased sharply during the 1980s, from 30% to 70%, saving an estimated 1.5 million lives annually.	• Nearly 3 million children die each year from immunisable diseases.
Food and nutrition	
• The per capita average calorie supply increased by 20% between 1965 and 1985.	• A sixth of the people in the South still go hungry every day.
• Average calorie supplies improved from 90% of total requirements in 1965 to 107% in 1985.	• 150 million children under five (one in every three) suffer from serious malnutrition.
Sanitation	
• 1.3 billion people have access to adequate sanitary facilities.	• Nearly 3 billion people still live without adequate sanitation.
Women	
• School enrolment rates for girls have been increased more than twice as fast as those for boys.	• The female literacy rate in the developing countries is still only two-thirds that of males. • The South's maternal mortality rate is 12 times that of the North's.

In greater numbers than ever before, people are moving across boundaries and continents

Skill formation, in addition to general education, promotes more productive uses of human capabilities. Cultivators in the Republic of Korea, Malaysia and Thailand — using modern technology — produced 3% more output for every additional year of schooling they had received. And the higher level of education of farmers in the Indian Punjab explains in part why their productivity is higher than that of farmers in the Pakistani Punjab. Investment in human capital thus increases people's productivity and enhances the chances of their employment — by raising the *potential* for future economic growth. Of course, if education does not create the skills demanded by society, it can lead to educated unemployment and considerable waste of human potential.

Migration

In greater numbers than ever before, people are moving across boundaries and continents in search of new opportunities — both economic and political. Expanded transport systems and communications networks have encouraged more and more people to leave their countries and settle elsewhere. They are more aware than previously of their deprivation — more aware of how their lives differ from those of people in other countries. And this drives them to search for the seemingly better life and greater opportunities across the border. If they had seen better opportunities at home, they might have preferred to stay. For many migrants the economic decision to leave is voluntary. For political and environmental refugees, however, there seldom is a choice.

Often well qualified, some migrants are highly trained specialists. They often leave for higher salaries and more job satisfaction. Some governments even see advantages in people leaving. Their remittances can be an important source of foreign exchange, helping to improve the balance of payments.

The brain drain hit Africa particularly hard in the 1980s. With a thin layer of qualified personnel to start with, the loss of even a few key specialists has had dramatic consequences. The brain drain from the more populous countries of Asia and from most Latin American countries is generally less dramatic.

In the early 1980s the number of economic migrants stood at around 20 million — and that of illegal migrants, generally less qualified than the officially registered ones, must be at least as high. So, perhaps 40 to 50 million people have moved in hope of a bigger share of the world's development benefits.

The traditional recipients of migrants from developing countries — Canada, Australia, New Zealand, the United States and the European countries — have adopted measures to limit the influx of migrants. The United States granted some 3 million people permanent immigration status in the first half of the 1980s, compared with 2.5 million in the five preceding years. Europe's foreign population has, for about two decades now, been around 10 million. And even in the Middle East region, immigration is stabilising.

South-South migration is growing because of the increasing restrictions on migration to the North and the increasing poverty in developing countries. The main recipients in Africa have been Côte d'Ivoire, Senegal, Ghana and Cameroon. The main countries of origin include such least developed countries as Burkina Faso, Mali, Guinea and Togo. Lesotho and Mozambique continue to be major suppliers of labour to South Africa.

Argentina, Venezuela and Brazil are about the only major recipient countries for economic migrants in Latin America, with the United States continuing to be the main destination by far. The main countries that export labour in the region are Mexico and Colombia.

In Asia the main releasing countries are Bangladesh, India, Pakistan, the Philippines, Thailand and the Republic of Korea — largely to the Arab states and the United States.

Popular participation and the NGO movement

Economic migration is one way for people to seek greater involvement in development. Popular participation in community affairs — economic, social and political — is another way, and in recent years it has gained in importance. Many community and other self-help organisations now assist people in exploiting their collective strength to resolve some of the challenges they face — their need for a road, a health centre, or an irrigation system, for education for their children, or for access to assets and credit.

Added to these community self-help organisations are a large and still growing number of nongovernmental organisations (NGOs) that typically work as intermediaries between people and governments.

Underpinning the NGO movement's growth are private initiatives by concerned citizens and the sponsorship of government. The NGOs' success in shifting the focus of development to people has in many countries moved them into a fully collaborative relationship with the state. Governments are beginning to realise that NGOs — small, flexible and with good local roots and contacts — often are much better suited to carry out the work of development than is a large bureaucratic machine.

One of the NGOs' big successes is in arranging credit for the poor. The poor traditionally stay poor because they have no assets and are seen as unworthy of even the smallest amount of credit. NGOs have changed this by showing that a joint-liability approach — with close contact and communication between debtor and creditor — can help boost repayment rates and open more credit opportunities for the poor within the official credit system. The NGOs have closely supervised, and provided advice to, borrowers — taking on the often very time-consuming functions that banks typically shy away from.

In Peru the Institute for the Development of the Informal Sector has established programmes to help small entrepreneurs and community groups gain access to credit. It provides bank guarantees for participants and arranges the technical and managerial advice and training they need to set up viable businesses.

Another NGO in Bangladesh, the Grameen Bank, provides innovative links between the government, commercial banks and outside donors on the one side, and landless entrepreneurs interested in borrowing but lacking collateral on the other. The Grameen Bank helps the landless organise into groups to secure loans, and most of its clients are women.

Other NGOs mobilising rural savings and making credit available to the rural poor include Rwanda's Banques-populaires, Zimbabwe's Savings Development Foundation, Ghana's Rural Banks and the Philippines' Money Shops (see boxes 4.2 and 4.3 in chapter 4).

The momentum of people's participation during the second half of the 1980s has done far more than prove that people can help themselves. It has contributed to a fundamental rethinking of the relationship between the state and the private sector. Policymakers now recognise that development can benefit from people's initiatives, and that these initiatives must be encouraged rather than stifled.

There is a growing consensus that the state must be strong and effective in creating an enabling framework for people to make their full contribution to development — to expand their capabilities and to put them to use — but that it should not undertake developmental functions that NGOs, entrepreneurs and people at large can carry out better.

Disparities and deprivation within nations

Every country has shared to a varying degree in the human progress over the past 30 years. But average improvements conceal considerable inequality within countries and mask the continuing severe deprivation of many people. The prevailing disparities also show the great potential for improving human development by distributing income better and by aggressively restructuring budget priorities.

This section focusses on the disparities between rural and urban areas, between males and females, and between rich and poor. Again, the lack of appropriate data hinders a systematic review. Use is thus made of special case studies to supplement the available cross-country data.

Average improvements conceal considerable inequality between rural and urban areas, between males and females, and between rich and poor

Rural-urban disparities

Two-thirds of the people in the developing countries live in rural areas, but in many countries, they receive less than a quarter of the social services for education, health, water and sanitation. For developing countries as a whole, people in urban areas have twice the access to health services and safe water and four times the sanitation services that people have in rural areas.

In many countries, rural-urban disparities reflect the distribution of income and the locus of power. These disparities, often high at lower levels of human development and per capita income, tend to narrow over time. But there are several exceptions to such a generalisation. Argentina has very high rural-urban disparities, despite its relatively high per capita income and human development. Tanzania, by contrast, has a fairly good geographical distribution of social services, even with its low income.

The following examples show how rural areas systematically lag behind urban areas in human development.

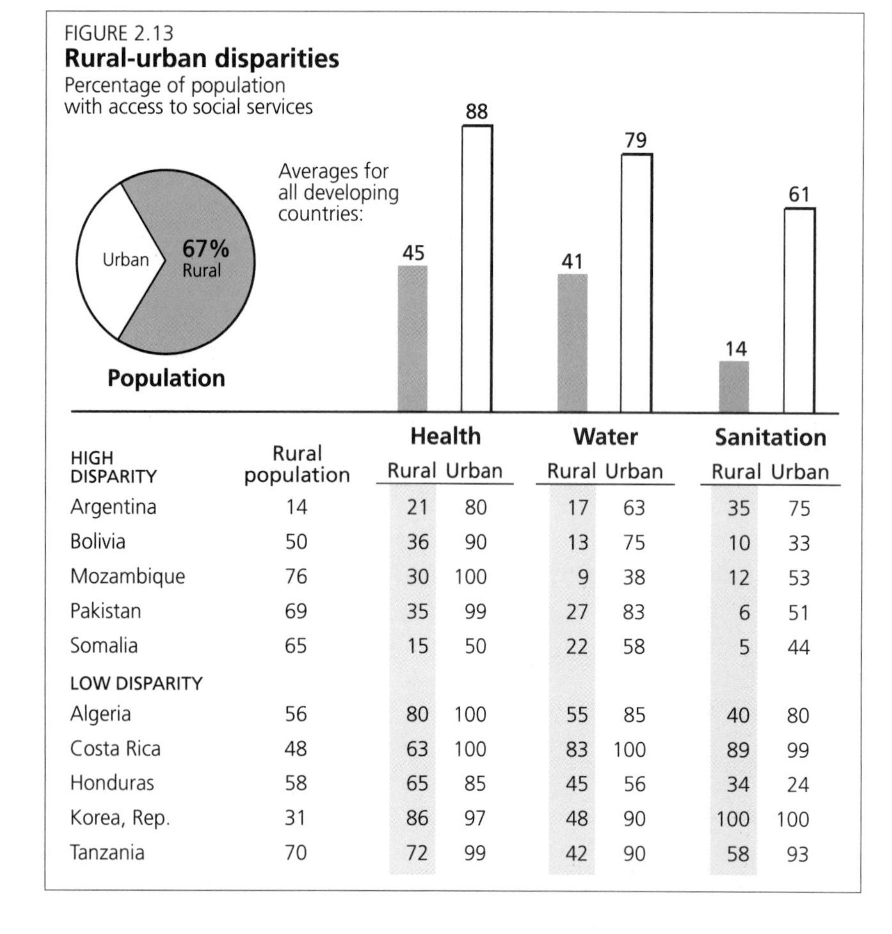

FIGURE 2.13
Rural-urban disparities
Percentage of population with access to social services

HIGH DISPARITY	Rural population	Health Rural	Health Urban	Water Rural	Water Urban	Sanitation Rural	Sanitation Urban
Argentina	14	21	80	17	63	35	75
Bolivia	50	36	90	13	75	10	33
Mozambique	76	30	100	9	38	12	53
Pakistan	69	35	99	27	83	6	51
Somalia	65	15	50	22	58	5	44
LOW DISPARITY							
Algeria	56	80	100	55	85	40	80
Costa Rica	48	63	100	83	100	89	99
Honduras	58	65	85	45	56	34	24
Korea, Rep.	31	86	97	48	90	100	100
Tanzania	70	72	99	42	90	58	93

• *Infant mortality*. For several Central American countries, infant mortality is generally 30% to 50% higher in rural than in urban areas. Costa Rica, Guatemala and Nicaragua narrowed some of the gap in the 1970s, but other countries have not been able to match urban progress in rural areas.

• *Life expectancy*. Rural Mexicans have a shorter life expectancy (59 years) than their urban counterparts (73 years).

• *Nutrition*. Data on the nutritional status of children in 31 countries show, without exception, higher rates of malnutrition in rural areas, 50% higher on average.

• *Literacy*. For selected countries in Africa and Asia, rural illiteracy rates generally are twice the urban rates — and for women in Latin America the rural rates are three times higher than the urban rates, and for men, four times higher.

• *Health facilities*. Access to health care is better in urban areas than in rural areas in every developing country. In some 20 developing countries the percentage of the population covered by health facilities in urban areas is more than twice that covered in rural areas. Even these figures understate the disparities since rural health facilities usually are simple clinics while urban facilities include hospitals with sophisticated equipment.

• *Water and sanitation facilities*. The rural-urban differences in the provision of water and sanitation are even greater. The coverage of the rural population is on the average less than half that of the urban population. In seven countries the proportion of rural dwellers with access to water was less than a fifth of that in urban areas. In Nepal access to sanitation facilities in urban areas was 17 times that in rural areas, and in Brazil the urban figure was as much as 86 times higher than the rural figure.

• *Income*. In most countries, urban incomes per person run 50% to 100% higher than rural incomes. The differences are particularly large in Africa. In Nigeria the average urban family income in 1978-79 was 4.6 times the rural. In Sierra Leone the average urban income was 4.1 times the agricultural income. And in Mexico urban per capita income was 2.6 times the rural. Rural-urban income differences remain wide, even after taking into account the differences in the cost of living between rural and urban areas.

To sum up, national data conceal large rural-urban differences, with rural areas performing systematically worse on the basic indicators of human development. Part of the reason is less access to social services, and part is lower income. Moreover, the rural and urban figures hide large dispar- ities within each area. These gaping disparities have major policy implications for restructuring the social spending of governments.

Female-male disparities

In most societies, women fare less well than men. As children they have less access to education and sometimes to food and health care. As adults they receive less education and training, work longer hours for lower incomes and have few property rights or none.

Both women and men shared the progress in improving the human condition from 1960 to 1980. In some fields women did even better than men, but substantial inequality remains. During the economic crisis of the 1980s, women had to bear a much greater cost of structural adjustment, and gender disparities tended to widen once again. Moreover, national data usually conceal the true extent of inequality between women and men (box 2.3).

Discrimination against females starts early. In many developing countries more girls than boys die between the ages of one and four, a stark contrast with the industrial countries, where deaths of boys are more than 20% higher than those of girls. And in 30 developing countries the death rates for girls were higher than or equal to death rates for boys, indicating the sociocultural patterns that discriminate against women.

The discrimination takes several forms. Young girls may not get the same health care and nutrition as young boys. In Bangladesh malnutrition was found among 14% of the young girls, compared with 5% of the boys. Families in India's rural Punjab spend more than twice as much on the medical care of male infants as on that of female infants.

The same neglect is evident in exceedingly high maternal mortality rates, mainly because health staff are in attendance for fewer than half the births. Maternal mortality rates were 1,000 or more (per 100,000 live births) in a few countries, and 400 to 1,000 in another 14 countries during 1980-84. In developed countries, maternal mortality rates rarely exceed 20 and are usually less than 10. No other North-South gap in human development is wider than that between maternal mortality rates, a symbol of the neglect of women's health in the Third World.

Gender inequality is reinforced in education. There still are 16 developing countries where female primary school enrolment is less than two-thirds that of males. And 17 developing countries have female secondary enrolments less than half those of males. For the developing world as a whole, the female literacy rate is now three-quarters that of the male. The gap has narrowed slightly in the last three decades, but much progress remains to be made.

The social dividend from female literacy tends to be very high. Higher female literacy is associated with lower infant mortality, better family nutrition, reduced fertility and

FIGURE 2.14
Female literacy and population growth

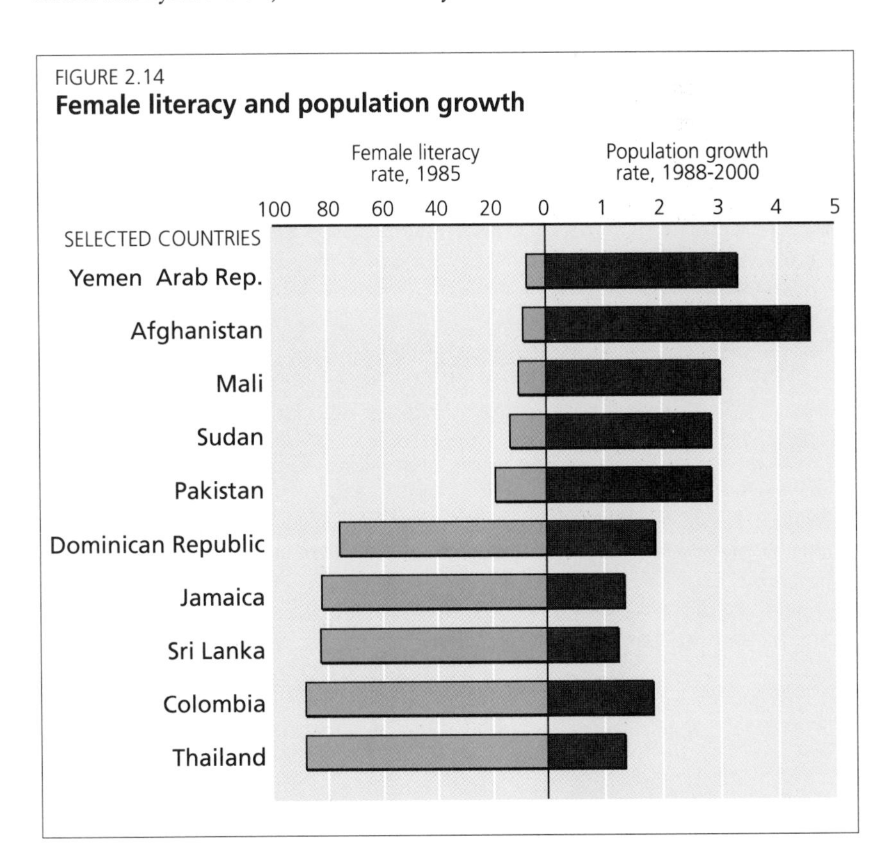

FIGURE 2.15
Female-male literacy disparities
Percentage of urban male

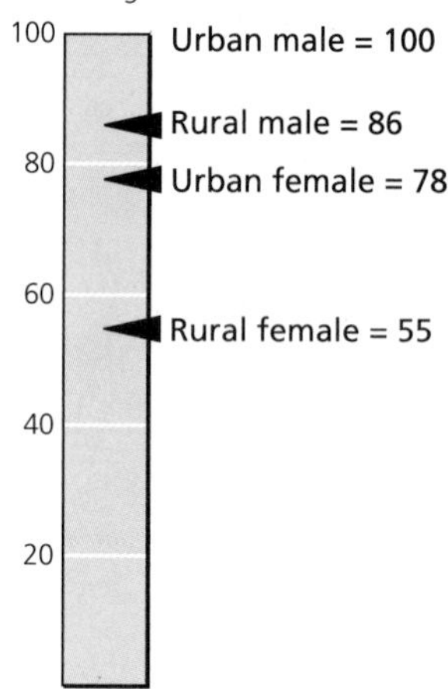

lower population growth rates. In Bangladesh child mortality was five times higher for children of mothers with no education than for those with seven or more years of schooling.

Better educated women also have smaller families. Colombian women with the highest education had four fewer children than women who had completed only their primary education. The continuing disparity in male and female education thus inflicts extremely high social and economic costs in the developing world.

Women typically work about 25% longer hours than men: up to 15 hours more a week in rural India and 12 hours more in rural Nepal. But their total remuneration is less because of their lower wage rate and their preponderance in agriculture and the urban informal sector, where pay tends to be less than in the rest of the economy. In urban Tanzania 50% of the women working are in the informal sector, in urban Indonesia 33% and in Peru 33%.

The persistence of female-male gaps in human development offers a challenge and an opportunity to the developing countries — to accelerate their economic and social progress in the 1990s by investing more in women.

In order to monitor progress towards the elimination of existing within-country disparities in human development, it would be desirable to have group and region-specific HDIs. How telling such indices could be is illustrated in technical note 4, which discusses the construction of a gender-specific HDI. Similar indices could be developed to monitor other disparities of special interest in a particular country, whether it be those between various ethnic groups, different geographical areas, rural and urban or rich and poor.

BOX 2.3

Women count — but are not counted

Much of the work that women do is "invisible" in national accounting and censuses, despite its obvious productive and social worth. The reason is that women are heavily involved in small-scale agriculture, the informal sector and household activities — areas where data are notoriously deficient.

But there is another aspect. Women's work, especially their household work, often is unpaid and therefore unaccounted for — processing food, carrying water, collecting fuel, growing subsistence crops and providing child care. For example, women in Nepalese villages contribute 22% to household money incomes, but when nonmarketed subsistence production is included, their contribution rises to 53%. It is estimated that unpaid household work by women, if properly evaluated, would add a third to global production.

Even when women are remunerated for their work, their contribution is often undervalued. In formal employment, women earn significantly less than men in every country having data. In the informal sector, where most women work, their earnings at times reach only a third (Malaysia) to a half (Latin America) of those of men.

Do women remain invisible in statistics because little value is attached to what they do? Apparently, yes.

Women have shouldered a large part of the adjustment burden of developing countries in the 1980s. To make up for lost family income, they have increased production for home consumption, worked longer hours, slept less and often eaten less — substantial costs of structural adjustment that have gone largely unrecorded.

The low value attached to women's work requires a fundamental remedy: if women's work were more fully accounted for, it would become clear how much women count in development. To do that requires much better gender-specific data on development. There is a need to redesign national censuses, particularly agricultural surveys.

Disparities between rich and poor

Income gaps and human development gaps are closely related in most developing countries, which is only natural since income is an important determinant of people's access to social services. In some cases, however, governments have changed this pattern through very active interventions with their social sector budgets. They did this by targetting their social spending and subsidies specifically on the poorer sections of society — and by reducing the appropriation of subsidies by higher income groups and vested power structures.

Two questions are of particular interest here. First, how do different income groups differ in terms of their human development? Second, who benefits from government social expenditures, which in many countries are said to aim at correcting the socioeconomic inequities resulting from the inequalities in the primary distribution of income.

Several studies show that the poor have very unequal access to social services and basic human development. For instance:

- In Brazil life expectancy in 1970 was barely 50 years for the bottom income group and 62 years for those with an income above $400. To put this in perspective, the expected life span of the poor in Brazil was no higher than the average in India, even though Brazil's average per capita income was about eight times that of India.
- In Mexico a person's life expectancy in

the lowest income decile was 53 years in the early 1980s, 20 years less than the average life expectancy in the top income decile.

- In Colombia infants in poor families are twice as likely to die as infants in the top income decile.
- In rural Punjab child mortality among the landless is 36% higher than among the land-owning classes.
- In a South Indian village the literacy rate in 1989 was 90% for Brahmins and 10% for people at the lower end of the caste hierarchy.
- In Zimbabwe child malnutrition was severe when the average family income was $51, mild at $168, and nonexistent at $230 and above.

Such evidence emphasises the need for careful monitoring of the beneficiaries of government spending to ensure that it reduces rather than perpetuates inequalities.

If the state provides the goods and services essential for human development free or at low cost — as in Sri Lanka in the 1960s and 1970s — it can reduce the handicaps the poor face. But the free or subsidised services may not reach many of the poor. That can happen — as it did in Egypt — where only urban food is subsidised or urban health services are provided. Information about social services may also be more accessible to the wealthier or better educated, who then manage to preempt the major benefit from such services.

Moreover, even free services have a cost. To gain access to health services or to attend school, people have to pay transport costs, and the time taken to use the services has an opportunity cost. That is why very poor families often keep their children out of school, especially at harvest time when farm labour is needed most.

Not enough research has been done on the distribution of social benefits by income group in developing countries, but scattered evidence shows that much social spending often goes for projects and programmes that subsidise the rich more than the poor.

- Hospital spending in Latin America, primarily benefitting the urban nonpoor, ranged from 64% of total central government expenditure on health in Guyana to 100% in El Salvador.
- In the Philippines in the early 1980s, annual subsidies to private hospitals catering to upper-income families exceeded the resources allocated to mass programmes (including malaria eradication and schistosomiasis) and to primary health care.
- In developing countries as a whole, tertiary education covered about 8% of the population but absorbed 73% of the education budget in 1973. The cost per student in tertiary education was 24 times that in primary education.

A major conclusion from all this evidence is that not all government spending works in the interest of the poor and that great care must be taken in structuring social spending to ensure that benefits also flow to them. The very rationale for government intervention crumbles if social expenditures, far from improving the existing income distribution, aggravate it further — an issue that is taken up at length in chapters 3 and 4.

Looking at all three types of deprivation, another major conclusion is that *poor rural women* in developing countries suffer the gravest deprivation. Many of them are still illiterate. Their real incomes have not increased and in some parts of the world have even fallen. Their births are still unattended by health personnel, and they face a high risk of death during childbirth. They and their children have almost no access to health care.

There are between 500 million and one billion poor rural women. For them, there has been little progress over the past 30 years.

For more than half a billion poor rural women, there has been little progress over the past 30 years

Reversibility of human development

Human progress during the 1960s and 1970s differed greatly from that in the 1980s.

In the late 1970s and early 1980s, very large imbalances had developed in the current account of the balance of payments in many developing countries. The non-oil developing countries had a combined deficit of $74 billion in 1980. Unlike the situation in much of the 1970s, there was no voluntary bank lending to finance the deficits. Voluntary lending dried up because the crisis was so widespread, affecting more than two-thirds of the countries of Latin America and Sub-Saharan Africa as well as several Asian countries.

The economies of most developing countries slowed down in the 1980s, except in Asia. Acutely affected by the crisis, they experienced a nearly continuous economic decline, and despite rigorous adjustment efforts, they were still showing severe imbalances at the end of the 1980s.

It is short-sighted to balance budgets by unbalancing the lives of the people

In 17 Latin American and Caribbean countries, per capita income fell in the 1980s. Average income per person in the region declined 7% between 1980 and 1988, and about 16% if account is taken of the deteriorating terms of trade and the resource outflow. Net investment per capita fell 50% between 1980 and 1985.

In Africa income per person declined more than 25% for the region as a whole, 30% taking into account the deterioration in the terms of trade. GDP did grow faster in 1985-87 than during 1980-84, but that growth was still slower than the growth in population, and incomes per person fell at roughly the same rate in countries with strong reform programmes as in countries with weak or no reform programmes. Investment fell more than 9% a year, and per capita consumption 1% to 2% a year.

Much of Asia, by contrast, was not very seriously affected. Between 1980 and 1986 GDP per capita rose 20% in South Asia and 50% in Southeast and East Asia, though some countries were badly hit, including the Philippines.

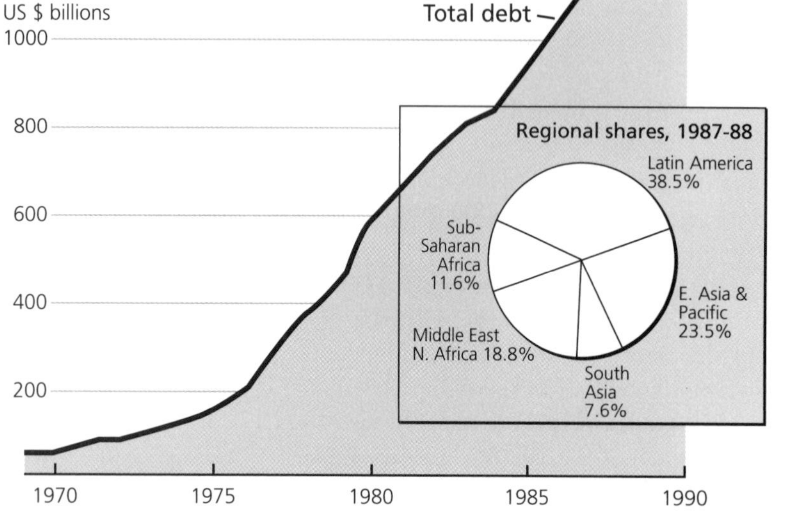

Evidence of the effect of these economic changes on social conditions is piecemeal because social data usually are not collected regularly at short intervals, or reported on systematically. Moreover, some social data — such as life expectancy — are generated by extrapolating past trends, until new empirical data, such as that from a population census, establishes a new trend. Few official statistics have thus begun to capture the effects of the 1980s' economic crisis on human development.

Judging from the piecemeal data that exist, many developing countries have had sharp breaks in their human development trends, and sometimes even reversals. Countries in Africa and Latin America suffered the most adversity.

In seven Latin American countries and six African countries, child malnutrition rose at some time in the 1980s. In two-thirds of the Latin American countries for which data are available, the progress in reducing infant mortality rates slowed or reversed — as it did in 12 of 17 African countries. Many households lost purchasing power and were left with incomes grossly inadequate to meet minimum food needs.

- In Ghana in 1984 even upper-level civil servants could only afford two-thirds of the least-cost diet to meet nutritional needs. A two-wage-earner household receiving the minimum wage could afford less than 10% of such a diet.
- In Uganda in 1984 an average-size urban family needed 4.5 times the minimum wage to meet its minimum food requirements.
- In Dar es Salaam in the mid-1980s, 58% of the women surveyed in low-income households reported that they had been forced to cut down from three meals a day to two, and 61% had reduced their consumption of protein-rich foods.
- In Jamaica in 1986 a family of four needed two to three times the minimum wage to have access to the minimum acceptable nutrition.

In many instances, high inflation, rising food prices, stagnant formal employment

FIGURE 2.16
Debt of developing countries

and curtailed government subsidies converged to push household incomes down. Latin America is estimated to have had 4 million fewer new jobs in 1980-85 than it would have had under previous trends — and its unemployment grew more than 6% a year. Africa had an annual increase in unemployment of 10% during the same period.

According to ILO estimates, wage earners have borne the brunt of the economic crisis, with real wages having been cut back severely. In Africa and Latin America, wage cuts of a third to a half were not exceptional. Between 1980 and the mid-1980s, real wages fell 50% in Peru and Bolivia, 30% in Mexico and Guatemala, and 25% in Venezuela. The share of labour income in the region's GNP declined by 25% between 1980 and 1987. In Africa, too, real wages fell more rapidly than income per person during the first half of the 1980s.

Rapid rises in food prices compounded the damage from falling real incomes. In many countries food prices rose faster than other prices because of reduced food subsidies, higher producer prices, decontrolled consumer prices and devalued currencies. Food subsidies fell from 1980 to 1985 in each of 10 countries examined in detail. Food price rises exceeded the general cost of living in five of six UNICEF case studies. And more than half the countries receiving World Bank structural adjustment loans had the availability of food per capita decline as a percentage of requirements from 1980 to 1987.

The declines in government spending on social services generally impaired human development in the 1980s. Social spending was not cut disproportionately more than total spending, but real government spending per person declined in around two-thirds of the countries of Africa and Latin America, in some cases considerably. Madagascar's real social spending per person fell 44% (during 1980-84), Senegal's 48% (1980-85) and Somalia's 62% (1980-86). In Zambia the real value of the drug budget in 1986 was a quarter of that in 1983, and only 10% of the budget was spent because of shortages of foreign exchange. In Bolivia central government health expenditure per person in 1984 was less than 30% of that in 1980.

The worsening of social conditions was far from uniform. Some countries protected the most vulnerable groups from the downward pressures. Zimbabwe, Botswana, Costa Rica, Chile and the Republic of Korea managed to adjust and protect the human condition, but these are countries that have done consistently well in human development (box 2.4). Moreover, many of them have well-established capacities for planning and managing national development.

The countries that protected vulnerable groups during the adjustments of the 1980s did so in a variety of ways.

- Some countries avoided excessively deflationary macroeconomic policies and thus managed to maintain incomes and employment. The Republic of Korea and

BOX 2.4

Adjustment with a human face in Zimbabwe

When Zimbabwe became independent in 1980, it launched a series of programmes in health, education and the productive sector to correct some of the large inherited racial inequalities and to improve the position of the poor. But imbalances developed, in part because of external shocks, and the government introduced a series of adjustment measures.

Some of the measures were orthodox: restraining the growth of credit, keeping wage increases below the rate of inflation, reducing subsidies, devaluing the currency and raising interest rates. Others were less orthodox: restraining dividend remittances, continuing import controls and adopting a more expansionary general stance than most policy packages approved by the IMF.

For much of the 1980s Zimbabwe failed to reach agreement with the IMF and adjusted on its own. The period of adjustment also coincided with an acute drought.

The government introduced measures to protect the most vulnerable population segments during the adjustment period.

- Credit and marketing reforms shifted resources to low-income farmers, whose share of credit from the Agricultural Finance Corporation rose from 17% in 1983 to 35% in 1986 and whose share of marketed maize and cotton rose from 10% to 38%.
- Expenditures on basic health and primary education increased rapidly. While the share of defence and administration in total government expenditure fell from 44% in 1980 to 28% in 1984, the share of education and health rose from 22% to 27%. Within the education budget, the share of primary education rose from 38% to 58% over the same period, involving a doubling of real per capita expenditure on primary education. A growing proportion of the rising health budget was devoted to preventive health care.
- Special feeding programmes were introduced, with a drought relief programme and a supplementary feeding programme for undernourished children. More than a quarter million children received food supplements at the peak of the drought.

Because of these efforts, the economic costs of adjustment did not become human costs. The infant mortality rate continued to decline, primary school enrolment rose at a rapid rate, and malnutrition did not rise despite the drought.

Without some end to the debt crisis, the impressive human achievements recorded so far may soon be lost

Zimbabwe adopted less deflationary adjustment policies than were typical.

- Some launched special employment schemes to maintain the incomes of low-income households. Chile undertook massive public works programmes, which were at one time employing as much as 13% of the work force. Zimbabwe diverted substantial amounts of credit to smallholder farmers.
- Some directed special nutrition support to the neediest. In Botswana and Chile, infants and children were carefully monitored, and food and other support was supplied as necessary.
- Some protected real expenditures on priority services in the social sector. Zimbabwe greatly increased spending on primary education and primary health care, cutting back on defence. Many countries supported low-cost and high-priority measures despite overall cutbacks in expenditure — and made progress in extending immunisation.

A general feature of the successful countries was their careful and systematic monitoring of human and economic variables. Good and up-to-date statistics on what was happening proved essential for appropriate and timely policy action.

Although many countries maintained their human development levels over this difficult period by redirecting resources towards priority areas — and indeed continued their progress in reducing infant and child mortality rates — it was clear that continuing economic decline would make such efforts increasingly difficult. Despite its economic problems, Jamaica maintained support for human development throughout the 1970s, but stabilisation programmes in the 1980s severely cut social spending, and there has been evidence of halts and even reversals in some human indicators.

Resumed economic growth is thus essential to allow the expansion of incomes, employment and government spending needed for human development in the long run. Without some end to the continuing debt and foreign exchange crisis in much of Africa and Latin America, the impressive human achievements recorded so far may soon be lost.

De-formation of human development

Human development is fragile. Economic slowdowns and their consequences — falling income, flagging employment, plunging wages and deep cuts in social spending — can quickly reverse progress.

This fragility is not limited to developing countries or to economic recessions. In the United States the number of homeless people has risen tremendously in the past years. And in the United Kingdom the distribution of income — whether original, disposable or final income — worsened during the 1980s, leading to a deepening of poverty.

Losses of human development gains may stem from the development path a country pursues, for development is far from unidirectional. Technological advancement has given a tremendous impetus to production and eased human life in innumerable ways. But it has also brought industrial pollution. The growing density of the transport network enhanced people's geographical mobility and access to developmental opportunities. It has also entailed environmental degradation.

The point is that development has desirable and undesirable effects. And people must be able to make informed choices about the weight they assign to the pros and cons. Is tobacco-smoking worth the risk of lung cancer? Is high speed on highways worth the deaths and disabilities it costs each year? What are chemical fertilizer's tradeoffs between increased agricultural production and polluted water resources? Such questions have no easy answers.

Many countries have seen more and more lives destroyed by rising crime, drug abuse, environmental pollution, family breakup and political turmoil. And now there is a new major threat to human life — the acquired immune deficiency syndrome (AIDS).

Development and crime

The relationship between crime and development is complex. Rapid socioeconomic change — often entailing dramatic consequences for people's lifestyles and the crumbling of traditional norms and values, but also sharp economic and social inequities—may lead to an increase in crime. Criminal activity, in turn, can further worsen the societal imbalances by destroying human lives and encouraging drug use. Perhaps worse, it makes people feel vulnerable and insecure, depriving them of dignity and optimism.

Property crime increases with higher levels of development. The relationship for other types of crime is less conclusive, but it is known that the developing countries' reported homicide and assault rates exceed those of developed countries, while the reverse is true for thefts and frauds.

For nine Western European countries the frequency of street crimes more than doubled between 1960 and 1980. From 1975 to 1980 the greatest increase was in drug crimes, which increased more than 10-fold globally, with increases for individual countries between 5% and 400% a year.

Crime apparently pays. Criminals are becoming more technically experienced and better organised, often with vast international operations and connections. The proceeds from organised criminal activity amount to billions of dollars, outstripping the GNP of many countries. But crime also imposes costs, for the frequent response to growing crime has been a large increase in police forces, in both developed and developing countries, syphoning off resources that could otherwise be available for development purposes.

The drug trade

The use of illicit drugs threatens the health and well-being of many millions of people in both developed and developing countries. Possibly even greater harm comes from production and marketing. The enormous illegal profits in producing and using countries criminalises society, corrupts law enforcers and brings political violence to countries and military conflicts between them.

More than 2 million people are directly employed in drug production and trade, which contributes much to the economies of drug-producing nations. Returns per hectare from growing narcotic crops in Latin America are 10 to 20 times those from legal crops. Yet the producers receive only a fraction of the street price, which is often as much as 120 times the production cost.

Drug abuse and trafficking defy measurement, but they are known to be increasing sharply. The cocaine seized between 1980 and 1985 increased more than fourfold, and the heroin sevenfold. WHO estimates that 48 million people worldwide regularly used illicit drugs in 1987—among them 30 million cannabis users, 1.6 million coca-leaf chewers, 1.7 million opium addicts and 0.7 million heroin addicts. The value of trade in illegal drugs exceeds that of world trade in oil and is surpassed only by the trade in arms.

The value of trade in illegal drugs exceeds that of trade in oil and is surpassed only by the trade in arms

Drug users are a third less productive than nonusers, three times more likely to be involved in accidents on the job and twice as often absent from work. Drug abuse during pregnancy means more miscarriages and infant deaths — and lower birth weights and mental achievements for the children that survive, with the babies of drug abusers often born as addicts. Intravenous drug takers also risk and promote the spread of AIDS.

Drug abuse imposes growing costs on drug users and their families, on governments for prevention, rehabilitation, medical and enforcement programmes, and on society for lost output and heightened violence. The United States alone spent $2.5 billion in 1988 for law enforcement against drug production and trafficking. Falling drug prices suggest, however, that these efforts are far from effective.

Attempts to control drugs have failed because the incentives for producers and traffickers and the demand pressures from consumers are far too strong. So, the battle continues to be lost at very heavy costs.

TABLE 2.5
Drugs seized worldwide, 1980 and 1985
(tons)

Drug	1980	1985
Cannabis herb	5,806	6,434
Cannabis resin	172	360
Cannabis liquid	1	1
Cocaine	12	56
Heroin	2	14
Opium	52	41

Environmental degradation

People should be able to live in a safe environment with clean water, food and air — and without undue health hazards from industrial wastes and other environmental degradation. The environmental hazards, already large, have been increasing over the past decades. These include health risks from the earth's warming, from the damage to the three-millimeter layer of ozone and from industrial pollution and environmental disasters.

Industrial pollution is foreclosing many opportunities for humankind

Some of the startling environmental disasters of the 1980s were:

- The leak from a pesticides factory in Bhopal, killing 2,500 people and blinding and injuring more than 200,000.
- The explosion of liquid gas tanks in Mexico City, killing 1,000 and making thousands homeless.
- The breakdown of the Chernobyl nuclear reactor, spreading radioactivity throughout Europe and greatly increasing future cancer risks.
- The warehouse fire in Switzerland that allowed chemicals, solvents and mercury to flow into the Rhine, killing millions of fish and threatening drinking water in Germany and the Netherlands.
- The 75,000 active industrial landfills in the United States, most of them unlined, allowing contaminants to leak into groundwater.

Such industrial pollution is accelerating the extinction of species and perhaps foreclosing many opportunities for humankind, especially in the medical field.

- At least 93% of the original primary forest in Madagascar has been destroyed, and about half the original species (numbering around 200,000) have been eliminated.
- In western Ecuador since 1960 almost all the forests have been destroyed to make way for banana plantations. As many as 25,000 species have been destroyed in the past 25 years.

Equally damaging, though less obvious, are the cancers, respiratory diseases, and diarrhoeal diseases from pollution.

- Only 209 of the more than 3,000 large towns and cities in India have even partial sewage facilities, and only eight have full sewage treatment. More than 100 cities dump untreated sewage, chemicals, and other wastes into the river Ganges. Three of every five people in Calcutta suffer from respiratory diseases related to air pollution.
- Deaths from lung cancer in Chinese cities are four to seven times the national average, with many of the deaths attributable to heavy air pollution from coal furnaces.
- In Malaysia the area around Kuala Lumpur has two to three times the pollution of major cities in the United States.
- In Japan air pollution reduces some wheat and rice crop production by as much as 30%.
- About 10,000 people die each year in developing countries from pesticide poisoning, while 400,000 suffer acutely.
- Diarrhoeal diseases from unsanitary facilities and dirty water kill an estimated 4 million children a year in developing countries.

Along with this deterioration, there has been some progress. Developed countries have tightened their regulations on pollution substantially. Air pollution in most developed cities has been declining. International action has been initiated on chlorofluorocarbons. Awareness of the environment's importance and the market's limitations in protecting it has greatly increased in recent years.

For the Third World, however, such progress has often been offset by pressures of population, poverty and urbanisation. The developing countries now exhibit the greatest increases in world pollutants, because few of them have the capacity to install, use and maintain environmentally benign technologies.

In absolute terms, the industrial pollution in the North is far greater than in the South. For example, 29% of the chlorofluorocarbons escaping into the atmosphere originate in the United States, 41% in Australia, Canada, New Zealand, Japan, and Western Europe, 14% in Eastern Europe and only 16% in developing countries. Acid rain is worst in central Europe, and about half the forest in Western Germany is already damaged.

According to the WHO the cost of measures to remedy environmental degradation and eliminate significant public health hazards is greater than the cost of prevention.

Refugees and displaced persons

Much human potential goes to waste because of forced migration — people compelled to abandon their homes and their assets because of political turmoil, military conflicts or ethnic strife. For such reasons, 12 to 14 million people had registered as refugees at the end of the 1980s.

The world has always known massive population movements. The Second World War and its aftermath displaced nearly 15 million people, including Germans, Hungarians, Poles, Czechs and Russians. The partitioning of the Indian subcontinent in 1947 uprooted more than 14 million people. More than a million Palestinians are displaced. Periodic wars and crises in various parts of Africa have also displaced millions — for example, Nigeria expelled a million people in 1983.

The refugee problem grew tremendously during the 1980s. Some 14 million people were displaced in 1988, compared with an estimated 8 million at the start of the decade. The Afghan war displaced about 5 million people, a third of the Afghan population. About 300,000 Somalis have fled to Ethiopia. Other growing groups of refugees include ethnic Turks from Bulgaria and Vietnamese boat people. And in Central America, 160,000 Salvadorans are scattered throughout South America (120,000 in Mexico), many Guatemalans are in Mexico and the United States, and many Nicaraguans are in Honduras, Mexico and the United States.

In addition to international refugees are the millions of people displaced in their own countries: 10 million in Africa in 1988, including 2.7 million in Uganda, 2 million in the Sudan and 1.1 million in Mozambique.

Added to these numbers are the numbers of environmental refugees, who rank today with political refugees: 12 to 14 million people who have abandoned their homes because of natural resource degradation and its sequels — drought, flooding, soil erosion, lost productivity, failed harvests and threats of hunger and death.

Changing household patterns

The traditional extended family has in many parts of the world been replaced by the nuclear family — typically two parents and their children. Accompanying the decline of the extended family has been the breakdown of the social security net and the support it provided to its members. In many countries, especially developing countries, the replacement systems — nurseries, health and unemployment insurance, and other social services — have not yet emerged. The uneasy transition has often been marked by considerable hardship, especially for the children, the elderly and the disabled.

Now, the nuclear family is itself breaking up in many countries and being replaced by single-person households and single-parent households. In the United Kingdom nuclear families of two parents with children accounted for only a quarter of the households in 1988. In the United States couples with children dropped from 44% of all households in 1960 to 29% in 1980, while one-person households rose from 13% to 23% and single-parent households from 4% to 8%. If the trend continues, only three of five young American families will be headed by a married couple in 2000. Trends are similar in other developed countries. The incidence of divorce has also been high in the North, and it appears to be on the rise in the South.

Poor women, in both North and South, are hurt most by these trends. Because

FIGURE 2.17
Refugees by region
Millions of people

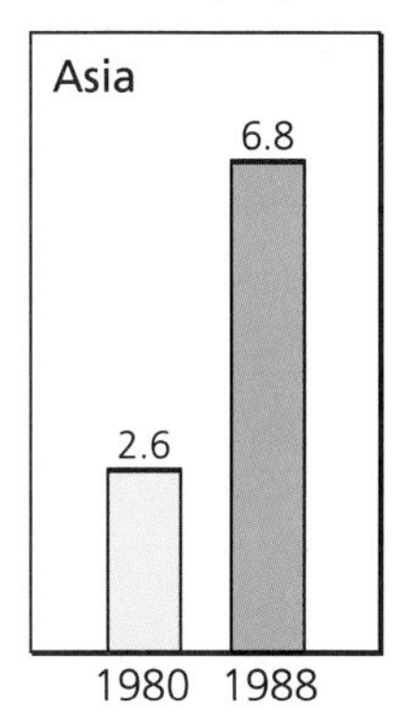

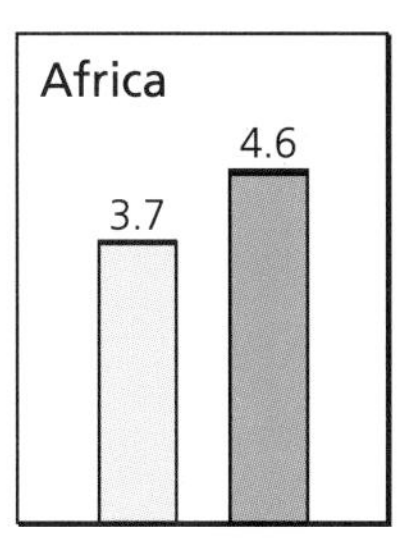

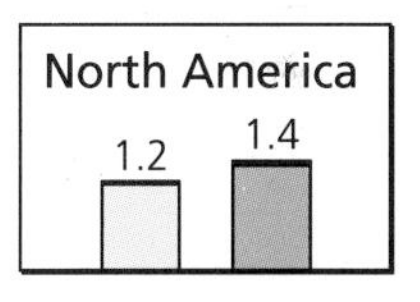

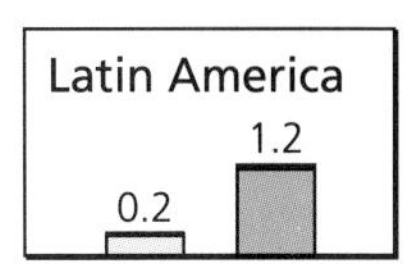

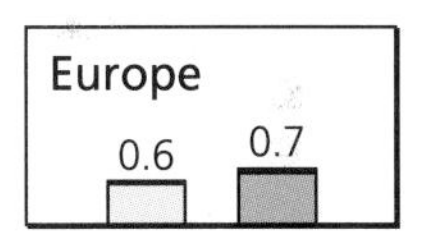

TABLE 2.6
Changes in the size of households in selected developed countries
(percentage of total households)

	One person			Couple with children			Single parent with children		
Country	1960	1970	1980	1960	1970	1980	1960	1970	1980
Canada	9	13	20	..	50	37	..	2	3
England and Wales	12	18	22	49	44	39	7	7	8
France	20	22	24	45	41	39	4	5	5
Germany, Federal Rep.	21	26	31	55	47	42	2	2	3
Netherlands	12	17	22	56	53	43	6	7	6
Sweden	20	25	33	37	30	25	3	3	4
Switzerland	15	20	27	48	45	41	5	5	4
United States	13	17	23	44	40	29	4	5	8

women are typically less qualified than men, they tend to go into lower paying jobs and have fewer opportunities to be upwardly mobile, leaving them less able than men to provide a decent living to their families. The increasing number of female-headed households has led to a feminisation of poverty.

Tropical diseases and the AIDS epidemic

A vast number of people in the developing world suffer from, or are threatened by, debilitating or fatal tropical diseases.

- Malaria is endemic in 102 countries and threatens more than half the world's population. There are 100 million malarial infections each year and about one million deaths.
- Onchocerciasis has infected nearly 18 million people and about 80 million people are seriously threatened by it. In many affected villages a third to a half of the adults have been blinded. Victims of the disease are concentrated in West Africa and in parts of Latin America and the Middle East.
- Schistosomiasis is endemic in 76 countries, with 600 million people at risk and 200 million people already infected.
- More than 90 million people are infected by filariasis, and an estimated 900 million people are at risk.

For onchocerciasis, the distribution of the drug ivermectin has in recent years been a striking breakthrough. For malaria, however, there has been little improvement in the numbers affected in the past 15 years. The situation may even have become worse, because there is significant underreporting for all tropical diseases.

The social consequences of tropical diseases are severe. In a village affected by guinea worm, for instance, agricultural productivity has dropped by 30%. In many cases, children are the most affected. In Sub-Saharan Africa malaria has caused more than 100,000 deaths of children under the age of one and nearly 600,000 deaths of children between the ages of one and four. Even when children survive, their subsequent growth and learning capabilities are often reduced.

Migrant workers are also at a great risk. Health workers have found a high incidence of malaria in the new settlements of the mobile and diverse population along the Amazon.

A decisive effort is thus required to advance research on the prevention and control of tropical diseases and to make drugs available to all people at risk.

AIDS emerged as a frightening threat to mankind only in the late 1970s. Between 5 and 10 million people apparently are infected worldwide, although only 133,000 cases were reported to the WHO at the end of 1988. Of the reported cases, about 68% were in the Americas, mainly in the North, 14% in Europe, 17% in Africa and 1% in Asia and Oceania. But these figures are gross underestimates because of the lack of diagnosis and reporting. The actual figures must be much higher, especially in developing countries.

Most AIDS cases are 20-40 year olds, the most productive members of the work force. In some cities in Africa, the rate of infection for this age group is thought to be 25%.

AIDS is likely to reverse many of the

BOX 2.5

The AIDS epidemic

The AIDS epidemic poses a serious threat to all countries, but it particularly affects developing countries that lack preventive health and social support services and that have a high incidence of infection. It adds burdens to debt, poverty, illiteracy, structural adjustment and other diseases.

The developing countries most affected include those in much of Central, Eastern and Southern Africa and a number of Caribbean countries, including French Guyana, Bermuda, the Bahamas, Haiti and Trinidad and Tobago. Rates of infection are also high for some subgroups in Brazil, Mexico and Thailand.

In its infancy, the epidemic has already induced sharp increases in adult, maternal and child morbidity and mortality rates in affected countries. Associated secondary epidemics of endemic developing country diseases, especially tuberculosis, are also occurring. National health budgets in many of these countries are inadequate, and health care systems are predominantly urban-centred with a curative orientation.

A distinctive feature of the epidemic is that — unlike famine, drought and poverty, which often claim the very young and the very old — AIDS claims those in the productive years and so also threatens the health of the economy.

Dependency ratios are increasing, and with per capita income on the decrease, there will be more dependents to feed with less. One study estimates that in 10 high-incidence African countries, more than 10% of the children have lost at least their mother to AIDS by the end of this decade.

As the epidemic intensifies, the already-limited social services and health insurance provided by governments or the private sector will be withdrawn because of high costs. Key sectors of the economy — including mining, transportation, defence and finance — may lose many of their trained workforce. Remittances from abroad, tourism and foreign investment could all be adversely affected. Infection rates in rural areas are increasing and will eventually reduce food and other agricultural production.

successes in reducing infant and child mortality and in raising life expectancy. It has been estimated that if 5% of the pregnant women in a typical African developing country are infected, the infant mortality rate would rise by about 13 per 1,000, an increase higher than the current rate in most developed countries.

The cost of caring for AIDS patients is imposing tremendous strains on health budgets. Public spending on AIDS research and education in the United States amounted to $900 million in 1988, and the cost of care was $50,000 to $150,000 per patient. Costs like these, if reproduced in developing countries, would soon absorb entire health budgets. Although the cost of care is much less in developing countries, the disease already is putting immense strain on budgets and taking resources away from other priorities. This pattern can only worsen as the disease spreads.

To sum up: the picture of human development needs qualification. Human progress does not take place automatically, and higher income is no guarantee for a better life. The problems of reversed or de-formed human development challenge both developing and developed countries, but they also underscore the centrality of human development as a continuing policy concern and priority. Development, even in countries at higher incomes, cannot afford to lose sight of its primary goal: the betterment of human life.

CHAPTER 3

Economic growth and human development

Economic growth is essential for human development, but to exploit fully the opportunities for improved well-being that growth offers, it needs to be properly managed. Some developing countries have been very successful in managing their growth to improve the human condition, others less so. There is no automatic link between economic growth and human progress. One of the most pertinent policy issues concerns the exact process through which growth translates, or fails to translate, into human development under different development conditions.

Typology of country experience

The human development experience in various countries during the last three decades reveals three broad categories of performance. First are countries that sustained their success in human development, sometimes achieved very rapidly, sometimes more gradually. Second are countries that had their initial success slow down significantly or sometimes even reverse. Third are countries that had good economic growth but did not translate it into human development. From these country experiences emerges the following typology:

- *Sustained human development*, as in Botswana, Costa Rica, the Republic of Korea, Malaysia and Sri Lanka.
- *Disrupted human development*, as in Chile, China, Colombia, Jamaica, Kenya and Zimbabwe.
- *Missed opportunities for human development*, as in Brazil, Nigeria and Pakistan.

The analysis of these country cases leads to several important conclusions. First, growth accompanied by an equitable distribution of income appears to be the most effective means of sustained human development. The Republic of Korea is a stunning example of growth with equity. Second, countries can make significant improvements in human development over long periods — even in the absence of good growth or good distribution — through well-structured social expenditures by governments (Botswana, Malaysia and Sri Lanka). Third, well-structured government social expenditures can also generate fairly dramatic improvements in a relatively short period. This is true not only for countries starting from a low level of human development but also for those that already have moderate human development (Chile and Costa Rica). Fourth, to maintain human development during recessions and natural disasters, targetted interventions may be necessary (Botswana, Chile, Zimbabwe and the Republic of Korea in 1979-80). Fifth, growth is crucial for sustaining progress in human development in the long run, otherwise human progress may be disrupted (Chile, Colombia, Jamaica, Kenya and Zimbabwe). Sixth, despite rapid periods of GNP growth, human development may not improve significantly if the distribution of income is bad and if social expenditures are low (Nigeria and Pakistan) or appropriated by those who are better off (Brazil). Finally, while some countries show considerable progress in certain aspects of human development (particularly in education, health and nutrition), this should not be interpreted as broad human progress in all fields, especially when we focus on the question of democratic freedoms.

The main policy conclusion is that economic growth, if it is to enrich human development, requires effective policy management. Conversely, if human development is to be durable, it must be continuously nourished by economic growth. Excessive emphasis on either economic growth or human development will lead to developmental imbalances that, in due course, will hamper further progress.

Policies for human development

Many factors influence the levels and changes in human development, ranging from aspects of the macro economy — which in turn are affected by developments in the international economy — to micro factors operating in individual households. Also important is at least one set of intermediate, or meso, variables: the level and structure of government expenditures and government programmes for the social sectors. Meso policies cover the whole range of fiscal policies, including those directly affecting the distribution of income, but the analysis here is confined to social expenditures. It can be broadened considerably through more research, particularly on the links between the level and structure of government expenditures and the distribution of income.

The main macroeconomic determinants of human development, together determining the levels and changes in household income, are initial levels and growth rates of income per capita and initial levels and trends in the distribution of income.

The main instruments of government for directly affecting human development levels are:

- *Across-the-board meso policies*: those for the provision of public goods and services in a way that does not discriminate among different social groups or regions, such as universal food subsidy systems, universal primary education programmes and nationwide immunisation programmes.
- *Targetted meso policies:* those for the provision of public goods and services to all members of particular target groups in the society, such as the food stamp programme for lower-income groups in Sri Lanka or a supplementary feeding programme that attempts to cover all malnourished children in a country.

Meso policies centre on health, education, potable water and other social services — usually provided by government — and can be measured by the shares of government budgetary expenditures in GNP or GDP. The level of meso policies can be described as low if government expenditures on the social sectors are less than 6% of GDP, moderate if they are between 6% and 10% and high if they are greater than 10%. Per capita public spending in the social sectors would be expected to rise with average per capita GDP. Richer countries may thus have higher absolute social spending per capita even if the level of their meso interventions, as defined here, is lower. Higher incomes can, therefore, have a positive impact on human development not only through ensuring high primary incomes but also by providing larger absolute resources to the government.

It is also desirable to distinguish different types of expenditures in each social sector, such as that on primary and tertiary education and on preventive and curative health care. Such distinctions describe the *structure* of expenditures within a particular social sector — and provide greater detail than the *allocation* of the total budget to different social sectors. A distinction between recurrent and capital expenditures can also be made.

There is no automatic link between economic growth and human progress

The literature provides fairly conclusive evidence of the association between different rates of success in human development and the relative importance given to different types of spending on social sectors. For example, spending on primary education and preventive health care is likely to lead to substantially larger improvements in human development than spending on higher levels of education and curative health care — at least at low initial levels of human development.

Meso policies can be well designed or less well designed, and their impact depends on their context. Government policies for universal primary education and for universal secondary education are across-the-board meso policies. But the former are more likely than the latter to be part of a well-structured package of meso policies where primary school enrolment ratios are still low.

Similarly, there are differences in targetted interventions. If substantial benefits accrue to nondeserving groups or do not accrue to deserving groups, the intervention is poorly designed. The balance between targetted and across-the-board policies also matters. Targetted interventions may be appropriate only under special circumstances, such as a temporary recession or extreme crisis, or only in countries that have the administrative capacity to manage efficient targetting. The circumstances should thus define the extent and duration of using targetted inter- ventions to protect or improve human development.

Social spending, directed towards the poor, must compensate for uneven income distribution

Meso policies become important when people's primary incomes, especially those of the poorest, are insufficient for them to obtain the goods and services needed to ensure a decent level of human development. Primary incomes are the disposable incomes of households from the normal workings of the economy. They often are insufficient in countries where incomes are generally low: even if the distribution of income is good, few people have primary incomes sufficient to ensure adequate human development. Primary incomes can also be insufficient where higher incomes are badly distributed: the incomes of some people may allow even developed country living standards, but for many others the primary incomes may be insufficient to meet their basic needs.

Well-structured meso policies are needed to compensate for the low primary incomes of important segments of the population. Where incomes are generally low but the distribution is good, well-structured across-the-board meso policies are likely to be appropriate. In countries with higher average income and good growth but skewed income distribution, some targetted interventions that favour the poorer segments of society may need to supplement the across-the-board policies. But even here, and especially in the long run, well-structured across-the-board policies — along with changes in the growth process — are likely to have the greatest payoff.

The patterns of human development described here are linked with differences in the relative roles of the macro and meso determinants and with differences in the relative roles of specific meso policies. These differences will become clearer in the following discussion of country experiences since 1960.

Indicators of country performance

Any assessment of human development would ideally use a composite measure, such as the human development index (HDI) that was introduced in chapter 1. But the HDI, now available for only one point in time, does not yet allow trend analysis. We could also consider several indicators separately — life expectancy at birth, mortality of children under five years of age, female and male literacy, and nutritional status, especially that of children. But good time series are also rare for many of these indicators.

A third option — the one chosen here — is to select an indicator that has fairly comprehensive time series data and that correlates closely with other indicators of human development. The under-five mortality rate meets both these requirements. Extensive empirical evidence suggests that reductions in the under-five mortality rate usually reflect improvements in nutrition — particularly that of pregnant women, infants and children — as well as achievements in education, especially female literacy. Estimates of life expectancy, in turn, are strongly influenced by under-five mortality rates, particularly in developing countries.

The long-run trends in under-five mortality rates thus provide a useful indicator of changes in human development. But these rates refer primarily to changes on only one side of the human development equation — the formation of human capabilities. They do not capture the use of human capabilities.

Sustained human development

Countries with durable progress in human development often started from very different initial conditions in 1960 and have at times followed quite different routes to sustain their success.

The Republic of Korea

The Republic of Korea has achieved human development through fast and equitable growth. For most of the people, primary incomes have grown enough to enable improvements in the human condition without significant government interventions. Social sector expenditures as a percentage of GDP have been relatively low.

Although Korea's economic management and the resulting growth and distribution are undoubtedly superior to that of most developing countries, its performance has not been consistently good. For example, its income distribution worsened during the 1970s, in part because skilled workers in the heavy and chemical industries, whose growth was emphasised during this period, earned far more than unskilled workers. In addition, income disparities between urban and rural areas, rather significant to begin with, increased further during the 1970s.

The main reason was the urban bias in the country's development strategy, particularly the concentration of resources in the capital city, Seoul. This bias meant that, despite remarkable growth, the distribution of income, while better than that in most developing countries, left much to be desired. Many Koreans were vulnerable to even temporary disruptions in the flow of primary incomes.

The disruptions came in late 1979 and 1980, when the country suffered negative growth for the first time in 20 years. Sparked by external shocks, the recession was also attributable to a bad harvest in 1980 and the political instability after the assassination of President Park in October 1979. Its causes could, moreover, be traced to less rigorous economic management during the second half of the 1970s when the government, spurred by the easy availability of foreign credit, embarked on an ambitious programme of investment in heavy and chemical industries.

The programme ensured a continuation of the remarkably high growth rates of the 1960s and early 1970s, but it also swelled the budget deficits, widened the trade gaps, and increased the external debt. The external shocks at the end of the 1970s were thus greater than they would have been under the more prudent and restrained growth strategy followed earlier.

The government's response in managing the economy — and in protecting the most vulnerable groups during and after the recession — provides useful policy lessons for human development. First, it embarked on a comprehensive programme of stabilisation, liberalisation and structural adjust-

TABLE 3.1
Under-five mortality and other basic indicators of human development

Country	HDI 1987	Under-five mortality rate (per 1,000 live births) 1960	1975	1988	Life expectancy (years) 1960	1975	1987	Adult literacy (%) Female 1970	Female 1985	Male 1970	Male 1985	Calorie supply as % of requirements 1965	1985
Sustained human development													
Korea, Rep.	0.903	120	55	33	54	64	70	81	91	94	96	96	122
Malaysia	0.800	106	54	32	54	64	70	48	66	71	81	101	121
Botswana	0.646	174	126	92	46	52	59	44	69	37	73	88	96
Sri Lanka	0.789	113	73	43	62	66	71	69	83	85	91	100	110
Costa Rica	0.916	121	50	22	62	69	75	87	93	88	94	104	124
Disrupted human development													
China	0.716	202	71	43	47	65	70	..	56	..	82	86	111
Chile	0.931	142	66	26	57	65	72	88	97	90	97	108	106
Jamaica	0.824	88	40	22	63	68	74	97	..	96	..	100	116
Colombia	0.801	148	93	68	55	61	65	76	88	79	82	94	110
Kenya	0.481	208	152	113	45	52	59	19	49	44	70	98	92
Zimbabwe	0.576	182	144	113	45	53	59	47	67	63	81	87	89
Missed opportunities													
Brazil	0.784	160	116	85	55	61	65	63	76	69	79	100	111
Nigeria	0.322	318	230	174	40	46	51	14	31	35	54	95	90
Pakistan	0.423	277	213	166	43	50	58	11	19	30	40	76	97

FIGURE 3.1
Sustained human development: country profiles

Average growth rates of GDP per capita

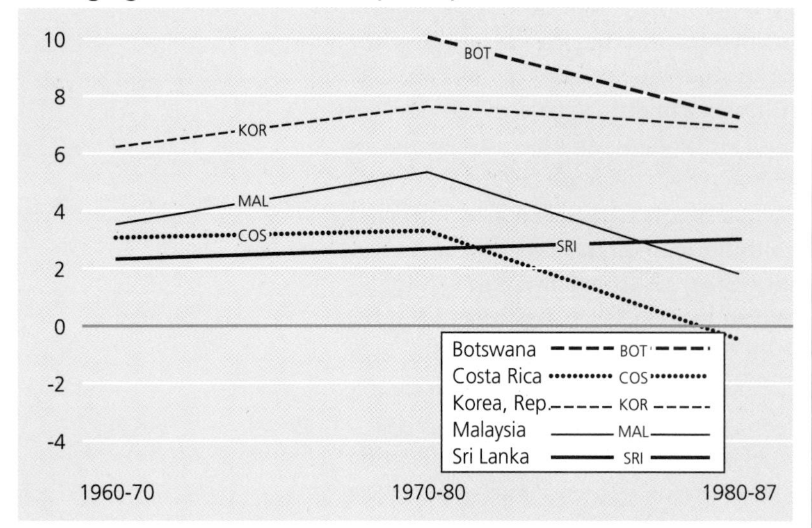

Under-five mortality rate

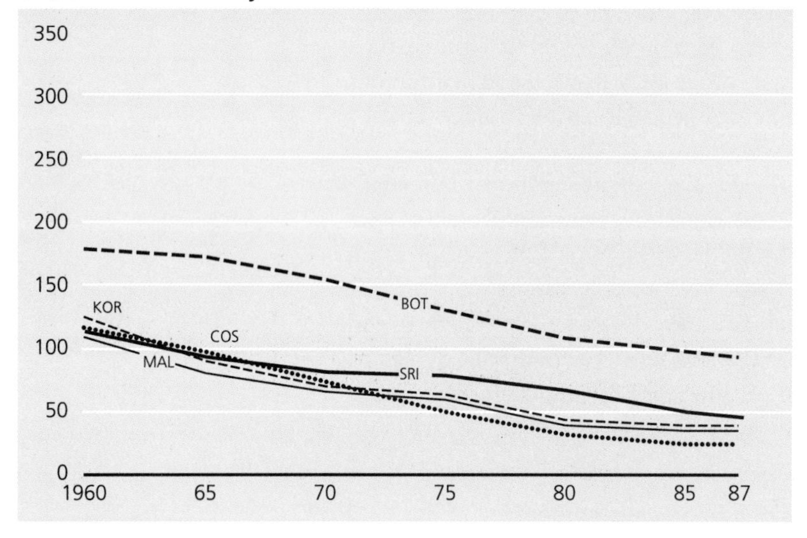

Social sector public expenditure, percentage of GDP

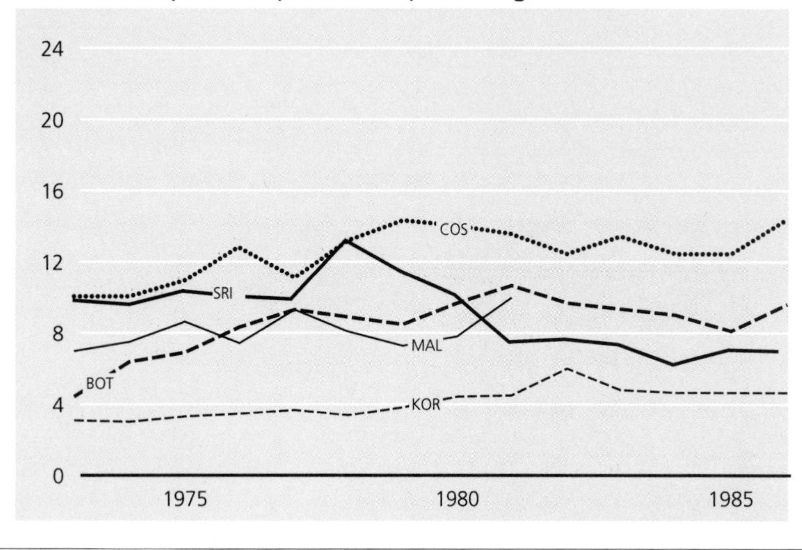

ment. Second, it introduced new social programmes and intensified existing ones.

The government deficit was drastically reduced, expansion in money supply was curtailed and inflation was brought under control. Many macroeconomic reforms in both internal and external markets were carried out in an economy that had gradually returned to more extensive controls in the 1970s after the substantial reforms of the 1960s.

In social expenditures, the coverage of population by health insurance was increased from a tenth in 1978, a year after its initiation, to almost a third by 1981 and to almost a half by 1985. In addition, a medical assistance programme for the lowest income groups was introduced in 1979. Members of poor families (depending on their income and ability to work) were entitled to free or subsidised medical care, especially maternal and child care.

In addition, the public works programmes to provide employment to the poor during crisis periods were temporarily increased during the recession. These programmes provided an estimated 9.4 million man-days of employment in 1980 alone. Direct income transfers were made to those who were unable to work and take advantage of these employment opportunities due to infirmity or old age. Moreover, the Livelihood Protection Programme, initiated in 1961, was expanded to benefit an estimated 2 million people in 1981, through grants of cereals and through cash for fuel and tuition expenses.

As a consequence of these effective meso interventions, the human development levels continued to improve even during the difficult years of 1980 and 1981, though at a temporarily slower rate. Meanwhile, the major changes introduced in macroeconomic policies restored price stability quickly. While the 1980s have not been easy years, the Korean economy has performed extremely well, promoting the human development of its citizens.

One important lesson from this experience is that countries with a generally impressive growth but a less impressive income distribution may require well-structured meso interventions, particularly targetted ones, during brief periods of slower growth. A second lesson is that to avoid lasting damage to the human condition, macroeconomic adjustments are needed to restore growth that has fallen off.

Malaysia

Malaysia's experience shows that growth alone is no guarantee of human development, but it also shows that human development is possible even under conditions of a fairly inequitable distribution of income — if effective meso policies are in place.

Malaysia in 1960 was a middle-income country with moderate human development and a reasonable income distribution, which subsequently deteriorated. The human condition nevertheless improved steadily, with the under-five mortality rate dropping from 106 in 1960 to 32 in 1988. Other indicators also confirm that there has been a steady and sustained improvement in the human condition.

Malaysia had good growth after 1960, though not as spectacular as that of the Republic of Korea. Per capita GDP grew by about 3% a year during the 1960s and accelerated to about 5% a year during the 1970s. Even in the difficult 1980s, Malaysia managed per capita growth of 2% a year, but the fruits of this growth were not equitably distributed. The Gini coefficient steadily increased between the late 1950s and mid-1970s — from 0.42 in 1958, to 0.50 in 1970 and to 0.53 in 1976. There has since been some improvement, but the distribution of income remains quite inequitable: the Gini coefficient in 1984 was still 0.48.

Malaysia's steady success in human development owes much to well-structured across-the-board meso policies. Public spending in the social sector averaged about 8% of GDP during 1973-81. That level of spending is not as high as that in Sri Lanka, which had lower income and poorer growth, but it also is not as low as that in the Republic of Korea, which had higher income and better growth.

Malaysia's meso policies were designed to benefit all groups in society, with special emphasis on the rural areas where the poorer people live. A detailed study of public expenditures in Malaysia found that this goal was actually being achieved. Central government expenditures per capita for education, health care, agriculture and pensions were highest in the rural areas in the early 1970s.

The distribution of primary incomes improved significantly after the effects of taxes and expenditures were taken into account. The ratio of secondary incomes — primary incomes plus the incidence of budgetary activities — to primary incomes declined steadily at higher incomes. For the lowest 10% of income recipients, the ratio of secondary to primary incomes was 1.5. This group thus received an additional 50% in "income" from government activities. Each of the lowest four deciles — the bottom 40% — had ratios of at least 1.20, while for the highest income decile the ratio was 0.93.

The Malaysian experience thus shows that steady improvements in human development are possible even in the context of good growth coupled with poor income distribution — if the benefits of meso policies can be distributed equitably.

To avoid lasting damage to the human condition, macroeconomic adjustments are needed to restore growth that has fallen off

Botswana

Botswana also translated the benefits of growth into human development through well-structured meso policies. Botswana started off as a low-income country with low human development and an uneven distribution of income. Botswana's human development is among the best in Africa, particularly Sub-Saharan Africa. Its under-five mortality rate fell from 174 in 1960 to 92 in 1988, still high but it started from a much higher level. The rate of reduction compares well with other successful countries over the past three decades. This success is also reflected in the remarkable progress in literacy. Moreover, Botswana has succeeded, unlike most African countries, in protecting the vulnerable groups during the adverse external circumstances of the 1980s.

An important factor in Botswana's steady and sustained improvement in the human condition is the exceptionally high growth since independence. Per capita GDP increased impressively at about 10% a year during 1965-80 and about 8% a year during 1980-87, a period when most African countries suffered negative growth. The high growth — based largely on minerals, especially on rapid growth in the production and export of diamonds — continued during the 1980s despite the drought (agriculture accounts for less than 10% of GDP). But since 80% of the population is rural and relatively poor, the drought threatened rural incomes and reduced the availability of food.

Data on income distribution are not available for Botswana, but unless the initial distribution was extremely inequitable, the high average growth rates are likely to have been accompanied by some growth in the income of even the poorer segments of the population, at least before the drought. Macro factors are thus likely to have contributed to the steady improvement in human development since independence.

Data on meso policies, not available for years before 1973, indicate a moderate but rising level during 1973-77 and a stable and fairly high level during 1978-86. Public expenditures in the social sectors rose from 4% of GDP in 1973 to about 9% in 1977 and for the most part remained between 9% and 10% thereafter.

Thus, while improvements in human development to the end of the 1970s may be attributed largely to growth and a moderate dose of meso policies, Botswana's success in maintaining the earlier achievements and protecting the vulnerable groups during the drought has been largely due to extensive meso policies, particularly targetted ones.

The government introduced comprehensive and substantial drought relief measures after 1982, two of which were particularly important:

- A public works programme provided employment on infrastructural projects, covering an estimated 74,000 workers in 1985-86 and replacing 37% of the income lost due to crop failure.
- Supplementary feeding programmes were launched for primary school children, children under five (for all in the rural areas and for the most malnourished in urban areas), pregnant and lactating women and tuberculosis patients. In 1985-86 there were an estimated 680,000 beneficiaries, nearly 60% of Botswana's population.

Funds were also provided to help repair water systems and to transport emergency water supplies to drought-stricken areas. Agricultural relief and recovery programmes assisted small farmers in clearing land and acquiring inputs, including free seeds. Livestock relief and recovery programmes assisted with vaccinations and feed and provided a guaranteed market for cattle.

The budgetary cost of the drought relief programme, which reached more than 70% of the population, was about $21 million in 1985-86, or 2% of GDP. Foreign donors contributed an equivalent amount. The total cost was thus moderate — showing that other poor countries could replicate the programme. Botswana also developed a system of nutritional surveillance and early warning to permit timely identification of problems and appropriate interventions, a system also likely to be replicable in other countries (box 3.1).

Botswana's achievements in human development have clearly been helped by the prosperity of the diamond industry and its impulse to growth. But it is also clear that meso policies for the provision of basic health and education facilities across the

BOX 3.1

Botswana's drought relief

A decentralised, cross-sectoral, and flexible monitoring network has helped Botswana respond quickly when drought affects villages or nomadic herders.

The system, instituted in the wake of the 1982 drought, is headed by an Interministerial Committee that has the decisionmaking power to channel resources quickly to drought-stricken areas. The Committee's information base is continually updated by an Early Warning Technical Committee that monitors rainfall, food supplies and reserves, agricultural conditions, and the nutritional status of children — and makes district-by-district recommendations for drought recovery assistance.

The technical committee is supported on the ground by the National Nutritional Surveillance System, which compiles monthly reports on the nutritional status of all children under five attending health facilities. It also gets weekly reports of rainfall at 250 recording points and monthly reports on agricultural conditions from Botswana's 120 extension district offices. The technical committee makes regular drought assessment tours to confirm and supplement the network's data.

Timely, local-level information is combined with top-level policy involvement to assure quick response. When health centres reported falling weights among children in 1984, the Interministerial Committee quickly provided supplementary feeding supplies for under-fives across the country. Further reports of malnutrition in the next year led to the restoration of full drought rations.

board — together with targetted policies to meet special needs during the drought — contributed much as well, particularly in protecting vulnerable groups.

Sri Lanka

Sri Lanka's experience can be divided in two phases: 1960-78 and after 1978. Modest growth characterised the first phase, with per capita GDP rising about 2.2% a year during 1960-70 and about 2.5% during 1970-80. But the distribution of income was fairly good, with the Gini coefficient of household income falling from about 0.45 in 1965 to 0.35 in 1973. After 1978, per capita GDP growth accelerated to more than 3%, but the distribution of income worsened. Estimates of Gini coefficients for 1978 and 1982 are comparable to those for the 1950s and early 1960s: above 0.45.

It can thus be said that Sri Lanka shifted from a regime of moderate growth with a good distribution of income (before 1978) to one of better growth with a poorer income distribution (after 1978). Throughout, however, the levels of income have remained relatively low. This has meant that, although growth was moderate and the income distribution good, substantial improvements in human development could not be achieved exclusively through the macro side, and the meso interventions had to be significant.

Indeed, Sri Lanka has a long history of social sector interventions dating back to the period before independence. As early as 1945, the government had extended free medical care to almost every part of the country and introduced universal free education up to the university level.

But its best known meso intervention is the nearly universal food subsidy introduced in 1942. That system persisted until 1979, with only occasional changes in the eligibility criteria and the quantities allowed. For example, the proportion of rationed rice to total rice consumed exceeded 70% at one time but declined to about 50% after 1966. As a proportion of the total calorie intake, rationed rice represented about 20% in 1970. The budgetary cost was substantial, varying between 15% and 24% of total public spending in the 1970s. In addition to the system of food subsidies, the strong interventions initiated in education and health before independence were maintained thereafter. All this is reflected in a high proportion of public expenditures on the social sector in GDP: about 10% during 1973-78.

In 1979, in the wake of the change in macro policy, the food subsidy programme gave way to a food stamp scheme: only households whose declared incomes were less than a specified level received food stamps, which could be used to buy basic foods from designated shops. This change was primarily intended to reduce the budgetary burden of government. The share of food subsidies in government expenditures dropped from 15% in the mid-1970s to about 3% in 1984, and the share in GNP dropped from about 5% to 1.3%. Overall, social sector expenditures declined from around 10% of GDP during 1973-78 to around 7% during 1980-85.

The relative reduction in social sector expenditures was countered, however, by a

BOX 3.2

Food stamps miss the target in Sri Lanka

In Sri Lanka, some of the poorest people lack access to their main staple — rice — despite a food stamp programme intended to help them. The main reason is lack of flexibility in programme design.

First, the shift from an across-the-board programme to a targeted scheme was introduced in 1979 in order to ensure that a relatively large share of the benefits from the government's food subsidies flows to the most deserving groups. However, inflation doubled the price of food between 1979 and 1982 — and halved the purchasing power of the food stamps, since their face value remained unchanged. As a result, the absolute amount of real income transferred to the poor was, at the end, considerably lower than before.

Second, after March 1980, no new applicants were accepted for the scheme. This disqualified all new-borns and families that subsequently suffered serious income losses. Meanwhile, many higher-income households continued to benefit from the scheme by underreporting their incomes.

The effects were quickly discernible. The national average daily calorie consumption per capita was virtually the same in 1981-82 as in 1979: just under 2,300 calories. But the per capita calorie consumption of the lowest decile fell from 1,335 calories to 1,181, and that of the second lowest decile from 1,663 calories to 1,558.

In contrast, the calorie consumption of the rich increased, mainly because they appropriated a greater share of the fruits of accelerated growth to more than compensate for the cutbacks in food rations.

The lesson conveyed by this experience is that the effectiveness of policy measures, especially that of targeted programmes, must be subject to continuous monitoring. This holds true, in particular, for policy measures being implemented under conditions of rapid socio-economic change — changing consumer or producer prices, flagging or expanding unemployment, and falling or rising wage levels.

better distribution of its benefits. In 1973 the middle-income groups benefitted most, but by 1980 the per capita benefits declined with rising incomes, and the poorest 40% of the population derived more benefit than other income groups from government spending.

The biggest change was in the distribution of benefits from education. The participation of low-income children in primary schools improved markedly in this period, so a greater share of the expenditure on primary schooling accrued to these groups. Moreover, the attempt to restrict food stamps to low-income groups managed to increase the *proportion* of benefits accruing to the poor. But the weaknesses of the new programme appear to have led to an *absolute* decline in some aspects of welfare, such as calorie consumption of the poorer segments of the population (box 3.2).

Promoting faster economic growth at the expense of equity can damage the invisible bond between the people and the government

Sri Lanka's experience thus suggests that, in a low-income country with a good distribution of income, well-structured across-the-board meso interventions can significantly improve human development. These policies proved vulnerable, however, to political and economic changes. In principle, the shift towards targetted interventions should have helped sustain improvements the human development, despite a worsening of the income distribution. But in practice, replacing across-the-board meso policies by targetted policies can worsen the position of some vulnerable groups.

An important lesson from Sri Lanka's experience is that promoting faster economic growth at the expense of equity — without effective social safety nets to protect human development, especially after a sustained period of good human progress — can damage the invisible bond between the people and the government and lead to considerable social and political turmoil. The question for governments in a similar position is whether, and to what extent, budgetary transfers are necessary if free markets fail to protect the poor adequately.

Costa Rica

The last example of sustained human development is particularly interesting because of its significant improvements in human development over a relatively short period. Costa Rica started as a middle-income country with an income distribution that was fairly moderate, at least for Latin America, and with a moderate level of human development. During 1960-87 the trends in the under-five mortality rates reflected human development approaching that in the developed countries. Notable, however, is the substantial improvement in the 1970s. The under-five mortality rate fell from 121 in 1960 to 22 in 1988, but much of that reduction came between 1970 and 1980, dropping from 76 in 1970 to 31 in 1980 — a more than 50% reduction in a decade.

Growth in the 1960s and 1970s was fairly good, with per capita income increasing more than 3% a year, but it turned moderately negative in the 1980s. The Gini coefficient declined from 0.52 in 1961 to 0.44 in 1971 but then returned to about 0.50 in 1977. Since then, there has again been some decline, but even at its lowest point in 1982 the Gini was still about 0.43.

Turning to meso policies, social sector expenditures expanded in Costa Rica. GDP was growing impressively, and the share of government expenditures in GDP was also increasing, from 18% in 1973 to about 25% in 1979. So, although the share of the social sectors in total expenditures remained stable, at the high level of more than 50%, the share in GDP rose from 10% in 1973 to 14% in 1979.

Social expenditures were also well structured. In the 1970s Costa Rica introduced major changes in its health strategies to ensure complete coverage of basic health services for the entire population. Under the first national health plan, launched in 1971, public resources for the health sector were increased, and efforts were made to increase the efficiency of their use. These programmes fell into two categories.

First, the strategy for primary health care was to extend the coverage of basically preventive services to people not previously served — through the rural health programme (begun in 1973) and the community health programme (1976). By 1980 water and sanitation services in rural and urban areas reached 60% of the population.

Immunisation campaigns against measles, diphtheria, pertussis and tetanus were launched, and sanitation activities (for potable water and sewage disposal) were intensified in rural areas. Community participation in health programmes was also encouraged.

Second, medical services were improved and systematically broadened, mainly by transferring the ministry of health hospitals (often poor in resources and frequently offering deficient services) to the Social Security System (CCSS). The CCSS doubled the number of centres offering outpatient services and tripled the amount of physician-hours between 1970 and 1980. There was thus an important restructuring of health expenditures: the number of hospitals fell from 51 to 37, and the number of outpatient installations rose from 348 to 1,150. Also by 1980, insurance coverage for illness reached 78% of the population. All these programmes paid special attention to regions with lower levels of human development.

The achievements of Costa Rica's public health programmes should not be considered in isolation. Public health had the political support of a government highly sensitive to social needs. It also had the economic support of the growth and prosperity after 1964. Improvements in educational attainments were important as well. The proportion of women who completed their primary education rose from 17% in 1960 to 65% in 1980, accelerating the decline in infant and child mortality.

Costa Rica shows that assigning a high priority to social sector expenditures, coupled with well-structured across-the-board policies, can dramatically improve the human condition despite only moderate growth and a poor distribution of income.

Disrupted human development

The countries in this category achieved success in human development, often dramatic success, but could not maintain it. Like the previous group of countries, they differ in their initial conditions and in the speed of initial progress before stagnation or reversals set in.

China

A low-income country with good income distribution, China dramatically improved its human condition through extensive, well-structured, across-the-board meso interventions (with some targetting) during a period of arguably moderate growth, roughly 1960-78. But even with good subsequent growth, reductions in the coverage of meso policies led to a stagnation or, by some accounts, even a reversal of these trends. Moreover, China's record is flawed by the absence of other vital human choices, including political and economic freedom.

China's achievements show up in the under-five mortality rates, reduced from 202 in 1960 to 98 in 1970 and more than halved to 43 in 1988. Other indicators tell a similar story.

There is some controversy over whether China has sustained its progress in the 1980s, a period of significantly faster growth in incomes. The dramatic reductions in the under-five mortality rates until 1980 seem to have slowed during the 1980s, even though the rates are still higher than those in the industrial countries.

A recent World Bank study suggests that China's earlier progress in improving the health of its people may have stagnated somewhat in recent years. For example, there are reports of substantial increases in the prevalence of schistosomiasis in certain regions of China. Although the evidence is not conclusive, it appears that China's achievements through the end of the 1970s may have slowed down considerably, if not reversed, on some fronts in recent years.

Widespread literacy and food programmes to help ensure adequate nutrition have been important in China, but the development of an effective health care system has contributed most to improving the human condition there. The Chinese health care system has many noteworthy features, some of them quite innovative (box 3.3).

- It strongly emphasises preventive health services over curative.
- It mobilises people to carry out preventive health campaigns.
- It delivers services even to remote rural areas.

Social expenditures must be restructured to benefit the many, rather than a few

FIGURE 3.2
Disrupted human development: country profiles

Average growth rates of GDP per capita

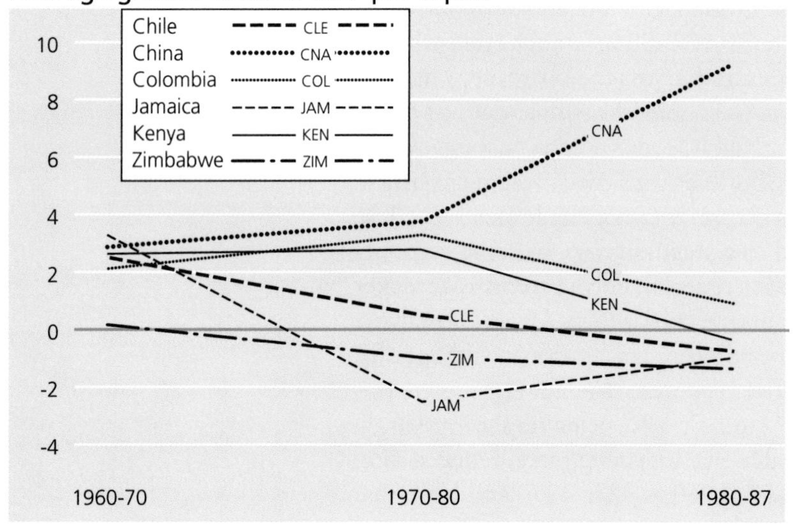

Under-five mortality rate

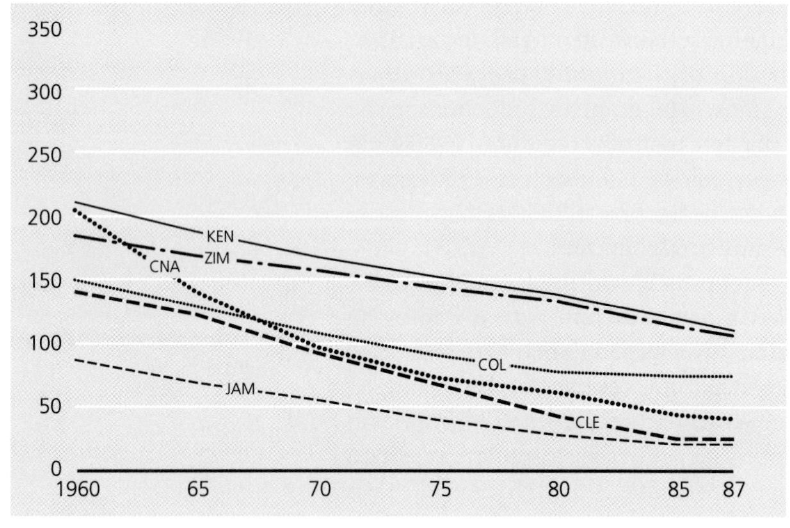

Social sector public expenditure, percentage of GDP

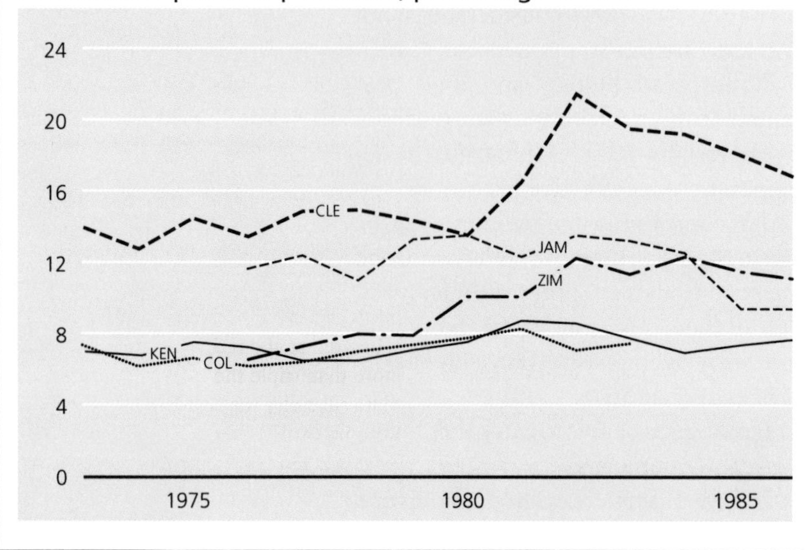

- It consumes a relatively large proportion of national resources.

China's advances in human development are also attributable to socioeconomic gains in meeting basic needs. China's approach to ensuring an adequate food supply to its citizens has differed from that in Sri Lanka. Food security was for many years built into the commune system, and the production brigades gave their members rations of basic foods in exchange for work. Communes sold grain or paid taxes on production to the state. The state could then guarantee food security to communities that, for some reason, were short of food and required relief grain.

Recent changes — the adoption of the household responsibility system in 1979 and the dismantling of communes in 1982 — radically altered the situation, with production now being liberalised and left more to communities and even individuals.

The recent economic reforms in China have also led to a collapse of the rural cooperative insurance system, removing the protection against the financial risks of ill health for the great majority of rural people. Those risks can be substantial because the Chinese health system recovers a high proportion of its costs: hospitals typically recover about three-quarters of their operating costs through user fees and drug sales. To put this in perspective, the costs per hospital admission average $36 for rural people, even though the annual rural income per capita is less than $100 in many regions. The costs are about twice as high ($75) for urban residents, but most of them are still covered by compulsory, state-subsidised health insurance.

The network of barefoot doctors apparently has been another casualty of the reforms, with rural health care coverage declining, county hospitals and rural clinics in financial distress, and private medical practice emerging again.

China's new "household responsibility system" reintroduced the concept of economic incentives for individual productivity. But the larger role for private and cooperative enterprises, the growth of piece work, and the establishment of liberalised enterprise zones — all part of the post-1978 reforms — also worsened the distribution of income across families and regions.

Although the events of 1989 may be another reversal, peasants and workers until recently were encouraged to produce privately for individual reward. The post-1978 reforms undoubtedly increased the production incentives, as reflected in accelerated growth, but they appear also to have hurt, probably unintentionally, the variables that contribute directly and indirectly to human development, slowing the earlier rates of progress. There is no rationale for neglecting social development in a period of accelerated economic growth.

Chile

Chile also saw its dramatic progress in human development falter. Chile started the 1960s as a middle-income country with a moderate distribution of income and a moderate level of human development, and then had its subsequent progress bring it close to developed country levels.

Like Costa Rica, Chile dramatically reduced the under-five mortality rate from 142 in 1960 to 26 in 1988, with much of the reduction coming in the 1970s. But unlike Costa Rica, which sustained its progress in human development on all fronts, Chile appears to have been less consistent.

Its calorie consumption per capita declined slightly between the mid-1960s and mid-1980s, and debate surrounds the trends in general living conditions since the mid-1970s. It has been suggested, for example, that the under-five mortality rate has declined despite a steady deterioration in overall living standards — reflected in sharp falls in real wages, worsening income distribution, rising incidence of certain diseases, deteriorating housing conditions and falling primary school enrolment ratios. It has also been suggested that poverty declined for the extremely poor but not for the poorer groups as a whole.

Chile's growth was moderate in the 1960s, flat in the 1970s and negative in the 1980s. Total government expenditures fell from 35% of GDP in 1973 to 30% in 1980. But the share of social sectors in total government expenditures, already very high at 40% in 1973, increased steadily to reach 50% in 1979 and average 60% in the 1980s. So, although the 1970s were a period of slow growth and declining government expenditures, social sector expenditures increased marginally through 1978, having been 14% of GDP in 1973.

Chile implemented across-the-board policies but targetted its health programmes on maternal and child care. In addition, a screening programme was established in conjunction with well-baby clinic check-ups (already reaching almost all infants) to detect and treat children suffering from malnutrition. This programme proved highly effective in protecting the most vulnerable groups during a period of economic instability.

Chile's experience shows that some human development indicators can improve dramatically — even during periods of poor growth — if well-structured across-the-board policies are combined with some

BOX 3.3

China's health care system

Shortly after the revolution China initiated campaigns to improve sanitation by eliminating the "four pests" (rats, flies, mosquitoes and bedbugs), to vaccinate against and cure infectious diseases, and to control the vectors of such major endemic disorders as malaria and schistosomiasis.

The keys to success were mobilising the masses, extending services to the remotest areas and making health services affordable.

Mobilising the masses. The Chinese tackled the "free-rider" problem that often hampers effective preventive health measures by making people responsible for them. According to some estimates, preventive measures accounted for less than 5% of the resources devoted to health. Mass mobilisation of surplus labour, especially during slack agricultural seasons, prevented the overtaxing of the health budget in a poor society — but achieved outstanding results.

Extending services to remote rural areas. The deprofessionalisation of health care providers through a mass cadre of barefoot doctors at the grassroots helped extend basic services into remote regions. Recent estimates show that while the number of western-style doctors per 100,000 population in China is two and a half times that in India, the number of village-level health workers was 4.5 times that in India.

Extensive health insurance coverage. During the early 1980s, financing came in almost equal amounts from three main sources: private outlays (32%), labour insurance (31%) and state budget expenditures (30%), with the residual financed by production brigades. Notable in this financing profile is the high proportion of expenditures mediated through insurance schemes, a reflection of extensive health insurance coverage. The coverage of insurance changed drastically after the economic reforms of the early 1980s. In 1981 about 70% of the population was completely insured. But there were considerable urban-rural differences in health spending. Urban expenditures, estimated at about $16 per person, were more than triple the rural expenditures. State subsidies for health in urban areas were almost 10 times those for rural areas — about $13 per capita compared with less than $1.50 per capita.

targetted interventions. It also shows that the targetted interventions may not help in maintaining *overall* progress if growth does not recover.

Jamaica

Targetted social spending may not help in maintaining overall progress if growth does not recover

Like Chile, Jamaica started the 1960s as a middle-income country with a moderate income distribution and a moderate level of human development. But instead of having dramatic improvements in the human condition, its progress was more uniform. And it had even less success than Chile in sustaining its progress during the 1980s. For example, the number of children admitted for malnutrition to the country's major children's hospital more than doubled between 1978 and 1985.

Jamaica's growth rates were very respectable in the 1950s, with per capita GDP increasing by nearly 7% a year on average, and still reasonable in the 1960s, with per capita GDP growing about 3.5% a year. Data on income distribution suggest, however, that a very inequitable distribution became even worse. The proportion of aggregate income going to the poorest 40% of the population, only 8.2% in 1958, fell to 7% in 1972, while the share of the richest 10% of the population increased from 43.5% to 50%.

No information is available regarding public expenditures in the social sectors before 1976, but data for 1976 onwards show a high level of meso interventions in the social sectors, particularly for 1976-80. Public expenditures in the social sectors ranged between 12% and 14% of GDP during 1976-80 and then fell to about 10% in 1985 and 1986. If social sector expenditures in the 1960s and early 1970s were comparable to those subsequently, the meso interventions before 1976 were quite important. It is thus likely that the steady improvement in human development during the 1970s may have been facilitated by respectable growth, complemented by meso policies that compensated somewhat for the skewed and worsening distribution of income.

Growth deteriorated, however, during the 1970s, with per capita GDP falling about 3% a year. There was some improvement during the 1980s, but growth rates remained negative. The effect was harshest on the poor, who suffered falling real incomes and unemployment. The government tried to maintain real wages and protect the vulnerable groups, with only partial success, through food subsidies, price controls and employment schemes. But with growing external and internal deficits — the current account deficit was more than $200 million and the government's budget deficit more than 15% of GDP in 1981 — these meso policies were difficult to sustain indefinitely.

After the change in government in 1980, government expenditures, including those on the social sectors, were cut as part of an adjustment programme, reducing the share of social sector public expenditures in GDP. With per capita real GDP falling during the 1980s, real per capita expenditures on the social sectors fell as well. Education expenditures per person under 15 are estimated to have declined 40% and per capita health expenditures by 33% between 1982 and 1986.

The government's attempts at targetted interventions achieved limited success. It introduced a food aid programme in 1984 to protect the most vulnerable — infants, school children, pregnant and nursing women, the elderly and the very poor people, together constituting about half the population. But the per capita benefits from the scheme fell short of the requirements. The adverse movements in the macro and meso determinants of human development slowed and in some cases reversed the rate of progress.

Colombia

Per capita GDP in Colombia grew moderately at 2.1% a year in the 1960s and 3.7% a year in the 1970s. The country did avoid a recession in the difficult 1980s, but its per capita GDP growth nevertheless slowed to about 1% a year.

Although modest, Colombia's economic growth in the 1980s made it possible for the government to maintain the per capita increases in social expenditures. The share of public spending on education in GNP moved up from 1.7% in 1960 to 2.8% in 1980, and that on health from 0.4% to 0.8% of GNP. But the adjustment programme adopted in 1984 reduced public expenditures, including the expenditures on social sectors. Social spending nevertheless continues to account for about one-third of total public expenditures.

Colombia's human development indicators mirror the overall economic situation. The country's income distribution improved in the 1970s and 1980s, with the Gini coefficient declining from 0.57 in 1971 to 0.45 in 1988. The under-five mortality rate fell from 148 in 1960 to 78 in 1980. Since then, the decline has been more modest — to 68 in 1988. There has also been a slowdown in the growth of real wages since 1987, but this decline appears to have come to a halt.

Although the slower growth in the 1980s would have required a compensatory increase in meso interventions, some elements of the government's earlier policy package — like its effective food stamp scheme — were discontinued, primarily for fiscal reasons. The economic adjustment policies appear, however, only to have slowed human progress, not to have reversed it. The challenge now is to convert the gains in economic growth that are expected from these policies into further improvements in human development.

Kenya

Human development in Kenya was for many years successful, despite difficult initial conditions — low income, low human development indicators and a rather uneven distribution of income. But the progress has slowed down in recent years.

A low-income country, Kenya had a reasonably good growth in the 1960s and in the 1970s, when its per capita GDP increased at about 3% a year. But like most African countries, it suffered negative growth in the 1980s, with per capita GDP falling about 0.9% a year. Detailed data on income are not available, but one estimate suggests an inequitable distribution. In 1976 the poorest 40% of households received only 9% of total income while the top 10% received 46%, representing 25 times the income of the poorest 10% of households.

The government's policies only partly offset the effects of bad distribution. Kenya's meso interventions have generally been moderate, with the share of social sector public expenditures remaining remarkably stable at 7% to 8% of GDP during 1973-86. Two-thirds of this spending was for education, and the rest mostly for health. Kenya's education system also benefitted from voluntary self-help (*harambee*) efforts. In 1970, for example, two secondary students in five were in unaided (mainly *harambee*) secondary schools.

In Kenya, therefore, the government's moderate efforts were supplemented by significant private involvement in the provision of social services, especially in education. This, along with a moderately good growth, contributed to Kenya's improvements in human development through the end of the 1970s. In the 1980s, however, the failure to increase the coverage of meso policies in the face of declining primary incomes and unequal income distribution appears to be associated with a deterioration in human development.

Zimbabwe

Human development in Zimbabwe, relative to the rest of Sub-Saharan Africa, has been very good. But Zimbabwe also suffered some stagnation after progressing from poor initial levels of human development.

Improvements in Zimbabwe have come despite steadily worsening growth since the 1960s, with per capita GDP falling about 1% a year in the 1970s and about 1.5% a year in the 1980s. Data on income distribution are being collected only now, but it is widely agreed that inequalities were significant before independence and have since been reduced by the redistributive policies of government — but only somewhat, leaving substantial inequality.

Zimbabwe's improvements in human development can therefore be attributed to social sector expenditures, which were moderate to high before independence in 1980. The country's experience since independence shows the difficulties of sustaining human development, even with well-structured meso policies, if growth remains negative for long periods.

After independence, the government gave greater prominence to social sector meso policies and restructured its social spending towards activities having a greater impact on human development, targetting those in need. These expenditures jumped to more than 10% of GDP after 1980.

Reversals in human development during adjustment periods can be avoided through careful policy management

At the time of independence, Zimbabwe inherited a highly inequitable health care system — reflected, for example, in the fact that 44% of publicly funded services went to sophisticated central hospitals that served only 15% of the population, while only 24% went to rural health services for the majority of the population. After independence, several measures were taken to redress these imbalances.

- Health care became free for those earning less than Z$150 a month, the vast majority of the population.
- The programme of immunisation against six major childhood infectious diseases and tetanus immunisation of pregnant women was expanded. The proportion of fully immunised children between 12 and 23 months is estimated to have increased from 25% in 1982 to 42% in 1984 in rural Zimbabwe and from 48% in 1982 to 80% in 1986 in Harare City.
- A programme for building hospitals and rural health centres was initiated — constructing 163 rural health centres by January 1985 and upgrading numerous rural clinics and provincial hospitals.
- A diarrhoeal control programme was launched in 1982, and a Department of National Nutrition was established — for nutrition and health education, for growth monitoring and nutrition surveillance and for supplementary feeding programmes for children.

In addition to these measures, most of which meet the criteria for well-structured meso policies in the health sector, there has been similar restructuring of education to increase the share of primary education in total public spending.

Although these improvements were not enough to prevent a slowdown of progress in human development, they may have prevented a reversal during the prolonged recession (box 2.4). Zimbabwe's experience shows that although it may not be possible to rely on meso policies alone to sustain progress in the face of poor growth, improvement in their structure can avoid reversals, at least in the short run.

Missed opportunities for human development

Brazil

Brazil failed to achieve satisfactory human development despite high incomes, rapid growth and substantial government spending on the social sectors.

An upper-middle-income country, Brazil had a per capita GNP of $2,020 in 1987. Except for 1980-87, when its per capita GDP grew at just over 1% a year, Brazil's growth has been quite good — with average annual growth of per capita GDP hovering around 3% in the 1950s and 1960s and rising to a very respectable 6.4% in the 1970s.

Central (federal) government expenditures in the social sectors ranged between 8% and 10% of GDP during 1973-86. As a percentage of total central expenditures, they remained at about 50% during 1973-79 but fell to 35% in 1986. Surprisingly, the level of meso policy interventions was quite high, even in comparison with countries at comparable income levels. Moreover, the social spending by state and local governments matched that of the central government. Total social expenditures by all levels of government and by the private sector are estimated to have constituted a quarter of GDP in 1986.

Despite rapid growth and substantial meso interventions, Brazil's human development record has been unsatisfactory. The under-five mortality rate was still 85 per 1,000 in 1988, almost twice Sri Lanka's and only slightly lower than Myanmar's, countries with per capita incomes amounting, respectively, to a fifth and a tenth of Brazil's. Life expectancy was 65 years in 1987, and the male and female literacy rates respectively were 79% and 76% in 1985.

These national averages hide significant

regional differences. In the poorer North-east, for example, infant mortality rates were more than twice those in the rest of Brazil in 1986 (116 compared with 52), life expectancy at birth in 1978 was only 49 years compared with 64 in the rest of Brazil, and child malnutrition was twice the national average.

There are two important reasons for such poor human development in Brazil. One is the extreme inequality of income distribution. The other is the inefficient targetting of public resources. The distribution of income in Brazil is among the worst in the world, with the Gini coefficient estimated at 0.60 in 1976, 0.56 in 1978, 0.56 in 1980 and 0.57 in 1983.

As indicated earlier, well-structured meso policies can compensate for a poor distribution of income and improve the human condition. This has not happened in Brazil because public resources did not reach the poor or improve the basic dimensions of human development. Substantial public subsidies were provided for "private" goods, usually consumed by the better-off sections of society, while "public" goods and services likely to have the widest impact on human welfare were neglected.

Brazil spends large amounts on social security (7.4% of GDP in 1986) and on housing (2.9% of GDP), with the benefits going disproportionately to the urban employed. Expenditures on social security may not have increased inequality since they are financed mainly by the beneficiaries, but a considerable amount of the expenditure in housing is for subsidies. Spending on health and nutrition, by contrast, had a lower priority: about 2.2% of GDP went for health at all levels of government in 1986.

In health, preventive programmes — such as immunisation, prenatal care and vector-borne disease control — are estimated to be about five times more cost-effective than curative programmes in reducing mortality. But an estimated 78% of all public spending on health goes to largely curative, high-cost hospital care, mainly in urban areas and especially in the urban South. This is in sharp contrast with the 87% of public health expenditure that Brazil allocated to preventive care in 1949, a

FIGURE 3.3

Missed opportunities for human development: country profiles

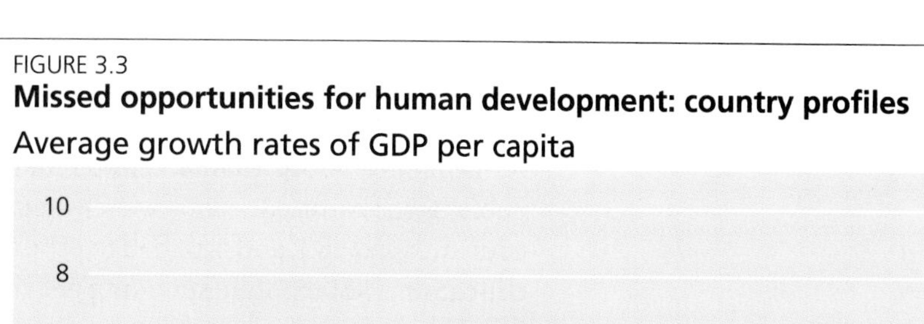

Average growth rates of GDP per capita

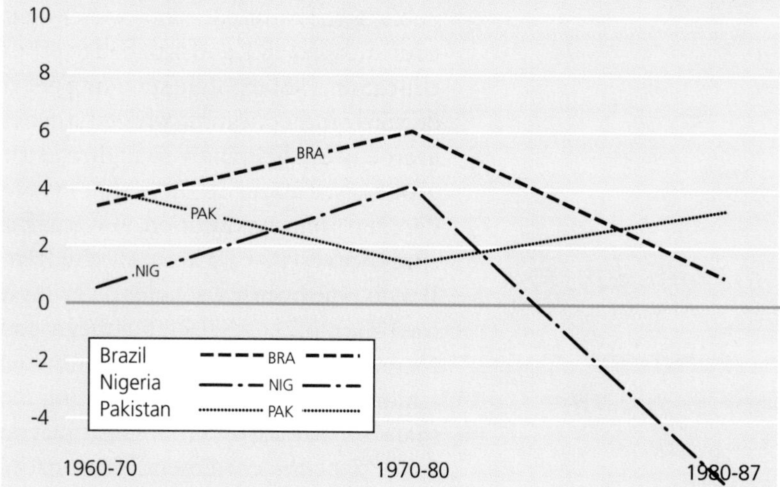

Under-five mortality rate

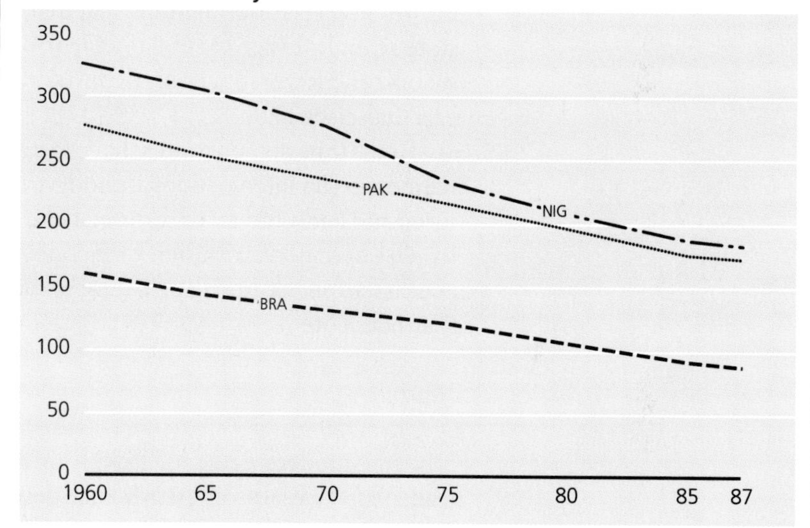

Social sector public expenditure, percentage of GDP

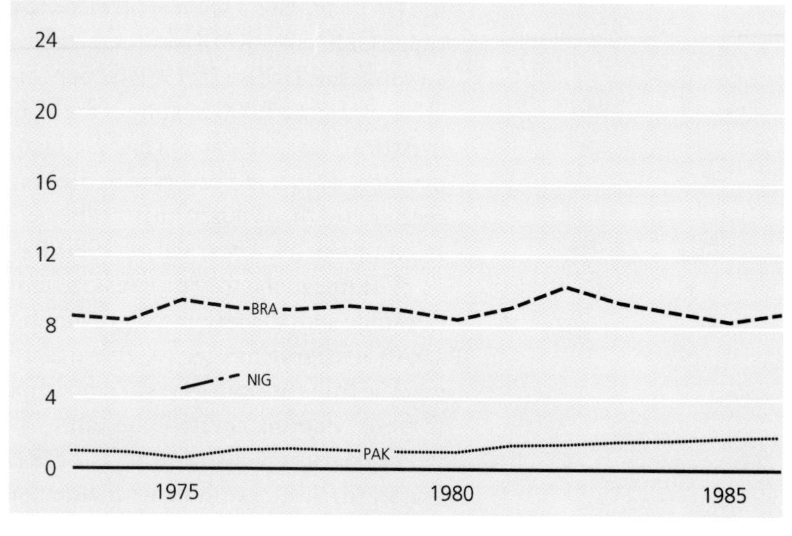

Substantial meso policy interventions, if poorly structured and badly targetted, cannot make up for an unequal distribution of income

share that fell steadily to 41% in 1961 and to a low of 15% in 1982 before rising to 22% in 1986.

Similarly, more than a quarter of all public spending on education went to higher education in 1983, and only half to primary education. Total public spending per student in higher education, where the benefits accrue overwhelmingly to higher-income groups, was about 18 times that in secondary and primary education. A World Bank study shows that 13% of all children in Brazil come from households receiving less than one minimum salary, but they account for only 1% of higher education enrolment. Children from households earning more than 10 times the minimum salary account for 48% of the enrolment but constitute only 11% of all children in the country. That is not the only inequity in the system. Spending per pupil is lower in municipal than in state schools, lower in rural than in urban schools and lower in schools in the Northeast than elsewhere.

Brazil thus demonstrates that substantial meso policy interventions, if poorly structured and badly targetted, cannot make up for an unequal distribution of income — even if the overall growth of income is more than adequate.

Nigeria

Nigeria's moderate rates of growth did not lead to substantial progress in human development. Its per capita GDP increased only 0.6% a year in the 1960s, partly as a result of the civil war. The discovery of oil led to per capita GDP growth of a very respectable 4% a year in the 1970s. In 1980 its per capita GDP of about $1,000 was one of the highest in Africa, classifying it as a middle-income country. This trend reversed in the severe recession of the 1980s, with per capita GDP falling about 5% a year during 1980-87.

The unsatisfactory progress in human development, despite rapid growth in the 1970s, can be attributed to several factors.

The fruits of rapid growth do not appear to have been distributed equitably. Evidence on income distribution is weak and scattered, but there is general agreement that the distribution was getting more unequal between 1960 and 1980, with the Gini coefficient for the late 1970s reported to be about 0.60.

Nor have the supplies of the goods and services that contribute to human development been adequate. The availability of food, for example, is estimated to have fallen by nearly a quarter between 1965 and 1975. The accompanying sharp rises in food prices suggest that food supplies did not keep pace with demand.

Detailed time series on the level and structure of social sector public expenditure are not available, and the IMF provides data only for some scattered years. But other evidence on per capita public expenditures in the health sector for 1964, 1970 and 1976 show very low levels both in absolute terms and in comparison with countries at similar incomes. For example, in 1976, total expenditure (current plus capital) was only about $1.75 per capita. By contrast, in 29 countries with a GNP per capita between $300 and $599, the per capita government expenditure on health exceeded $2 in 18 countries and $6 in 11 countries.

The bias in public spending towards curative services was also heavy. For example, in the second five-year plan (1970-74), 80% of federal capital expenditure was earmarked for teaching hospitals and urban areas. Lagos, with about 4% of the population in 1970, had more than 90% of all registered medical practitioners in 1973, 67% of all state hospitals and clinics and 72% of all private clinics. This strong bias towards curative care in urban areas meant that only a small proportion of the rural population had access to medical services. One estimate suggests that only 25% of Nigerians, most of them in urban areas, had health coverage in 1975.

Education received a higher priority than health in the national plans. In 1977, for example, education absorbed more than 40% of the recurrent federal budget and 55% of the recurrent state budgets, but these figures conceal the neglect of primary education. Although universal primary education was a major objective in the mid-1970s, the structure of the government's education spending has not reflected this priority. Primary education received less than 20% of public current educational expenditure in 1981, among the lowest ratios in Africa.

A systematic analysis of the distribution of the benefits of public expenditure in 1977-78 concluded that the federal government's capital expenditure was unambiguously pro-rich in both urban and rural sectors, although the distributional incidence of federal recurrent expenditure among urban and rural households was rather proportional and tended to maintain the status quo of income distribution. At the upper end of the income distribution, however, there was a tendency for benefits to rise as a proportion of income. So, the structure of public expenditure in Nigeria did nothing to compensate for the maldistribution of income.

Nigeria thus provides a clear example of failed trickle-down — of missed opportunities for human development. Rapid growth did not significantly improve the human condition because of basic flaws in the growth process and the failure to restructure meso policies to compensate for them.

Pakistan

Pakistan's GDP per capita rose almost 4% a year in the 1960s. Although the rate of growth dropped to 1.6% a year during the 1970s, it became respectable once again during the 1980s, increasing at about 3.5% a year during 1980-87. The distribution of income has been moderate.

But the country's human development has been unsatisfactory, particularly when contrasted with that of Sri Lanka, whose growth before the 1980s was fairly modest and whose per capita income has been broadly similar. In 1987, life expectancy in Pakistan was only 58 years, much lower than Sri Lanka's 71 years and even below the average of 61 years for low-income countries, among which Pakistan counts as one of the richest. Likewise, its under-five mortality rate — 277 in 1960, compared with 202 in China — was still 166 in 1988, compared with 43 in China.

Pakistan's performance on other important basic indicators of human development leaves much to be desired. Its adult literacy rate in 1985 was strikingly low at 30%, with large gender disparities — female literacy was 19%, male literacy 40%. And its gross primary school enrolment ratio was still only 40% in 1987. Pakistan is still far from universal primary education, something Sri Lanka has already achieved and China is pursuing. Again, the gender disparities are wide: in 1987 fewer than a third of Pakistani girls were enrolled in primary schools, compared with half the boys.

This dismal performance despite respectable rates of growth and a moderate income distribution can be explained by the failure of meso policies. Although growth has been good, Pakistan is still a low-income country. This low income implies that primary incomes — even if equitably distributed, which they are not — are insufficient, on their own, to permit the bulk of the people to acquire the goods and services needed for a decent life. Pakistan thus needs well-structured meso policies to promote human development — policies that have been gravely deficient.

There are several examples of failed trickle-down — of missed opportunities for human development

Several factors explain the failure of rapid economic growth to translate itself into satisfactory human development. Education and health are provincial responsibilities, but the provinces lack adequate financial resources — and a major decentralisation of financial powers from the federal government to the provincial governments in accord with the 1973 constitution has been pending with the National Finance Commission since 1974. There is also a serious imbalance between military and social expenditures — an imbalance that grew much worse in the 1980s as military expenditures rose five times while public sector development expenditures only doubled.

Pakistan spends a very small part of its budget on the social sectors — and a large and growing part on the military, preempting scarce resources that could otherwise be earmarked for education and health. Only 2.2% of Pakistan's GNP went for education and health in 1986, compared with 6.7% for military expenditures. Military spending was three times the spending on education and health. Even adding the fairly considerable spending by the provincial governments on education and health does not challenge the overall conclusion: Pakistan spends too little of its GNP on social development.

Moreover, a large part of the limited social expenditures goes to lower-priority activities. Of the public current expenditures on education, 24% was for tertiary education in 1985, compared with 7% in Sri Lanka in 1986, and only 40% was for primary education. There appears to be a similar bias towards lower-priority activities in health spending, but some recent policy changes are steps in the right direction. A nationwide immunisation programme was financed by postponing the construction of an expensive urban hospital. Education spending was tripled in the last four years. And a special tax was levied on all imports to finance additional spending on education.

Pakistan's overall experience shows that inadequate social spending and poorly structured meso policies can prevent a low-income country from improving the human condition even if there is rapid economic growth with a relatively moderate distribution of income.

CHAPTER 4

Human development strategies for the 1990s

The 1990s are shaping up as the decade for human development, for rarely has there been such a consensus on the real objectives of development strategies. The UN Committee for Development Planning summarises this emerging consensus best: "In the 1990s people should be placed firmly in the centre of development. The most compelling reason for doing so is that the process of economic development is coming increasingly to be understood as a process of expanding the capabilities of people."

Any development strategy for the 1990s will combine a number of objectives: among them, accelerating economic growth, reducing absolute poverty and preventing further deterioration in the physical environment. The departure from earlier development strategies lies in clustering all these objectives around the central goal of enlarging human choices.

The economic slide in the Third World in the 1980s, particularly in Africa and Latin America, must be reversed in the 1990s — and the accelerating economic growth must be used to advance the cause of human development. Average per capita income growth in the two worst affected regions actually declined in the 1980s. Higher levels of investment and sensible economic management are needed to increase annual growth rates to between 2% to 3% in the 1990s.

Besides expanding human capabilities and creating a conducive environment for their optimum use, development strategies must attend to people who live in absolute poverty and need special government support to reach an acceptable threshold of human development. On current projections, the number of people in absolute poverty is likely to increase from more than 1 billion to around 1.5 billion by the end of the century. The sharpest increase is expected in Africa, from around 270 million to some 400 million. A central objective of future development strategies must be to *reduce* the number of people in poverty in each country by the year 2000.

There is also a growing consensus that the objective of safeguarding the natural environment must be integrated with future development strategies, recognising that environmental problems of developed countries differ from those of developing countries. In developing countries, poverty often causes environmental damage — deforestation, soil erosion, desertification and water pollution — and environmental damage reinforces poverty. Thus, the environmental priorities of developing countries often concern natural resources, especially water and land.

In the developed countries, by contrast, affluence may cause different environmental problems — waste, carbon dioxide emissions, acid rain. Thus, their concerns are often with air pollution, which has reached alarming global proportions.

Any well-conceived development strategy must respect the differing perceptions of developing and developed nations on environmental issues and reflect the different stages of their development. In addition, the development process should meet the needs of the present generation without compromising the options of future generations. However, the concept of sustainable development is much broader than the protection of natural resources and the physical environment. It includes the protection of human lives in the future. After all, it is people, not trees, whose future options need to be protected.

This chapter deals with the policy measures that could accelerate progress in human development in the 1990s.

Policy measures for priority objectives

Growth with equity

Growth with equity is the best recipe for accelerated human development

The wealth of analysis delving into the "causes of growth" has produced three conclusions.

- The investment rate is an important determinant of growth, but there is considerable uncertainty about how much extra growth comes from more investment. To sustain growth, countries should aim to maintain the investment rate at 15% to 20% of GDP. (Countries cutting their investment in real terms — many in Africa and Latin America in recent years — will thus find it very difficult to sustain their growth.)
- Even more important is the rate of technical change, associated with science, technology and the development of human capabilities. So, promoting human development is important not only in itself but also as a critical input to the growth process.
- The policy environment is important for the efficient use of investment resources and for adapting to changing world conditions in ways that permit sustained growth.

As with growth, the literature on the determinants of income distribution is vast and complex, but it has produced two general conclusions about the better distribution of primary income that does so much for improving human development.

- Good asset distribution, which for developing countries usually means good land distribution, is important. A study of alternative development strategies over the past 30 years found that a good distribution of primary income was invariably associated with fairly equal land distribution. Countries that have had a land reform — China, the Republic of Korea and the Democratic Republic of Korea — have reduced poverty and inequality quite considerably. Most countries that have not — such as Brazil and the Philippines — continue to have large numbers of people living in poverty, even when they have managed high rates of economic growth.
- Rapid expansion of productive employment opportunities is essential for spreading incomes throughout the population. In mixed economies, such an expansion comes through rapid, labour-intensive growth, as in the Republic of Korea. In socialist countries, the state's ownership of most assets, accompanied by employment policies that typically secure jobs for all productive members of the labour force, tends to generate a good primary distribution. But such countries have often sacrificed efficiency for equity.

Growth with equity is the optimal combination for generating good macro conditions needed to achieve human development objectives. Despite much controversy on the appropriate policy environment, there is a modicum of agreement that the essentials for equitable growth comprise (i) sensible and flexible use of prices to reflect opportunity costs, (ii) the opening of market systems, (iii) supportive policies towards investment, technology and human resources and (iv) policies for distributing assets and expanding productive employment opportunities — with the appropriate mix tailored to individual countries.

Meeting the needs of all

A well-structured set of meso policies, especially needed where the distribution of primary income is poor, involves two features to ensure that the benefits reach the deprived. First is across-the-board provision of basic services, generally desirable for basic health and education. Second are targetted schemes directed towards deprived groups, such as income support and some food subsidies.

Well-structured meso policies usually require a mix of the two. The across-the-board provision of services alone may be enough in countries with good macro policies and especially a good distribution of primary income. The targetted schemes can be important where macro policies result in a skewed primary distribution and meso policies are needed to compensate, though here too some services need to be provided across the board.

Since social sector expenditure policies are reviewed later in this chapter, the focus here is on the provision of privately supplied goods, such as food. Policies to ensure that everyone has access to enough food may consist of:

- *Income support schemes.* Public works employment schemes have been successful in Chile and in India, in the Maharashtra drought relief scheme. Alternatively, direct cash payments can support the income of households in extreme poverty — a policy common in developed countries but less suitable in developing countries where the number of households involved is much greater and the administrative machinery weaker. Some countries, such as Chile, have nevertheless targetted cash support successfully to needy households.
- *Food subsidies.* An alternative or complement to income support schemes is to hold down food prices through food subsidies of various kinds (box 4.1).
- *Special nutrition programmes.* Such programmes can cover particular segments of the population — such as providing free school lunches in primary schools, which have the added advantage of encouraging school attendance and improving concentration at school — and at those identified as being nutritionally in need. Chile and Botswana have used these programmes to combat extreme malnutrition.

Tackling disparities

Disparities within countries are one of the biggest obstacles to improving the human condition. To reduce rural-urban disparities, the proportion of resources allocated to rural areas must increase, and even more important the decisions about priorities and resource allocations should be made locally. Such decentralisation of the decisionmaking for the allocation of public goods may be one of the most important ways of reducing rural-urban gaps.

Female-male disparities have to be tackled at several levels. Laws need to be changed to provide equal access to assets and employment opportunities. The institutions that provide credit and disseminate technology need to be restructured to reach many more women. Reforms are also needed to bring about full female participation in political, bureaucratic, and economic decisionmaking at every level. And traditional biases in the household against the young, especially females, must be blunted. In all this, the equality of women's access to education is essential. In addition, specially targetted programmes should, where appropriate, support the health and nutrition of young women. For example, every country should provide health care for women during pregnancy and birth.

The reallocation of social infrastructure is important in reducing the disparities between rich and poor. Measures are needed to encourage greater use of health and educational facilities by low-income groups — say, through nutritional support programmes in health clinics and school feeding programmes. And where access to education is limited, it is important to ensure admission by merit, rather than by connection, as is common.

BOX 4.1

In defence of food subsidies

Food subsidies can do much to stabilise food prices, transfer income to the poor and maintain political and social stability.

- They contributed the equivalent of 16% of the purchasing power of low-income families in Sri Lanka (at their peak).
- They increased consumption of the poorest 15% of the urban population by 15% to 25% in Bangladesh in 1973-74.
- They contributed about half the income of low-income families in India's Kerala state in the late 1970s.
- They accounted for about 16% of the income of the poorest quintile of the population in Egypt in the early 1980s but only about 3% of the richest quintile.

Often effective in transferring income to the poor in societies where overall income distribution is quite uneven, food subsidies have established an essential social safety net in many poor societies at not too great a cost (generally 1% to 2% of GNP). They have often compensated for the lack of social security schemes existing in industrial nations. The subsidies have, moreover, created an invisible bond between the poorer masses and the government. But when that bond is broken without creating an alternative social safety net, the political and social violence can cost far more than the subsidies.

The design of food subsidies always demands care. The budgetary burden should be kept manageable. Incentives for food production should not be discouraged. Targetting should ensure that benefits reach the poor to make the programme cost-effective. To reduce costs, the subsidies need to be targetted towards low-income households — by subsidising foods mainly consumed by low-income households or sold in low-income areas.

Rather than disapprove of food subsidies across-the-board, policymakers should devote their energy to designing food subsidy packages that redistribute income efficiently without hurting the efficiency of resource allocation.

Encouraging more participatory development

This Report places people at the centre of human development — as the agents and beneficiaries of the development process. People's needs and interests should guide the direction of development, and people should be fully involved in propelling economic growth and social progress.

Participatory development starts with self-reliance, which means people being able to take care of themselves. To stress people's economic, political and social self-reliance is not to argue against state intervention in human development. On the contrary, greater participation of all people in the development process depends on carefully designed government policies and programmes. But government interventions in support of human development should also encourage private initiative in the broadest sense — including that of private entrepreneurs, that of nongovernmental organisations (NGOs) and other community-based and self-help organisations, and that of people as individuals or households.

Welfare measures are an important aspect of policies for the poor, but the long-term solution to poverty requires more developmental measures. The poor have to find access to the means and opportunities for bringing them into the mainstream of development. Today, many policies and programmes for the poor adopt such an enabling strategy. Vocational training and other types of skill formation are important elements of these strategies, as is the provision of credit to the poor. NGOs have done much to make credit programmes work (boxes 4.2 and 4.3).

The effectiveness of some NGOs in programmes that require close and direct involvement with people has been a major factor in increasing the collaboration between governments and NGOs in many developing countries.

Promoting private initiative

Governments throughout the world have increasingly come to recognise that the private sector can and should play an important role in the development process.

Four major policy areas are generally recognised as essential to private sector development in the developing world:

- The creation of a proper enabling environment for private sector development, including new legislation and regulations supportive of private sector growth.
- Privatisation, especially of productive functions more efficiently performed by the private sector.
- The development of micro, small, and medium-size enterprises through such mechanisms as small-scale credit schemes, volunteer executive programmes and venture capital.
- The improvement of public sector management and private sector management training.

What is required is a smaller but more effective public sector capable of creating an enabling development framework and guiding private investments into priority areas for human development. The role of the public sector should be confined primarily to building economic infrastructure and to providing social services. Successful development will depend on the right strategy mix — on putting in place a policy

BOX 4.2

Rural banks in Ghana

Ghana's rural banks, supervised by the central bank, serve areas that other financial institutions ignore. There are 106 such banks — independent and run by the community — providing places to save and to make loans, principally to small farmers and the owners of cottage industries.

Growth has been extremely rapid — from 148,000 cedis in 1977 to 862 million cedis in June 1985, and from 802 savings account holders in 1977 to 221,000 at the end of 1984.

This growth shows that rural peasants and village dwellers will save if they have confidence in the bank, find it convenient and have ready access to their savings.

Since their start, the rural banks have lent 554 million cedis, almost all from local savings, with loans averaging 12,000 to 18,000 cedis ($200 to $300 at the 1985 exchange rate). By the end of 1984, there were 32,000 borrowers.

What accounts for this success?

- Mobilising local initiative is vital. The directors of the banks are local leaders committed to their community's development. In each community, a broad base of community members are shareholders.
- Funds mobilised locally are used locally, in sharp contrast to many banks that funnel rural savings to the cities.
- Loan approvals are based on the producer's reputation in the community, not on abstract guidelines and collateral requirements that eliminate most potential borrowers. The owners of the smallest farms and smallest businesses are the best candidates for loans.
- Administrative costs are kept low by using simple and standard procedures and recruiting staff from the community.

The potential for savings in rural Ghana is considerable. Local savings represent well over 90% of the loan fund from the first year of operation.

package that combines private and public sector strategies in the interest of people-oriented development.

Appropriate strategies and sequencing

The combination of policies appropriate for a country depends in large part on the level of income per capita, on the achievement in human development and on the distribution of assets and income. Five categories of countries can be differentiated by the combination of these conditions, with policy combinations suggested as appropriate for each category (box 4.4).

Although well-designed meso policies can compensate for a poor income distribution, they cannot substitute for the economic growth needed to finance the meso policies over the long run. Policies to sustain or restore economic growth are thus critical for *every* category of country.

As countries progress, their changing conditions will call for new policy combinations. Governments also face the problem of how best to sequence actions in the social sectors when each of them has major deficiencies. The limited evidence suggests that where choices have to be made, primary education should be given priority, with the provision of low-cost health interventions coming a close second. This issue of sequencing demands further research to define "production functions" for various aspects of human development and to identify social returns to different types of social expenditures over time (box 4.5).

Policies for adjusting countries

The policies appropriate for adjusting countries do not differ in their fundamentals from those applicable more generally. The major difference is that many adjusting countries are suffering *declines* in government expenditures and in per capita incomes. Moreover, in debt-ridden countries the resources available for the social sectors are further limited by the need to devote a high proportion of the budget to interest payments. These countries have particular difficulty in securing a favourable macro environment for human development and must thus pay special attention to well-structured meso policies. Because of their very tight budgets, they need to focus on low-cost programmes — to keep down the cost of across-the-board interventions — and to rely more on targetted schemes.

Debt-ridden adjusting countries have the greatest difficulty in securing the resources for improving human development — and the greatest need, since continuing downward pressures on human development will further undermine their long-run growth prospects. Moreover, living conditions in many adjusting countries are becoming intolerable.

In addition to their focus on meso policies, it is particularly important for these countries to restore equitable growth. For this, they need greater international financial support — to give them more time to adjust, to combine adjustment with growth and to protect and promote human development.

The discussion here assumes that the overriding objective of governments is the improvement of human development for all their people. As is well known, however, the reality is often otherwise. Governments are subject to many pressures from interest groups. Moreover, their objectives are often complex and multidimensional: to stay in power, to serve particular interest groups and sometimes to enrich themselves. That is why the political will and institutional

BOX 4.3

Money shops in the Philippines

The money shops of the Philippine Commercial and Industrial Bank make small loans — between $125 and $1,250 — to market stall-holders at a reasonable profit. The key to the money shops' success is their closeness to — and close interaction with — borrowers.

The first two money shops opened in 1973. By 1979 there were 70 of them scattered throughout the Philippines. Either in or on the fringe of private or public markets, the money shops meet the short-term (30 to 60 day) credit needs of commercial customers. The shops also accept deposits.

The shops usually operate in fairly large urban markets, because there must be 400 businesses in the market site to justify placement. Facilities are extraordinarily simple, often a wooden stall large enough to accommodate four employees.

For stall-holders to qualify for the minimum loan of $125, they must have a daily sales volume of $7.50 to $8.75 and profits of 10% to 50%. The terms are 14% interest for one year plus a 2% monthly service charge, with daily repayment. The recovery rate has been 98%.

The Philippines' experience demonstrates that it is possible to reach fairly modest income groups through commercial banks at market rates of interest if loans are properly targetted.

BOX 4.4

Different strategies for different contexts

Different countries have followed different strategies in translating economic growth into human development. They can be classified according to their initial socioeconomic conditions and the policy packages which they adopted, leading them to successful human development. Such a classification helps identify combinations of policy measures appropriate to different development contexts.

Category I countries have been facing the greatest difficulties. They had low income, low human development and uneven income distribution. Many countries in this group suffer from further handicaps. Some are debt-ridden adjusting countries and have experienced particular difficulties in establishing good macroeconomic conditions and increasing the resources needed for human development. Some depend primarily on the export of one commodity whose international prices can fluctuate violently. Nevertheless, the successful ones among these countries — for example, Kenya and Zimbabwe — have improved human development by adopting strong, well-structured meso policies.

Category II countries are those that started out with low income, a moderate income distribution and modest levels of human development. Many of them have — as do several category I countries — civil strife and war deflecting scarce resources from human development. International support for human development should be focused on category I and II countries.

Category III countries, despite their initially low incomes, have significantly better prospects than those in categories I and II. Having already achieved moderate human development, partly because their income distribution was not too uneven, they needed continued growth and the expansion of appropriate across-the-board meso policies to accelerate progress. Sri Lanka and the Republic of Korea followed this strategy successfully.

Category IV countries have low levels of human development despite moderate income levels, which could have placed better performance within easy reach. Judging from past experience, these countries need to improve their primary income distribution and compensate for their poor income distribution through much stronger and better-structured meso policies. Malaysia is a middle-income country that raised human development levels in this way.

Category V countries should — if they follow good meso policies while maintaining their generally good growth and distribution performance policies — attain high levels of human development within a few years. Costa Rica and Colombia provide examples of what such countries can achieve.

Country strategies for human development

Initial conditions	Priority policies	Countries that have successfully implemented these policies
Category I		
Low income Uneven income distribution Low human development	Improve growth and distribution Increase share of social spending Target social subsidies and programmes	Botswana Kenya Zimbabwe
Category II		
Low income Moderate income distribution Low human development	Improve growth Maintain distribution Increase share of social spending Across-the-board meso interventions	China
Category III		
Low income Moderate income distribution Moderate human development	Improve growth Maintain distribution Increase share of social spending Across-the-board meso interventions	Republic of Korea Sri Lanka
Category IV		
Middle income Uneven income distribution Low and moderate human development	Maintain growth Improve distribution Increase share of social spending Target social subsidies and programmes	Malaysia
Category V		
Middle income Moderate income distribution Moderate human development	Maintain growth and distribution Increase share of social spending Across-the-board meso interventions	Chile Colombia Costa Rica Jamaica

capacity to follow the most appropriate strategy are often lacking.

The political resistance can be great when reductions are proposed in social expenditures benefitting primarily powerful and favoured groups. The political resistance can be even greater when it comes to pruning military expenditures or undertaking land reforms. By contrast, the potential beneficiaries of change generally have little voice or political muscle.

As this Report has emphasised, however, success in human development is widespread and possible even in poor countries. Moreover, deeper analysis of the human condition in each society will by itself exercise considerable pressure for change. And often, a suitable climate has to be created for any major change.

Setting global targets for human development

The global targets that the international community sets at world conferences and during UN General Assembly debates must be seen more as desirable objectives — as indications of the direction that development should take — than as carefully calculated projections of what is feasible and realistic. Several global targets have been set during the past three decades, many of direct relevance to human development. Some have been broad: health for all by the year 2000. Others more precise: under-five child mortality rates in all countries should be 70 or halved, whichever is the lower, by the year 2000.

Setting global targets for socioeconomic progress for each decade has its advocates and its critics. The advocates point out that the adoption of global targets creates a conducive environment and political pressure for their serious pursuit nationally and internationally. The critics maintain that the global targets have no price tag, are not differentiated according to different country situations, are not accompanied by concrete national and international plans for implementation — and that any link between national progress and global targets is only incidental. Instead of taking extreme positions, it is far more productive to ensure that global target-setting for the 1990s is more realistic and operational.

Quantified global targets for the year 2000 do exist for some of the key indicators of human development examined in this Report.

- Complete immunisation of all children.
- Reduction of the under-five child mortality rate by half or to 70 per 1,000 live births, whichever is less.
- Elimination of severe malnutrition, and a 50% reduction in moderate malnutrition.
- Universal primary enrolment of all children of primary school age.
- Reduction of the adult illiteracy rate in 1990 by half, with the female illiteracy rate to be no higher than the male illiteracy rate.
- Universal access to safe water.

BOX 4.5

Priority research agenda for human development

In preparing this first *Human Development Report*, it became obvious that considerable research is needed in many areas before policy recommendations can be offered with confidence. The following topics emerge at the top of the policy agenda for such research.

Data collection. Far fewer resources are invested in collecting information on human development than in generating conventional economic data. As a result:

- National data on life expectancy, literacy, and child malnutrition are available not year-by-year but only through periodic household surveys or decennial censuses. This makes it difficult to estimate the effects of changing conditions and especially to pick up any deterioration in human development quickly. A permanent programme of representative national household surveys should provide regular monitoring of the human condition.
- For most indicators, only countrywide data are available, with little information on the various indicators for income or population groups, or for urban and rural areas, or even for major geographical regions. Data on absolute and relative poverty are also often lacking. All this information is essential for appropriate policy design and for assessing the effectiveness of policies.

A production function for human development. Little is known about how inputs relate to outputs in human development — for example, about what combination of health services, education and nutrition support bring about the best improvements in child mortality. Yet without this knowledge, governments have difficulty in identifying cost-effective and efficient policies. A comprehensive survey of microstudies in the sociological, economic, medical, biological and public health fields would be a useful first step in developing production functions for human development. Evidence from Western Europe, Japan, and the most successful developing countries would also help increase knowledge about the optimal sequencing of policies towards the social sectors.

Financing and restructuring the social sectors. More research is needed on alternative strategies and methods of financing the social sectors, including general tax reforms, local finance and the use of fees. In addition, the studies should be carried out on the potential for restructuring within the social sectors.

Decentralisation and local government. Research is needed as well on the forms and impact of decentralised decision-making — and on mechanisms for inducing effective participation, especially among groups usually unorganised and with little influence, such as poor women and the landless. The role of the private sector and NGOs in promoting human development also needs further study.

It is necessary to ensure that global target-setting for the 1990s is more realistic and operational

A possible measure for the realism of these targets is the rate of past progress in particular countries (see the annex tables).

Immunisation. If developing countries maintain their past rate of progress, most of them could achieve full immunisation coverage of their children by the year 2000. The main exceptions requiring additional effort are Niger, Mauritania, Mozambique, Yemen Arab Republic, Liberia, Ghana, Côte d'Ivoire, Papua New Guinea, Libya and Mongolia. But even in these countries, the target may be attainable since past achievement rates are only a crude yardstick.

There have recently been major improvements in the quality of vaccines and vaccination technology. In addition, more people recognise the importance of immunisation, and more trained medical personnel are available for the implementation of vaccination programmes. Because of these advances, the coverage of child immunisation increased sharply from 30% in 1981 to nearly 70% in 1988, saving an estimated 1.5 million lives annually by 1988.

The growing problem of AIDS cautions against complacency, however, for its spread may make parents more hesitant to bring their children for vaccination. The solution, altogether manageable, lies in taking the necessary sanitary measures, using the new self-destruct syringes and embarking on public information campaigns.

Child mortality. Reducing the child mortality rate in all countries by half is ambitious in the light of the last three decades' achievements. The annual rate of reduction required to reach this target is higher than what most countries have achieved, particularly in Africa. At past rates of progress, 23 countries — almost all with low human development — will not attain the target until after 2050.

The child mortality rate differs from the immunisation rate in that it is an outcome rather than an intervention. Immunisation requires vaccines, health personnel and other inputs that can be defined in a rather precise manner. But child mortality is the result of many factors, some controllable and others less controllable. Pushing the rate up are malnutrition from food shortages during droughts, increasing poverty resulting from overall economic stagnation, and the spread of AIDS. Pulling it down are higher female literacy, cleaner water, better sanitation, broader immunisation and so on. The under-five mortality rate thus defies precise prediction, and its reduction requires especially careful planning and monitoring.

Malnutrition. One significant target for ensuring the survival and development of children in the next decade is to eliminate severe child malnutrition and to halve moderate malnutrition by the year 2000. Most countries will require annual reductions in malnutrition of around 5% to 7% to meet the global target — not overly high and within the reach of properly targetted nutrition policies and programmes. The cost of such programmes is generally small, and the payoff large. Moreover, attaining some of the health and education goals for the 1990s will greatly contribute to reducing child malnutrition.

Primary school enrolment. Perhaps the most important human development target for the year 2000 is to ensure universal primary school enrolment for boys and girls. Although past performance could be calculated for only some developing countries, universal primary school enrolment may indeed be a reality by 2000. Some countries will have to increase their enrolment rates sharply: Mozambique, Malawi, Rwanda, Haiti, Saudi Arabia and Nicaragua. Others will have to reverse recent declines: Mali, Somalia, Tanzania and Morocco. But if primary school enrolment is universal by 2000, literacy rates will rise dramatically throughout the developing world in the early part of the next century. This most valuable of investments in human development will help unleash the creative energies of the four-fifths of humankind living in the developing world. The Declaration and Framework for Action adopted at the World Conference on Education for All held in Jomtien, Thailand, on 5-9 March 1990, holds out the promise of more rapid progress in this area as it has moved to the top of the policy agenda for the 1990s.

Adult literacy. The global target for the year 2000 is a 50% reduction in today's adult illiteracy rates, with a sharper increase in female literacy so that the gender gap in literacy disappears. Several countries may achieve the target for male literacy if they maintain or somewhat accelerate their past progress. The real problem is in such populous countries as India, Pakistan, Egypt and Sudan, which despite their very low literacy rates have allocated only modest amounts to education and thus expanded literacy at a painfully slow pace over the past three decades. The effort required for female literacy is even greater.

Is the global target for literacy realistic? The answer is difficult because there are many uncertainties. Male-female disparities and role differentiation are deeply rooted in the socioeconomic traditions of many countries, and breaking away from these traditions will take time. An encouraging note, however, is the stronger developmental role of NGOs in recent years and the explosion of an informatics revolution. With their support and involvement, faster improvements in literacy might be possible than those in the past, especially if their efforts augment rather than replace government programmes.

Safe water. Safe drinking water is available to 700 million more people than in 1980, and sanitation to 480 million more, yet the majority of people in the developing world still lack these basic necessities. Even so, the target of providing safe drinking water to all by 2000 is attainable for most developing countries if they maintain past rates of progress. But capital investment must go hand in hand with human resource development to ensure the proper maintenance and repair of the infrastructure, especially for water. And development expenditures have to be matched by appropriate provisions for recurrent expenditure to finance the required technical personnel, including village maintenance brigades. Wrapped up in this issue of budget allocations are questions of user fees, community financing and other aspects of resource mobilisation.

Realism of targets. Appraising the realism of developmental targets is complex because developmental conditions and challenges vary so much from country to country. Past achievements are only a rough indicator of future progress. New technological breakthroughs might accelerate future achievements, but societal values may need to change before such objectives as female literacy can become fully attainable. Moreover, much individual and institutional capability remains to be built before such goals as clean water and sanitation for all can be achieved. And the natural calamities of drought and floods may throw countries back in their efforts to eradicate hunger and malnutrition. The initial steps often are the easiest, with the path becoming more treacherous later on. Nor is staying on track assured: the 1980s have seen several reverses in social progress, and maintaining past progress may become difficult unless economic performance improves significantly.

Another important issue is whether financial resources will be adequate for implementing the programmes to reach these targets. The financial costs for attaining various global targets generally have not been worked out, however, either globally or for individual countries. Without such costing, realistic planning and analysis can hardly begin.

UNESCO and the UNDP recently prepared a rough calculation of the cost of achieving universal primary school enrolment by 2000: $48 billion over the next 10 years, or between $4 and $5 billion if cost-effective methods are used. This figure implies that countries will have to increase their current budgetary outlays for primary education by about 50% during 1985-2000, an average annual increase of nearly 3% (compared with 1.7% a year over 1975-85). This does not sound overly ambitious. After all, this amount equals only two days of military spending in the industrial countries, or one week of military spending in the Third World, or about 2% of annual debt servicing by developing countries. However, the implications for some countries, especially those in Africa and others in the least developed category, need to be carefully watched. Their budgetary outlays for education may have to more than double at a time when their per capita GNP is projected to be flat or declining.

The operational feasibility and overall credibility of global targets in human devel-

Universal primary school enrolment by the year 2000 will require added spending of $5 billion a year

opment will increase considerably if four criteria are met.

- The number of global targets should be kept small to generate the necessary political support and policy action for their implementation. The international agenda is already crowded, and having too many targets diffuses policy attention.
- The implications for human and financial resources must be worked out in detail, country by country, *before* fixing any global targets — to ensure that the targets are realistic.
- Different targets should be fixed for different groups of countries, depending on their current state of human development and past rates of progress.
- National strategies for human development should bridge national planning and global target-setting, for without national development plans the global targets have no meaning.

Comprehensive human balance sheets must be in all national development plans

National plans for human development

The first step in preparing a human development plan is to draw up an extensive inventory of existing human resources and skills, of people's health, education and nutrition, their absolute and relative poverty, their employment and underemployment and their progress in the demographic transition. That inventory should also describe the prevailing disparities between females and males and the distribution of social services between urban and rural areas and among different income groups. And it should capture the cultural ethos, ideological aspirations and real motivations of the people.

In other words, a serious attempt should be made to prepare comprehensive human balance sheets for the first chapter in every national development plan, relegating the usual national income accounts — GNP, exports and imports, saving and investment, and so on — to technical annexes. Obviously, it is possible to plan for people only if more is known about them and if they are not reduced to mere abstractions.

The second step is to identify priorities. Existing conditions should be compared with the objectives to be achieved under the plan. Feasible objectives can be estimated either by observing what other countries in broadly comparable circumstances have already achieved or by using national data to calculate maximum possible improvements given the resources likely to be available. The difference between feasible objectives and initial conditions provides a first indication of possible priorities: the greater the difference, the larger the scope for improvement and, possibly, the higher the priority for policy.

The third step is to rank the priorities according to the people's preferences. There can be no presumption that two countries with equally low human development and equally sparse resources would attach equal importance to each component of human development. Adjusting plan priorities to reflect country preferences implies, of course, that countries have effective mechanisms for ascertaining people's preferences. The effectiveness of these mechanisms often depends on the democracy and decentralisation of political and economic systems and the encouragement of participatory development.

The fourth step is to translate these priorities into specific goals for primary indicators, such as life expectancy, literacy, and nutrition. The goals would be refined in the light of the programmes and targets for specific policy instruments and contributory indicators, such as access to health and education services and access to clean water. A country may have as a goal a five-year increase in life expectancy (a primary indicator), but it normally will have to link this goal to contributory indicators (such as the availability of health services and calories) and policy instruments (such as investment in food production and health services) to enable planners to transform the general objectives into precise guidelines for action.

There usually are several roads towards achieving a given human development goal. Life expectancy can be increased by greater access of the population to safe drinking water, by broader coverage of immunisation for children, by expanded primary health care facilities, by generally improved nutrition or by a combination of all these measures. Because resources are scarce, policymakers have to choose from among the various programmes. The various measures thus compete with one another, but they can also be complementary — that is, the presence of one programme can enhance the effectiveness of another. For example, a school feeding programme to improve the nutrition of children or a programme for rapid increases in female literacy could increase the effectiveness of a programme to expand primary health care.

The task of the development planner is to achieve as great an improvement as possible in a primary indicator with the resources available or, put differently, to achieve a given improvement in a primary indicator at the lowest possible resource cost. This requires knowledge about the functional relationship between a primary indicator and the various measures that can affect it and knowledge about the complementarity among policy instruments — knowledge that often is lacking.

Ideally, the planner should know the full range of available alternatives and the cost of each. For example, health services could be provided by expanding curative facilities at technologically sophisticated hospitals, by expanding preventive services and small clinics or by a combination of the two. The cost of these different approaches is likely to vary considerably.

Present knowledge about the "production function" for different components of human development is rudimentary. Serious research is clearly needed in this area. Meanwhile, planners have to get on with their job and do the best they can — perhaps using knowledge from other countries, particularly those with similar resource endowments and incomes but above-average progress in human development. Examining successful experiences can tell much about the instruments most effective in achieving particular goals. One major objective of future Reports will be to summarise such practical experience.

The fifth step in human development planning is to match the cost of planned human development programmes with the available resources. Ethiopia (with a per capita income of $130 and a domestic savings rate of 3%) and Congo (with a per capita income of $870 and a domestic savings rate of 21%) are in the same broad category of countries with low human development. Yet what is feasible in the Congo clearly might not be in Ethiopia. It is important, therefore, for planners to identify accurately the volume of resources that can be made available for human development by reallocating domestic expenditures and by raising additional resources.

Priorities are bound to differ among broad groups. For countries with low human development — almost all in Sub-Saharan Africa and low-income Asia — policymakers will aim at rapid improvement in all the primary indicators. Priorities are likely to be high for child immunisation programmes, primary health care (especially in rural areas), primary school enrolment (especially for girls), safe drinking water, food production and its distribution to malnourished groups, and programmes to reduce population growth.

Priorities in the countries with medium human development can be much more selective than in the first group. Their primary health care and access to safe drinking water (particularly in urban areas) are quite good. School enrolment rates, except in the Middle East and some Latin American countries, are high. Calorie deficiencies are negligible, except in some Latin American countries. Such countries can thus focus their human development programmes on reducing internal gaps: rural-urban, female-male and poor-rich. They can also focus on some objectives beyond the primary indicators, such as secondary enrolment, better quality education, improved food distribution, adequate shelter and a cleaner physical environment.

Human development planning can be done only at the national level, but many governments in the developing world are still not fully equipped to undertake such exercises entirely on their own. International agencies can provide the necessary technical expertise and assistance at the request of developing country governments for formulating their human development plans. The United Nations system, in particular, must assume a major responsibility in human development in the 1990s, since

Ideally, the planner should know the full range of available alternatives and the cost of each

FIGURE 4.1
Declining investment rates
Percentage of GDP

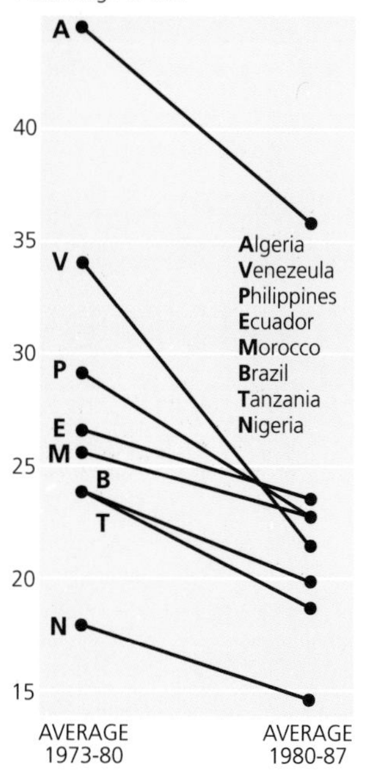

UN agencies are already dealing with individual social sectors and concerns. What is needed is to bring their expertise together at the country level for integrating human development in the overall macroeconomic framework. It is encouraging to note that the UNDP has launched an initiative along these lines in association with other UN agencies.

Financing human development

For several countries, the proportion of government expenditure on education and health fell between 1972 and 1987. Doubling the present proportion of such spending is a conservative estimate of what is required.

The rate of investment declined precipitously in several African and Latin American countries in the 1980s and must be raised to at least its level in the 1970s. An increase in overall resources is absolutely necessary in most developing countries — from special taxes, user fees and community contributions. In addition, budgetary resources are going to have to be reallocated within and across sectors, especially from the military. The international community has an important role in all of this, starting with a return to net positive transfers of resources from the developed countries to the developing countries.

FIGURE 4.2
Declining expenditure on health and education
Central government expenditure, percentage of GNP

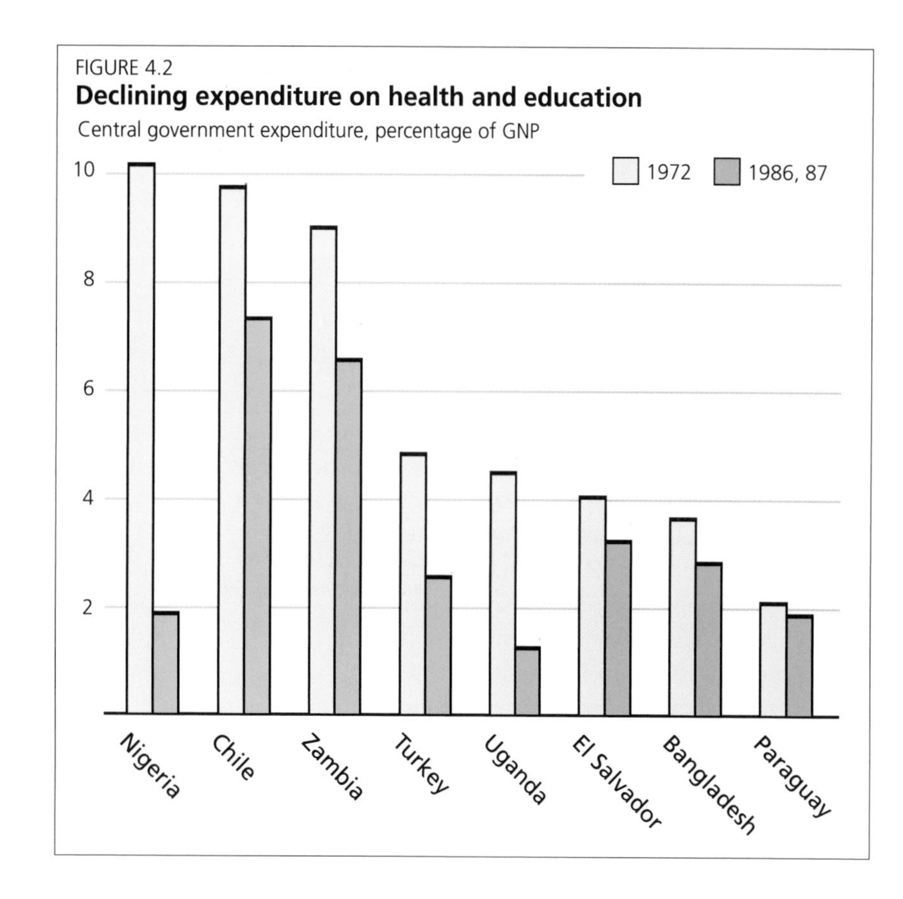

Many countries of Africa, Latin America, Asia, and North Africa and the Middle East need to begin by restoring their investments in health and education to the levels of the 1970s. Each of these regional groups would also need to increase food production by at least 3% a year, and that will require a steady and stable rate of investment in food production of upwards of 3% of their GNP. Investments in water supply in these regions will also have to be substantial.

In Asia the current rate of overall investment is adequate in middle-income countries. Except in the Philippines, the rate of investment in all major countries in this group was no lower in the 1980s than before. Some major countries in North Africa and the Middle East experienced declines in their rates of investment. In these countries a restoration of the rate of investment to higher levels will be necessary to enable the reallocation of resources to human development.

The countries of Latin America and the Caribbean experienced the sharpest reduction in public expenditure on health and education in the 1980s. They thus need to plan for a large increase in public spending on these two sectors. Their investments in agriculture will also have to be high, and considerable investment is required for expanding rural water supply — so it will not be possible to attain the region's goals for human development unless the general availability of resources increases dramatically in Latin America. This issue of limited resources is intimately linked with a satisfactory resolution of the debt crisis.

Social expenditures in many countries have to compete with other urgent claims on generally scarce budgetary resources, making the setting of budget priorities a formidable task. The discussion now turns to the possibilities for mobilising additional domestic resources, for restructuring priorities within the social sectors and for restructuring priorities between sectors.

Raising additional resources

Taxes. Tax reform — particularly reducing tax avoidance and tax evasion through closing loopholes and simplifying procedures — can raise extra resources for the social sectors. Ghana and Jamaica have raised their tax revenue considerably through their improved tax collection efforts. Other countries have introduced special taxes to finance social expenditures. Brazil created a special fund for its programmes in health, nutrition and education and for the support of small-scale farmers, financed by a 0.5% surtax on sales and a 5% surcharge on corporate income tax. Pakistan levied a 5% *Iqra* (literacy) surcharge on its imports to finance high priority projects in education. Extra taxes can also be levied on luxury items and on goods that are detrimental to good health, such as cigarettes and alcohol.

Discriminatory user fees. Governments are also turning more to discriminatory user fees as a source of revenue. Across-the-board user charges can seriously reduce the participation of the poor in development, such as the school attendance of children from low-income families, which respond more to prices than the better-off families do. The user charges introduced at primary schools in some francophone countries in Africa amount to between 7% and 15% of the average annual income of an inhabitant of these poor countries. User fees in the health sector have also been found to reduce the use of health services by poorer people, especially preventive services.

At the same time, however, the share of private health expenditures in the total is quite high in several developing countries: 31% in Brazil (1981), 41% in Jordan (1982), 50% in Zambia (1981) and 70% in Thailand (1979). The share of private secondary school enrolments in developing countries is similarly high: 41% in Tanzania, 54% in Burkina Faso and 60% in Kenya.

Could the apparent willingness of better-off people in society to pay quite high fees for private education and health care — and the obvious inability of poorer people to afford even minimum fees — be skilfully woven into a differentiated system of user charges?

People's willingness to pay, and hence the feasibility of introducing additional user fees, depends on what they expect the additional resources to be used for. They certainly will be more prepared to pay if they can assume that the resources will be invested in quality improvements or expanded coverage, such as opening new facilities in areas not served.

The user-fee system with a focus on the poor should have two objectives. One is to initiate distinct improvements in social services to increase the preparedness of the better-off to pay for services. The other is to ensure that the poorer sections of society pay nothing more than nominal fees, particularly for primary education and primary health care. A specially designed health

BOX 4.6

Singapore's Mediservice scheme

Faced with mounting medical costs, Singapore's Ministry of Health launched the Mediservice scheme in 1983. In addition to promoting individual responsibility for maintaining good health, the scheme aimed at building up people's financial resources to give them means to pay for medical care during illness.

Compulsory savings for medical care regularly set 6% of earnings aside in a personal Medisave account, and withdrawals are permitted for hospital charges and some outpatient procedures, such as minor surgery.

Medisave does not cover general outpatient treatment, which in Singapore is affordable. Nor is it intended to cover long-term chronic illnesses, since other modes of care already provided through subsidised government programmes and by voluntary bodies play a major role.

First implemented in all government hospitals from April 1984, Medisave is an added source of financing for families' medical expenses. This shift in public cost-sharing has freed government tax revenues to improve the public health services, especially preventive and chronic health care.

BOX 4.7

Sharing health costs in the Republic of Korea

The indigent in the Republic of Korea — those aged over 65, the disabled, children under 18 without parents or with parents over 60, and people residing at welfare facilities — get free medical care. People who are better off but still have low incomes — less than about $50 a month in 1985 — and subsistence farmers have to pay 20% of the inpatient fees, except in Seoul where they pay 50%. The selection of eligible persons is made once a year.

The scheme enables qualifying Koreans to receive primary health care at private clinics designated by the Ministry of Health and Social Affairs or from health centres and community health practitioners. They are referred to secondary and tertiary hospitals if necessary. And about half the country's medical facilities are designated for this purpose.

Although there have been occasional delays in payment because of the shortage of local government funds, the scheme has benefitted 3.3 million people — 600,000 indigents and 2.7 million low-income people, nearly 8% of the population.

insurance scheme in Singapore is making higher user charges more acceptable to the better-off sections of society (box 4.6). And a user-fee system in the Republic of Korea shows that such a system can protect the lower-income groups and still work efficiently (box 4.7).

Voluntary community contributions. In a troubling number of developing countries, the services for health, education and other physical infrastructure are rapidly deteriorating because of the lack of maintenance and repair, reflections of scarce budgetary resources. Teachers are abandoning their jobs because salaries are declining or payments have become irregular. Health personnel have to work without medicines and other consumables. To stop this trend, government officials (often with foreign donors) are turning to local people for voluntary contributions to maintaining local services. In such self-help schemes, villagers offer their free labour for construction and maintenance work, contribute food for government personnel and pay for drugs and other services.

The experience with these self-help schemes has been mixed. Sometimes they have worked well without leading to inequities, as in Senegal (box 4.8). In other instances they have created second-class institutions and perpetuated inequalities, as with many of Kenya's *harambee* schools. Great care must therefore be taken in designing community self-financing.

BOX 4.8

Community financing in Senegal

Community financing has persuaded the Senegalese to take much more responsibility for their own health. To give just one example, the system of self-management in 1983-84 raised funds equal to 80% of the budgetary appropriations, excluding those for staff, to the Ministry of Public Health.

Managed by the people and financed by patients' contributions, the system covers all hospitals, health centres, health posts, and village health huts. The charge is 26 cents per adult and 13 cents per child for treatment in hospitals and health centres, and half that for health posts.

Health committees, which administer the receipts, include representatives of every health hut in the village, and they are gradually learning the intricacies of management. Of the receipts, 60% go for drugs, 30% for birth attendants and community health workers and 10% for operational expenses.

Village life is organised around the health hut. The council of elders, the rural council, the mothers' committees, and the health committee all meet under the palaver tree to discuss problems of health, hygiene, and cleanliness. They consider how to replace their stock of drugs, how to collect receipts, and how to pay, in money or in kind, the community health worker they have chosen.

With 90% of the villagers using the health huts, they waste far less time, money and effort than before, when they had to go to a distant health post.

Reallocating budgetary resources within sectors

Countries with successful human development have often restructured government expenditures within the social sectors from low priority to high priority uses — say, from curative to preventive health or from tertiary to primary education. They have also maintained the expenditure on lower priority areas by finding new forms of finance for them, such as loan schemes for tertiary education and insurance schemes for higher level health care.

Normally, social expenditures need 25% to 30% of the total development allocations to maintain a proper balance between economic and social progress. And in countries with a legacy of human neglect, the allocations for social sectors may have to be even higher.

Some of the highest payoffs can come from shifts within the social sectors. Choices have often to be made between primary health care facilities for all or expensive hospitals for a few, between highly subsidised university education or free universal primary education, between sites and services for slums or middle-class urban housing, and so on. Examples of critical imbalances in social expenditures abound.

- Many countries have high dropout rates in their primary schools, sometimes above 70%. Investments to keep children in schools — say, through providing free school lunches or organising new drop-in schools, as in India — could thus pay a high dividend.
- Most developing countries invest too little in technical education (with secondary technical enrolment less than 2% of many countries' total secondary enrolment), so they often produce large numbers of generalists whose skills are not in high demand.
- The mismatch between supply and demand for certain skills leads to considerable unemployment among the educated, even when there is a shortage of skills.
- Primary education generally receives proportionately less public funds than do

FIGURE 4.3

Critical imbalances in social sectors

Excessive school dropouts

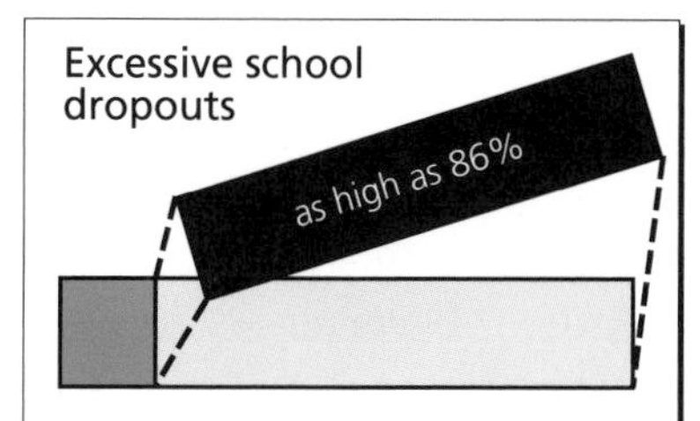

(% of grade 1 enrolment not completing primary school)

Lao PDR	86
Haiti	85
Yemen Arab Rep.	85
Chad	83
Central African Rep.	83
Bangladesh	80
Brazil	78
Myanmar	73
Nepal	73
Madagascar	70
Industrial countries	**11**

Army of generalists

less than **2%** in many countries

(secondary technical enrolment as % of total secondary enrolment)

Zimbabwe	0.1
Sri Lanka	0.4
Bangladesh	0.6
India	1.2
Niger	1.3
Morocco	1.3
Malaysia	1.3
Mauritius	1.4
Myanmar	1.4
Tanzania	1.5
Industrial countries	**23.0**

Educated unemployed

Educated unemployed highest rate- 28%

(as % of total unemployed)

Syria	28
Yemen, PDR	15
Indonesia	14
Peru	13
Guatemala	12
Singapore	9
Korea, Rep.	8
Uruguay	8
Panama	8
Venezuela	7
Industrial countries	**13**

Inverted educational pyramid

82% of expenditure for secondary & higher

18% for primary education

(primary education expenditure as % of total education expenditure)

Nigeria	17
Liberia	18
Uganda	20
Cuba	21
Venezuela	25
Mexico	27
Ghana	29
China	29
Algeria	29
Singapore	29
Industrial countries	**34**

Nurses per doctor

as low as 0.4 (1 nurse per 2.5 doctors)

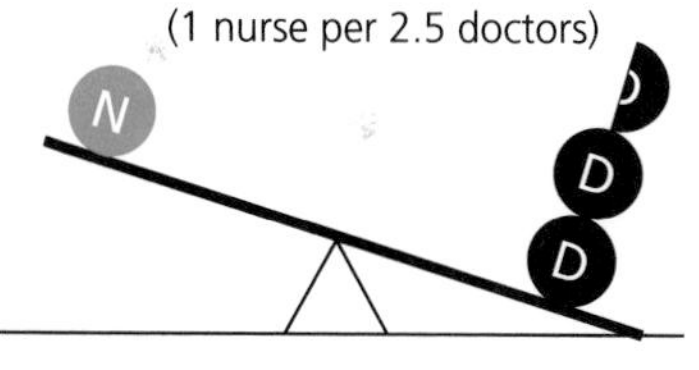

Argentina	0.4
Pakistan	0.6
Bolivia	0.6
China	0.6
Bangladesh	0.7
Jordan	0.9
Syria	0.9
Brazil	0.9
Iraq	1.0
Egypt	1.0
Industrial countries	**4.2**

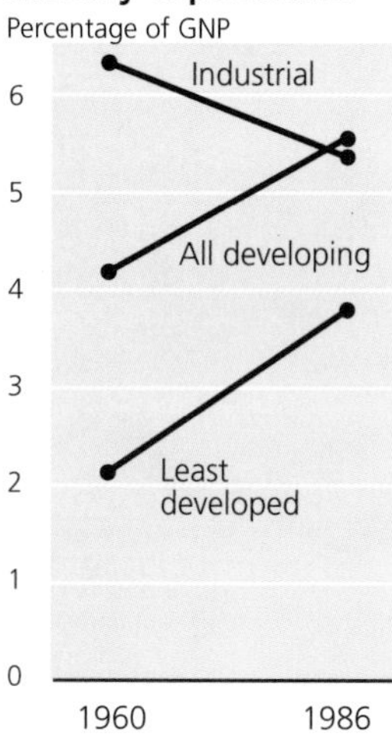

FIGURE 4.4
Military expenditure

college and university education, which in many cases could be more self-financing.

- Many health systems lack paramedics (especially nurses), obliging doctors to perform the tasks of their support staff.

The efficiency of social spending can be greatly improved by identifying such imbalances and taking steps to correct them. Since social budgets are going to remain limited in relation to demand — even under the most optimistic assumptions about the future — more effort must go into squeezing the maximum social progress from limited resources.

Reordering budgetary priorities across sectors

Many countries spend a high proportion of their budgets and GDPs on defence, offering great potential for switching resources towards the social sectors. Where there is less scope for such shifts, other sectoral reallocations may be possible. Spending on inefficient parastatals can be cut, some government activities can be privatised and development spending can be made more efficient. Care must be taken, however, not to divert resources required for the maintenance of economic infrastructure that is so important for sustaining growth. In addition, successful negotiations on debt reduction would release some of the vast resources that have been going to interest payments for use in the social sectors.

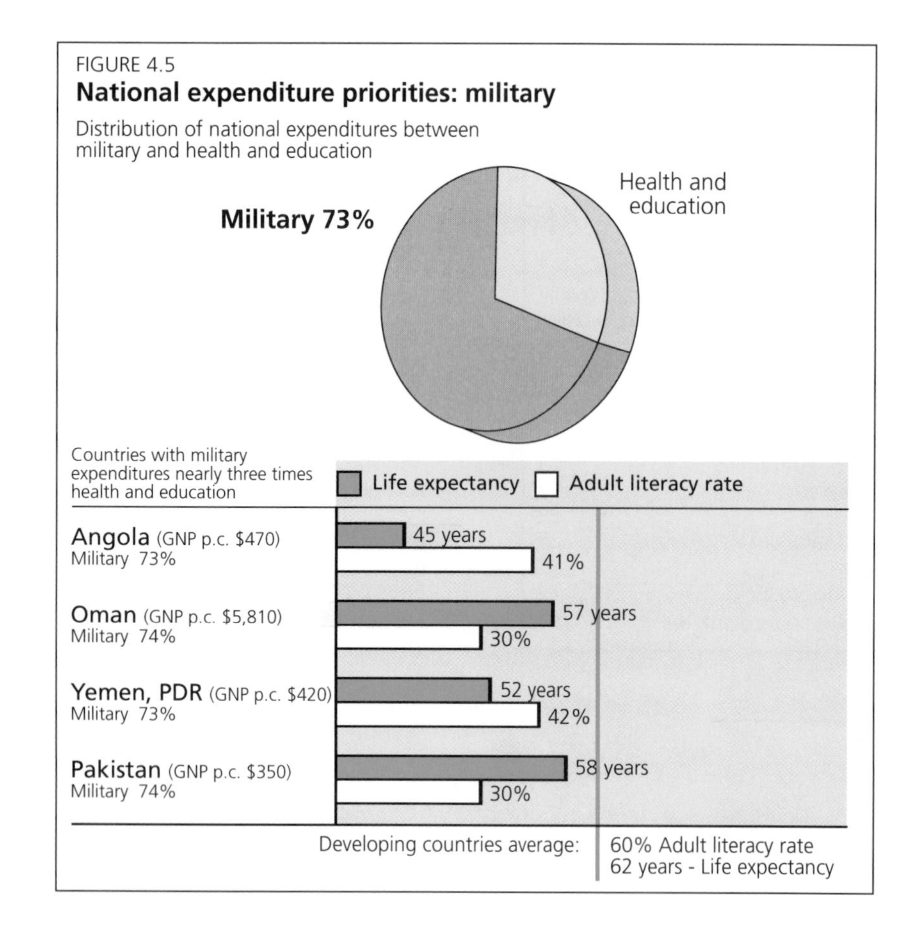

FIGURE 4.5
National expenditure priorities: military

The rapid rise of military spending in the Third World during the last three decades is one of the most alarming, and least talked about, issues. It continued even in the 1980s despite faltering economic growth in many developing countries and despite major cutbacks for education and health.

Military expenditures of the developing countries have increased 7.5% a year during the past 25 years, far faster than military spending in the industrial countries (table 4.1). Their total expenditures multiplied nearly seven times — from $24 billion in 1960 to $160 billion in 1986 — compared with a doubling for the industrial countries. And of the incremental growth of nearly $500 billion in annual global military expenditures between 1960 and 1986, nearly 30% was additional spending by the developing countries. As a result, the share of developing countries in global military expenditures rose from 7% in 1960 to 19% in 1986.

Whereas the industrial countries reduced the share of their GNP allocated to the military from 6.3% in 1960 to 5.4% in 1986, the countries of the Third World increased their share from 4.2% in 1960 to 5.5% in 1986. Most astonishing of all, the least developed countries nearly doubled the percentage of their GNP spent on the military — from 2.1% to 3.8%.

It is worth recalling in this context that average per capita income of the developing countries is only 6% of that of the industrial world. And at 1985 levels of GNP per capita, annual military costs represented nearly 160 million man-years of income in the Third World, three times the equivalent military burden of industrial countries. Obviously, the poverty of the people of the developing world has been no barrier to the affluence of their armies.

The sharp increases in military expenditures have not only preempted diminishing budgetary resources and squeezed social services and economic growth. They have also consumed considerable amounts of foreign exchange. Arms imports by developing countries skyrocketed from $1.1 billion in 1960 to nearly $35 billion by 1987, or three-quarters of the global arms trade. According to the World Bank, military debt is more than a third of the total debt for many large developing countries. What is frightening about all these figures on military expenditures is that they may well be underestimates, since few governments reveal their true military spending.

In the developing countries the expenditure on the military is more than that for education and health combined, compared with just over half in the industrial world. Even in the least developed countries, spending on the military is almost equal to that on education and health combined. Over 25 developing countries spend more on the military than on education and health, sometimes more than twice as much, have several times more soldiers than teachers and spend around 6% or more of their GNP on defence. There are eight times more soldiers in the Third World than physicians.

Among developing countries with the highest shares of military expenditures are some of the poorest and least developed countries — Angola, Burundi, China, Myanmar, Sudan, Uganda, the People's Democratic Republic of Yemen and Zaire. Despite the more than 800 million people in absolute poverty in South Asia and Sub-Saharan Africa, South Asia spends $10 billion a year on the military, and Sub-Saharan Africa $5 billion.

There obviously is great potential for increasing human development budgets if military expenditures are not allowed to increase, or are even reduced, during the 1990s. The total military expenditure of the Third World is estimated at almost $200 billion. If past trends continue, it would increase between $15 billion and $20 billion every year during the 1990s. Any reduction in past trends would thus release considerable resources for social programmes.

The emerging detente between the superpowers demands a candid reassessment of past military spending and an open dialogue about future options. If human development programmes are to be more liberally accommodated in future budgets, the international community can encourage this through four specific actions:

- The ratio of military to social expenditures should increasingly be accepted as one of the criteria for external assistance.
- Ceilings should be negotiated not only for development debts but also for military debts.
- Arms shipments to developing countries should be discouraged, especially since pressures for such shipments are likely to increase as defence industries face major production cuts in the 1990s.
- Major world powers should be urged to

TABLE 4.1
Rising military expenditure in the Third World

Region	Billions of 1984 dollars		Percentage of GNP		Annual percentage growth 1960-86
	1960	1986	1960	1986	
World	345	825	6.0	5.4	3.4
Industrial countries	321	666	6.3	5.4	2.9
Developing countries	24	159	4.2	5.5	7.5
Least developed countries	0.5	3.4	2.1	3.8	7.5

FIGURE 4.6
National expenditure priorities: social sectors

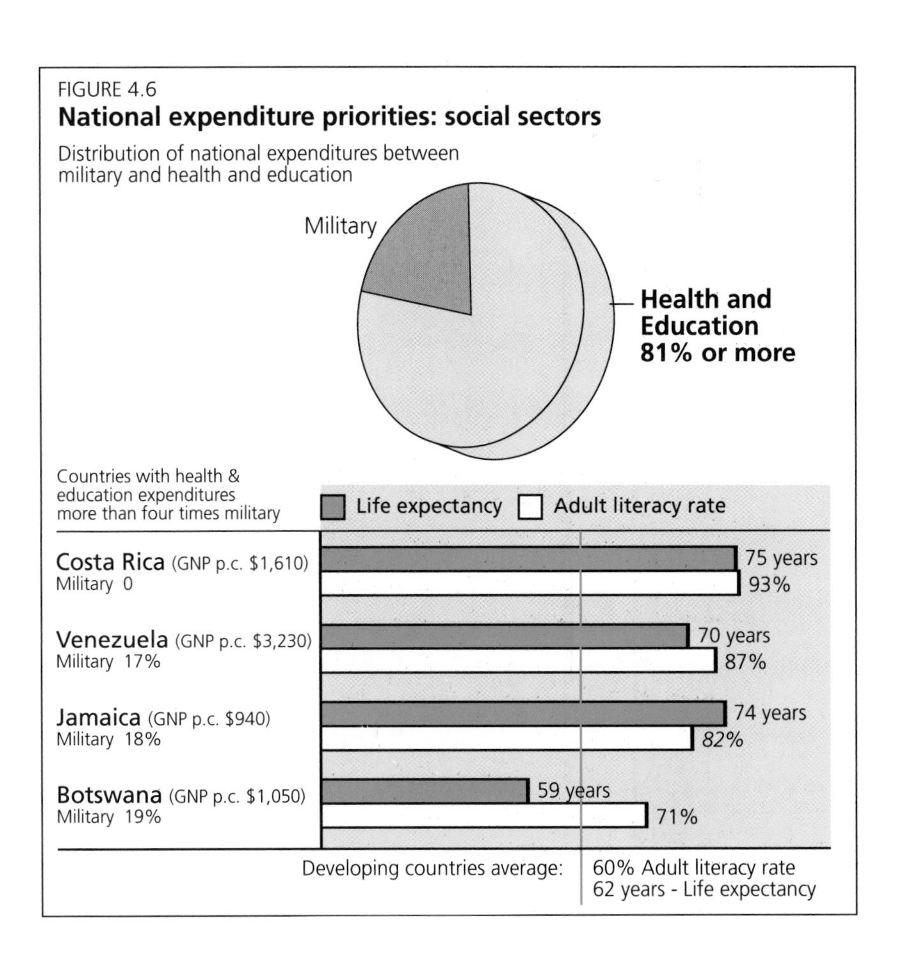

promote peaceful development in the Third World by defusing regional tensions (particularly in Southern Africa and the Middle East) and ensuring global security and economic justice.

The responsibility of industrial countries, and other major arms exporters, for the rising military expenditures in the Third World must also be noted. Defence assistance budgets of developed countries have often increased even when net economic assistance has declined. Defence industries in the industrial world have often aggressively sought willing clients in the Third World, offering soft credits and on occasion even illegal gratuities. Many developing countries have served as convenient battlegrounds for the cold war rivalries between the superpowers. To put this in perspective, the military budget of just one superpower currently exceeds the combined military spending of the Third World by 50%.

The 1990s offer a unique opportunity for a substantial reduction in military expenditures in all nations. The question is whether such a reduction will release substantial resources for the real peace effort — the attack on human deprivation — a question that must receive very serious consideration in the coming decade.

External environment for human development

Although the battle for human development must be fought in the developing countries, a favourable external environment can help considerably. But that environment has been far from favourable in the 1980s. The net transfer of resources to the developing countries has turned negative — from a positive flow of nearly $43 billion in 1981 to a negative flow of nearly $33 billion in 1988. Primary commodity prices have reached their lowest level since the Great Depression of the 1930s. Foreign debts of developing countries have crossed $1.3 trillion and now require nearly $200 billion a year in debt servicing (box 4.9). Never before have the developing countries faced such difficult external circumstances as those in the 1980s — and that has been a major cause of the setbacks in human development during this decade.

An urgent task for the international community is to restore a favourable external environment in the 1990s and to assist in the implementation of essential human goals. The first priority must be a return to positive net transfers of resources to the developing countries, possible only if a satisfactory solution is found for the debt crisis. But so far the debt problem of the highly indebted nations has resisted solution — despite suggestions for debt reschedulings, debt swaps and debt reductions. The basic problem is that no solution is without costs. If developing countries are to receive meaningful debt relief, either the governments or the commercial banks of the rich nations must incur the cost.

The impasse persuades many analysts that a new debt refinancing facility should be created under the auspices of the IMF, the World Bank or both. Such a facility could be funded by a new creation of special

TABLE 4.2
Military expenditure as a percentage of education and health expenditure

	Share of 1984 GNP allocated to		Military expenditure as % of education and health expenditure
	Military	Education and health	
All developing countries	5.5	5.3	104
Least developed countries	3.8	4.1	92
All industrial countries	5.4	9.9	55

TABLE 4.3
Soldiers or teachers

	Armed forces as % of teachers
Some of the worst examples	
Ethiopia	494
Iraq	428
Oman	275
Chad	233
Yemen, PDR	200
Pakistan	154
Some of the best examples	
Costa Rica	0
Mauritius	10
Côte d'Ivoire	13
Ghana	14
Jamaica	20
Brazil	24
All developing countries	68
Least developed countries	121

drawing rights (SDRs) if the United States reduces or eliminates its trade and budget deficits (see box 4.10). The cost of refinancing the developing countries' debt would thus be transferred imperceptibly to the entire international community.

Such a new international debt facility would be a significant improvement over today's disarray. The facility could allocate limited financial resources with more focus — to maximise the debt reduction. It could help promote the establishment of a set of commonly accepted international rules — to replace the sporadic, complicated and often inequitable ad hoc process that now prevails. Above all, a concerted international approach would help promote broader understanding that the resolution of the debt crisis is a shared responsibility, and that it would, if successful, benefit the entire world.

The world trading environment also requires concerted international action. Protectionist pressures have continued to increase during the first three years of the Uruguay Round of multilateral trade negotiations. There has also been a move towards regional trading blocs: the EC in 1992, the U.S.-Canada bilateral trade agreement and the Pacific-rim trading bloc around Japan. Unless the trading interests of the developing countries are protected through careful negotiations, there is a real danger that these countries, especially the least developed countries, will be marginalised.

Progress in the Uruguay Round has already been considerable for services, tropical products and the functioning of the GATT system. Among the issues still to be resolved are agriculture, and whether the EC will agree to a time-bound commitment to at least partial liberalisation; textiles, and whether developed market economies will agree to a time-bound commitment to return to GATT rules of bound tariffs, the elimination of nontariff restrictions, and most-favoured nation treatment; safeguards, and whether coverage is to be selective or universal, and what the role of multilateral surveillance is to be; and intellectual property rights, and whether negotiations will be broad or confined to trade.

It is vital that the parties to the Uruguay

FIGURE 4.7
Reversing resource flows
North to South net transfers (US$ billions)

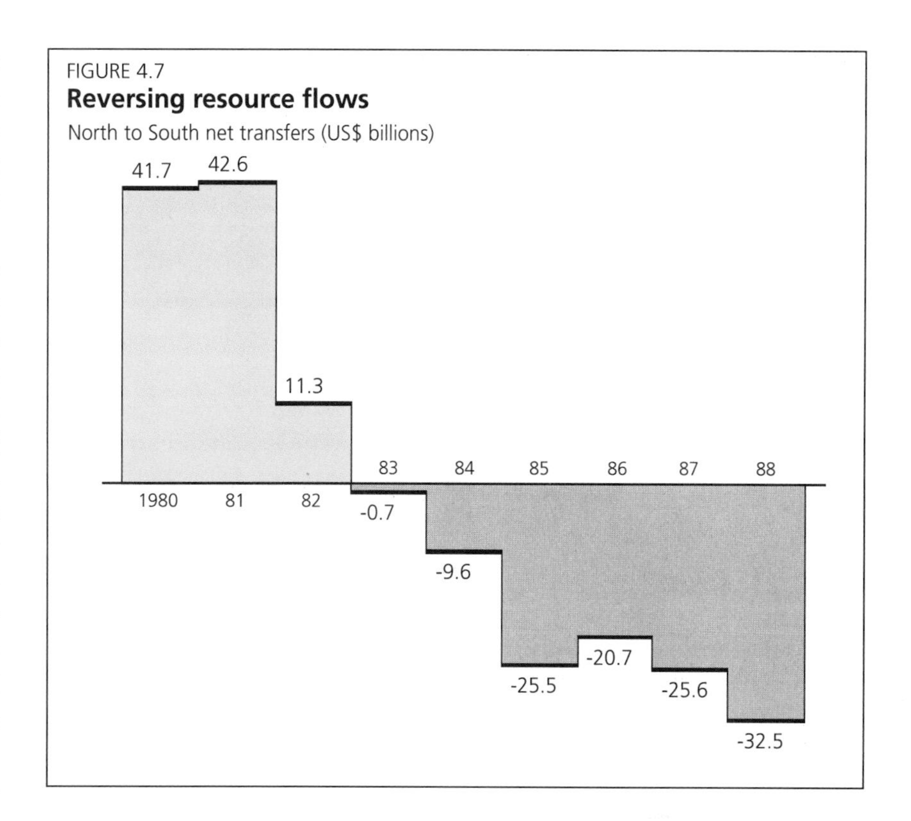

BOX 4.9

The lingering debt crisis

Developing nations owed foreign creditors $1.3 trillion at the start of 1989, just over half their combined gross national products and two-thirds more than their annual export earnings. The annual interest obligations on this debt are approaching about $100 billion and, with amortisation, the annual total debt service bill is now approaching $200 billion.

The debt servicing bill is so steep that only four nations of 21 in Latin America are up-to-date on payments to private creditors. Despite their effort to maintain their standing with multilateral lenders, eight countries are in arrears to the World Bank and 11 to the IMF.

The mounting debt burden of third world countries has reversed the North-South resource flows. According to a United Nations study, a sample of 98 developing countries transferred a net amount of $115 billion to developed countries between 1983 and 1988. Capital flight of billions of dollars, particularly from Latin America, made the situation much worse.

The World Bank regards 17 countries as having a serious debt problem: Argentina, Bolivia, Brazil, Chile, Colombia, Costa Rica, Côte d'Ivoire, Ecuador, Jamaica, Mexico, Morocco, Nigeria, Peru, Philippines, Uruguay, Venezuela and Yugoslavia.

In the highly indebted countries, per capita GDP declined by 1% a year on average during the 1980s, while countries that escaped the debt crisis had annual increases of 4%. For the 17 highly indebted countries (most of them in Latin America), gross capital formation fell by 40% in real terms between 1982 and 1985. In Africa capital formation fell from 21% of GDP at the start of the decade to less than 16% in 1988. And as mounting evidence from around the world suggests, the debt crisis and its consequences have also taken a heavy toll on human development.

A UNICEF study concludes that in the 37 poorest countries — many of them also weighted down by debt — health spending per head has fallen more than 50% during the last decade, spending on education more than 25%. And in some of the more indebted countries, infant and child mortality rates have risen. There is also accumulating evidence of falling employment, substantial cuts in real wages and deteriorating social indicators in the indebted nations.

Round settle these remaining issues before the stipulated time limit at the end of 1990 so that developing countries can benefit from a liberal and expanding trade environment in encouraging economic growth and human development. It is also essential that developing countries prepare themselves for the trading challenges of the 1990s, particularly in the newly emerging global services economy (box 4.11).

The donor community should also consider various ways of assisting human development strategies more directly during the 1990s.

First, there has been a steady decline in the official development assistance (ODA) allocated to education, health, nutrition and family planning — from more than 17% of total ODA in 1978-79 to less than 15% in 1988. This trend must be reversed. It is not enough to suggest that resources are fungible. External resources finance a major part of development expenditures in many countries, and it matters greatly how those resources are earmarked. Donor signals are also important in persuading reluctant finance ministers to devote a larger share of domestic resources to social spending, which entails large recurring expenditure.

Second, donors can help in the formulation and implementation of human development strategies in the 1990s by offering new conditions for cooperation. They could give explicit support to human development — for example, by specifying that human development programmes should be the last (not the first) to be reduced in an adjustment period or by making clear that external assistance would be reduced if a country's military expenditure exceeds its social expenditure.

Third, technical assistance should be carefully scrutinised to ensure that it builds human capabilities and institutions in the developing countries. The record is not reassuring. Africa now gets $4 billion a year in technical assistance — about $7 per person. Yet, human development and national capacity-building in this region have progressed only reluctantly. For example, Tanzania currently receives an annual amount of some $300 million in technical assistance. Much of that goes into the salaries and travel of foreign experts (while the country's total civil service budget is $100 million). Could not some of this $300 million be spent more profitably on building institutions and human capabilities in Tanzania? The ever-increasing technical assistance belies the claim that its purpose is to build national capacities and phase itself out, at least in its present form.

A sobering thought in concluding this discussion concerns the shifting demographic balance of the world. As chapter 2 showed, the population of the developing countries, 69% of the world total in 1960, is expected to increase to 84% by 2025 while the share of industrial nations will decline correspondingly from 31% to 16%. If the future generations in the developing world cannot improve their conditions through liberal access to international assistance and trade, will there not be an even greater compulsion to migrate in search of economic opportunities? When human beings are educated, means of communication are

BOX 4.10

Proposal for an international debt refinancing facility

There are at least three essential elements for a satisfactory solution to the debt crisis.

First, the commercial lenders should not be made to appear to be carrying the risk of present illiquidity or potential insolvency. It is not important that they are paid back immediately — but they must have the assurance of getting back what is due to them, or at least a mutually agreed part of the debt.

Second, the developing country debtors can pay only what their trade surplusses allow them to pay. But large trade surplusses are neither possible nor desirable in the long run, as they will require a massive transfer of resources from poor to rich nations. There is no real alternative to a substantial debt reduction, with governments and commercial bankers in the developed countries bearing the cost.

Third, to find a long-term institutional solution, it is absolutely necessary to set up an international intermediary that can reach country-specific settlements within the framework of an international consensus on the debt crisis.

The basic task of an international debt refinancing facility will be to:

- Stretch out maturities.
- Reduce interest costs to a defined ceiling of export earnings.
- Arrange substantial debt reduction.
- Apportion adjustment costs between the indebted nation and its external creditors.
- Protect new lending levels.
- Ensure expanding export markets.
- Reverse declining resource transfers.
- Reach a new equilibrium in the balance of payments at a higher, not a lower, level of production and employment.
- Ensure improved policies of domestic economic management in the indebted nations.

The logical umbrella for such a facility is joint IMF-World Bank sponsorship. The facility will have to operate on two fronts — improving both the internal policies and the external environment of indebted nations. It may need to reactivate the dormant mechanism of SDR allocations to provide the necessary resource backing for its debt reorganisation efforts.

easily accessible, and the wave of human freedom is bringing down international barriers, what will hold back another tidal wave of international migration? The answer lies in enabling the developing countries to improve human choices so that their people do not have to seek an option abroad because of sheer despair and deprivation.

Implementing human development strategies

Making informed policy choices

Informed choices about development — by policymakers and people generally — are critical to human development. Those choices depend on expanded and improved data gathering and policy analysis.

Today's systems of social statistics need considerable strengthening in all developing countries to improve the coverage, reliability and disaggregation of data, especially by gender, income group and geographical area. Such strengthening is also needed to enhance the speed, regularity and timeliness of gathering, analysing and disseminating information to interested users.

Many countries need to organise detailed and nationally representative household surveys. The strengthening of national capacity in this area has been the main concern of the UN National Household Survey Capability Programme, launched in 1980. Such surveys are also being supported within the context of the Social Dimensions of Adjustment Project, sponsored by the UNDP, the World Bank, the African Development Bank and other agencies and bilateral donors. Given the need for low-cost data collection, the rapid rural assessment method is being used increasingly as a shortcut in studies interested in the living conditions of particular population groups or in particular local problems. The method is also being used in urban studies.

Data and information about micro-level behaviour have to be complemented by information on how macroeconomic policy measures affect people's lives. Conversely, modelling exercises should also look at the consequences that positive and negative social trends can have on economic variables. As chapter 2 concluded, poverty has a high price, especially in terms of wasted human potential and lost productivity. Discrimination against females also has a high price, as does environmental degradation. Making these high costs explicit can help muster the political will needed to overcome today's rather narrow focus on income expansion and the preoccupation with shorter-term concerns.

Targetting policy measures and social programmes

Countries can economise on scarce budgetary resources and enhance the effectiveness of human development programmes by tailoring them to the specific needs and interests of the intended beneficiaries. As the discussion in chapter 3 demonstrated, the need for targetting is great if a country has low income and a fairly inequitable income distribution. Targetted interventions usually adopt one of two approaches or a combination of both. Some interventions aim at being *inclusive* — ensuring the widest possible coverage of the identified beneficiaries within a given budget. Other interventions aim at being *exclusive* — limiting the access to the benefits to a well-

BOX 4.11

The new global services economy

The technological breakthroughs of the past two decades — particularly in informatics — have transformed traditional services. Human skills are now the most important input into modern banking, finance, advertising, communications, business management and public administration.

Services today are the dominant part of the world economy. They generate nearly 70% of the GNP and employment opportunities in the industrial nations. But they still lag behind in the developing countries, contributing 48% to GNP and 18% to employment.

Between 1970 and 1980 the trade in services increased by an average of 19% annually, reaching $435 billion in 1980. It is estimated that trade in services may reach nearly $1 trillion by 2000.

Most developing countries are still net importers of services. Their net deficit in services increased from $4 billion in 1970 to $58 billion in 1980. The rapid expansion of trade in skill-intensive services offers a tremendous opportunity to developing countries — if only they can impart new knowledge and skills to their people.

The emergence of a new global services economy shifts comparative advantage more in favour of people than natural resources. Developing countries have a majority of the world's people but still only a small share in global trade in services. The opportunities are there, however, for developing countries willing to organise themselves for the modern service sector of the 1990s.

defined segment of the population.

Targetting has generally been applied to food subsidies and nutrition programmes — with considerable success. In Tamil Nadu, India, a feeding programme covers more than 17 million people. All children aged six months to three years are weighed either at a community health centre or at their parents' home. Those found to be underweight receive supplementary feeding for 90 days. If their condition does not improve during that time, they are referred to a health programme. The Tamil Nadu programme is thus open to all children. But through weight-monitoring, it is focused on the neediest. The rather labour-intensive screening of all children of the defined age group has contributed much to the programme's success. Botswana's experience has been the opposite. Lack of technically qualified staff, and therefore ineffective screening, allowed many healthy children also to benefit from the government's feeding programme.

Targetting of social policies and programmes must match the management capacity in a country

The lesson from these and similar experiences is that tight targetting can work if the technical and administrative capacity is in place. Otherwise, looser targetting is preferable. For example, school lunch programmes in Brazil and Jamaica are targetted to school children, but they allow access to all qualifying children irrespective of their economic status.

For food subsidies, targetting has sometimes restricted the beneficiaries to certain income groups (through ration cards or food stamps) and sometimes subsidised commodities consumed more by the poorer people in society (cassava, sorghum and pulses). Faced with a tightening budget, Sri Lanka tried the first route, Morocco the second. Often the transition from across-the-board to targetted subsidies proved politically difficult (because previous beneficiaries were reluctant to forgo their entitlements) and administratively unmanageable.

Targetted programmes have normally absorbed between 1% and 10% of total government expenditure, or between 0.2% and 2% of GNP. A careful study of targetted schemes and different country experiences leads to six broad guidelines:

- The targetting of government interventions, particularly food subsidies, must carefully balance the economic and political feasibility of the scheme.
- Given the powerlessness and vulnerability of the poor, meeting their interests might require allowing the participation of some not-so-poor groups who are more capable of asserting themselves politically.
- The approaches selected for targetting must match the management capability in the country. Tightly targetted programmes, typically more difficult to manage, may ultimately become more expensive than loosely targetted programmes.
- A geographical approach to targetting is sometimes feasible if the poor are concentrated in a particular area.
- A combined approach to targetting — geographical targetting combined with targetting by commodity for food subsidies or with weight monitoring for nutrition interventions — has proven highly effective in many instances.
- Targetting must be used to move beneficiaries out of target groups to avoid dependence on government interventions. Nutrition interventions and food subsidies should be linked to income-generating programmes and to incentive schemes for relinquishing entitlements, particularly for beneficiaries who are not in need but have political clout.

Selecting cost-effective technology

Since financial resources are often very limited, planners must focus on priority tasks and find the most cost-effective methods to achieve their social objectives. The greater the cost-effectiveness of human development programmes, the higher will be the pressure on political leaders and finance ministries to provide the needed resources — and the greater will be the mileage from each resource unit.

The recent rekindling of interest in cost-effective technology has been triggered not only by the financial crisis of the 1980s but also by the finding that low-cost technology in many instances is not only cheaper but better. There are examples of such technologies in all sectors: oral rehydration technology and breast-feeding in health, improved wood-saving stoves in energy, or rain-harvesting techniques in agriculture.

Linked to the issue of appropriate technology is the growing concern of governments about the use of local development inputs — equipment, supplies and expertise. Several countries have made progress in developing their domestic capacity in pharmaceuticals, such as Argentina, Brazil, Mexico, the Republic of Korea, India and Egypt (box 4.12). Other countries produce weaning foods and other basic consumer goods locally.

Conclusions

The analysis here of some of the key issues in planning, financing and implementing human development strategies in the 1990s points to four main conclusions.

First, given the derivative but abiding significance of goods and services in expanding human options, countries must broaden the commodity base for national prosperity. But how can economic growth promote human development? The link is not automatic. It results from deliberate and effective public action that enables people to participate in, and benefit from, the process of development — to develop their individual capabilities and put them to the most creative and productive use.

Second, public action is often necessary to supply social services and make them available to the entire population. This applies particularly to education and health services, including water supply and sanitation. Better distribution of food and shelter may also require public intervention if the income distribution is skewed towards a few and if the vast majority are denied their essential needs. A major task of government is to correct the distribution of incomes and assets through income transfers and the widespread distribution of public goods for human development. To be avoided, however, are situations where the more powerful capture a disproportionate share of the public social services. The distribution of social services is not neutral with respect to income groups. Targetted income transfers can help in reaching the poorer beneficiaries.

Third, human potential will be wasted unless it is developed — and used. Economic development should create a suitable environment for the use of human talents. It should match the human dexterity the society needs with the human skills that are cultivated. But national production must also expand to make good use of human potential. One lesson of development is that judicious use of markets — without a plethora of inefficient controls — often creates an enabling environment for good use of individual talent and potential. But even with efficient incentives and fast general expansion, there will be those who, for one reason or another, may not be able to earn even a minimum of satisfactory income, have even a minimum of adequate nutrition or acquire even a minimum of relevant education. Guarantees of public support and suitable social safety nets are then needed. Further, the chronically deprived and dispossessed must be brought up to a threshold of human development to enter the mainstream of economic growth. But then it is time for governments to step aside — because freedom to participate in the market according to one's talents and preferences is the best vehicle for produc-

BOX 4.12

A cost-effective strategy for essential drugs

Most developing countries now have drug legislation and essential drug lists. But many do not appear to have the capacity to enforce either one of them.

Only a quarter of 104 countries surveyed had a well-defined drug policy, and only a fifth had a proper system in place for the procurement and distribution of drugs. The result was wasteful, inflated expenditure on drugs in many countries.

The WHO List of Essential Drugs mentions 250 items, but many more than that are usually on the market. In India, for example, around 25,000 drugs are available. But expert opinion says that most common diseases could be treated with few basic drugs: chloroquine, acetylsalysilic acid (aspirin), paracetamol, ferrous salt and penicillin.

These drugs could be procured for all the population in the developing world at 2% of the current spending. If the basic drugs list had 30 items, the cost would be equivalent to 7% of the current spending.

A more rational drug policy, coupled with improved procurement and distribution systems, could thus save much in public sector budgets and trade balances.

Greater local production can often bring further economies in drug spending. Developing countries were importing 41% of their requirements in 1984. Local drug production could bring down their costs in many cases.

China produces 90% of its needs. And advanced drug manufacturing capacities exist in Argentina, Brazil, Mexico, the Republic of Korea and Egypt. But in most of these countries, multinational corporations have an overwhelming share of the market: 30% in Egypt, 50% in Argentina, 78% in Brazil and nearly 100% in some African countries.

tive use of human capabilities.

Freedom, therefore, is the most vital component of human development strategies. People must be free to actively participate in economic and political life — setting developmental priorities, formulating policies, implementing projects and choosing the form of government to influence their cultural environment. Such freedom ensures that social goals do not become mechanical devices in the hands of paternalistic governments. If human development is the outer shell, freedom is its priceless pearl.

CHAPTER 5: A SPECIAL FOCUS

Urbanisation and human development

Cities in the developing countries today present many contrasts. They contribute to human development — as well as constrain it. They are centres of affluence — as well as concentrations of poverty. They bring out the best in human enterprise — as well as the worst in human greed. They contain some of the best social services available in the country. But they are also host to many social ills — overcrowding, unsanitary living conditions, drug addiction, alienation, social unrest and environmental pollution.

What should be the perspective on urbanisation? For human development strategies, what makes sense is to reinforce the creative and productive capabilities of the cities and to overcome their many social ills. That is the urban challenge in the developing world.

To meet this challenge, there are four main items on the agenda for managing cities in the 1990s. The first is to decentralise power and resources from central government to municipalities. The second is to generate municipal revenues so that cities can pay their own way. The third is to develop strategies to meet the needs for housing and for urban infrastructure — and to target special assistance to poor communities and weaker groups. The fourth is to improve the quality of the urban environment.

Urbanisation in the developing countries

The growing concentration of people in cities is by now almost exclusively a developing country phenomenon (box 5.1). The urban population of the developing countries, now 1.3 billion, is expected to grow by nearly another billion in the next 15 years.

According to recent UN projections, the developing world's rural population will reach an upper limit by 2015 and beyond this point all future population growth will be concentrated in urban areas. By 2015 half the developing world's people will live in urban areas.

Growth rates are expected to be fastest in Africa, where the urban population may double between 1985 and 2000. Absolute growth will be greatest in Asia, where cities will gain another 500 million inhabitants during the same period.

Most of this growth — two-thirds of it in many Asian and Latin American cities, but less than one-half in many African cities — will be the natural increase of populations already in the cities. The remainder will come from rural-urban migration, the incorporation of villages into expanding ur-

BOX 5.1

The urban explosion

This is the century of the great urban explosion. In the 35 years after 1950, the number of people living in cities almost tripled, increasing by 1.25 billion. In the developed regions, it nearly doubled, from 450 million to 840 million, and in the developing world, it quadrupled, from 285 million to 1.15 billion.

In the past 60 years the developing world's urban population increased 10-fold, from around 100 million in 1920 to close to 1 billion in 1980. Meanwhile, its rural population more than doubled.

- In 1940 only one person in eight lived in an urban centre, and about one in 100 lived in a city with a million or more inhabitants.
- In 1960 more than one person in five lived in an urban centre, and one in 16 in a city with a million or more.
- In 1980 nearly one person in three was an urban dweller, and one in 10 lived in a city with a million or more.

The population of many of Sub-Saharan Africa's larger cities increased more than sevenfold between 1950 and 1980 — Nairobi, Dar es Salaam, Nouakchott, Lusaka, Lagos and Kinshasa among them. During these same 30 years, populations in several other third world cities — Seoul, Baghdad, Dhaka, Amman, Bombay, Jakarta, Mexico City, Manila, São Paulo, Bogota and Managua — tripled or quadrupled. In-migration has usually contributed more than natural increase to their growth.

This growth has been far beyond anything imagined only a few decades ago — and at a pace that is without historic precedent.

FIGURE 5.1
The ten largest cities, 1960

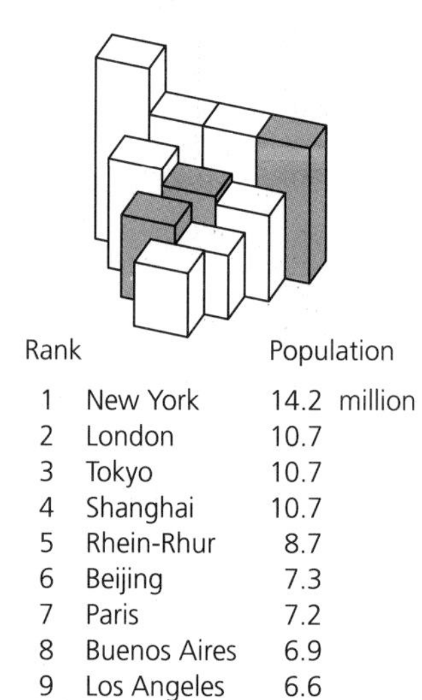

Rank		Population
1	New York	14.2 million
2	London	10.7
3	Tokyo	10.7
4	Shanghai	10.7
5	Rhein-Rhur	8.7
6	Beijing	7.3
7	Paris	7.2
8	Buenos Aires	6.9
9	Los Angeles	6.6
10	Moscow	6.3

The ten largest cities, 2000

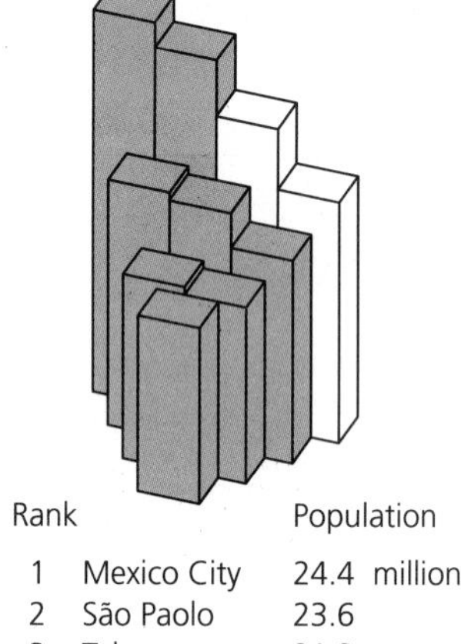

Rank		Population
1	Mexico City	24.4 million
2	São Paolo	23.6
3	Tokyo	21.3
4	New York	16.1
5	Calcutta	15.9
6	Bombay	15.4
7	Shanghai	14.7
8	Tehran	13.7
9	Jakarta	13.2
10	Buenos Aires	13.1

■ in a developing country
□ in an industrial country

ban municipalities and the changing definition of settlements from rural to urban as they reach a given size.

In 1960 just three of the world's 10 largest urban agglomerations were in the developing world, and only one, Shanghai, had more than 10 million people. By 2000 there will be 18 cities with more than 10 million people in the developing world, and eight of them will be among the 10 largest cities in the world.

Most spectacular is the expected growth of cities with more than 4 million inhabitants. In the 1960s there were 19 such cities, nine in the developing world. In 1980 there were 22 in developing countries. By 2000 there will be 50 — and by 2025, 114 of a total of 135 in that year.

Cities and human development

Modern transport and communication link the world's major cities closer in a global network for the exchange of goods, services, knowledge and expertise. Leading national development efforts, the big cities generate a significant share of their countries' GDP.

- Lima, with 27% of Peru's population in 1981, accounted for 43% of the GDP.
- Manila, with 13% of the Philippines' population the same year, produced 33% of the GDP.
- Lagos, with 5% of Nigeria's population in 1978, accounted for 57% of total value added in manufacturing.
- Greater São Paulo, with 10% of Brazil's population in 1980, contributed 25% of the net national product in that year and more than 40% of the total value added in manufacturing.
- Port-au-Prince, with 14% of Haiti's population, generates 40% of the national income.

The metropolises of the Third World clearly generate the resources needed for their proper management.

The urbanisation of poverty

Major population movements to cities are also shifting the main burden of poverty to urban areas. There were around 40 million urban households living in poverty in 1980, compared with 80 million rural households. By 2000 the number of poor urban households living in absolute poverty is projected to increase by 76% to 72 million, and that of poor rural households to fall by 29% to 56 million.

The poverty of urban dwellers is apparent in recent studies by the Economic Commission for Latin America and the Caribbean: 22% of the people in Panama City were poor in 1983, 25% of urban Costa Ricans in 1982, 64% of the people in Guatemala City in 1983 and 45% of the people of Santiago, Chile, in 1985.

Urban economies cannot absorb all the rural poor. The persistent problem is that attempts to tackle urban poverty directly — by creating jobs and providing public services unavailable in rural areas — simply attracts more of the rural poor, and their migration wipes out any gains.

Urban poverty was once thought to be the product of underemployment, not unemployment, but evidence from Sri Lanka, Malaysia, Colombia, Côte d'Ivoire and Tunisia shows that open urban unemployment has grown since the mid-1970s and is highest among the poor. Urban poverty is also linked to economic fluctuations: the conditions for the urban poor worsened during the economic crisis of the 1980s after improving or at least stabilising during the 1970s.

Malnutrition and disease in the cities

Malnutrition appears to be spreading in third world cities. In Colombia, Costa Rica, Guatemala, El Salvador, Tunisia and Morocco, rural diets are superior to those of the urban poor, particularly in calorie intake. And in several countries there are more severely malnourished children in low-income urban areas than in rural areas.

Darkening this grim picture, urban health statistics often underestimate the seriousness of disease and malnutrition in poor neighbourhoods.

- In Manila the infant mortality rates are three times higher in the slums than in the rest of the city, rates of tuberculosis are nine times higher, diarrhoea and anaemia are twice as common, and three times as many suffer from malnutrition.

- In Bombay the prevalence of leprosy in one slum was 22 per 1,000, compared with 7 per 1,000 for the city as a whole.
- In Singapore the incidence of hookworm, ascaris and trichuris among squatters was more than twice the incidence among flat dwellers.
- In Abidjan tuberculosis was six times as prevalent in the more deprived areas of Abidjan as in rural Côte d'Ivoire.
- In Dakar a third of the peri-urban population sampled had ascaris, compared with only three cases in 400 in rural Senegal.
- In the large slums of Haiti's Port-au-Prince, more than 20% of new-borns die before their first birthday, and another 10% before their second, rates almost three times those for rural areas.

The urban poor may have higher nominal incomes than the rural poor, but their real incomes are seldom better. The reason is that the social services of government, while generally favouring urban areas, seldom reach the urban poor. Housing for the poor is considerably worse in the urban areas, as are environmental conditions, water supply and sanitation. Few governments have effective programmes to reduce urban poverty, leaving the urban poor to fend for themselves — to provide their own shelter, to find their own work in the informal sector and even to grow their own subsistence and market foodcrops.

The quality of the urban environment

None of the cities in the developing world can afford the infrastructure of developed mega-cities. They contend with poor drainage systems and the risk of flooding. Very few have complete water or sewerage systems, and the lack of water and sewerage is particularly severe in slums and squatter settlements (box 5.2). Most of the developing world's cities have unreliable electricity or telephone connections, congested and badly maintained roads and grossly inadequate public transportation.

The economic downturn of the 1980s made the difficulties in supplying and maintaining urban infrastructure even worse. Structural adjustment policies squeezed budgetary allocations for social sectors and impeded the ability of cities to service basic needs. In Dar es Salaam the per capita spending on urban services fell 11% a year between 1978 and 1987. In Nairobi the capital spending on water and sewerage fell from $28 per capita in 1981 to $2.50 in 1987, and maintenance spending from $7.30 to $2.30 — for an average annual decline of 28% for capital and maintenance spending combined.

TABLE 5.1
Projected increases in urban population in major world regions, 1985-2000

Region	Urban population (millions) 1985	Urban population (millions) 2000	Absolute increase (millions)	Percentage increase
Africa	174	361	187	108
Asia	700	1,187	487	70
Latin America	279	417	138	49
Oceania	1.3	2.3	1	77
Developing countries	1,154	1,967	813	70
Industrial countries	844	950	106	13
World	1,998	2,917	919	46

BOX 5.2

Poor water supply and sanitation in larger cities

Four large cities highlight the obstacles to providing water and sanitation services in today's economic and social climate:

Dakar, Senegal. In the early 1980s only one small inner-city area had facilities for the removal of household and public sewerage. Only 28% of urban households had piped water connections, while 68% relied on public standpipes and 4% bought water from vendors. In Pikine, on the outskirts of Dakar, an average of 700 persons relied on each standpipe, and in one neighbourhood, there was only one standpipe for every 1,500 persons.

Calcutta, India. Some 3 million people live in shanty towns and refugee settlements without potable water. They endure serious annual flooding and have no way to dispose of refuse or human wastes. Another 2.5 million live in underserviced older areas. Piped water is available only in the central city and in parts of adjoining municipalities. Sewerage connections are limited to a third of the former colonial core, and the poor maintenance of drains allows periodic clogging and accentuates flooding.

Karachi, Pakistan. Potable water is piped or transported from the Indus River, some 160 kilometres away, and is available for only a few hours a day in most areas. A third of the households have a piped water connection, and most of the more than 2 million people in squatter settlements must either use public standpipes or buy water from vendors at inflated prices.

Bangkok, Thailand. A third of the people lack access to piped water and rely on water bought from vendors or neighbours. Piped water in housing, commercial and industrial developments comes mainly from deep wells, which when polluted must be dug deeper. Pumping from the water table is causing the land to subside, making flooding more severe. There is no piped sewerage system (the cost of installing one is more than $1 billion), and human wastes are generally disposed of through ill-maintained septic tanks and cesspools.

Large disinvestments in urban services foreshadow grave consequences for urban environments and the productivity of cities. In Tanzania, for example, many working days are lost each year to the intermittence of water and electricity supplies, reducing business tax revenues and increasing consumer prices. Similar productivity losses over the past two decades in Tanzania are the result of underinvestments in public transport.

The housing problem

The formal housing sector rarely produces more than 20% of the new housing stock in third world cities. The rest is produced informally, with various degrees of illegality, ranging from the unlawful occupation of land to the pervasive neglect of building codes, infrastructure standards, zoning restrictions and regulations for land use and subdivision. During the early 1980s, just one formal housing unit was added to the total urban housing stock for every nine new households in the low-income developing countries. The great bulk of the growing urban population is thus being housed in unauthorised informal settlements.

Overcrowding is serious in inner-city slums, where the supply of unauthorised housing has been severely curtailed, and where rent control legislation discourages new rental units. In Kumasi, Ghana, three of every four households have only one room to live in. Restricted supply has led to similar overcrowding in many other third world cities as well. Half the population of Calcutta, a third of the urban population of Mexico City and most people in urban Africa live in such conditions. Families alternate with other families in their use of the same dwelling unit at different hours of the day.

Controlling the invasions and encroachments by squatters has meant that many urban poor must rent accommodation or take the more expensive, and often less accessible, option of buying land in illegal subdivisions. Renters usually are poorer than owner-occupiers in the informal settlements, and many of them would rather live in cheap accommodation (and invest back in their villages) than invest in a house in the city.

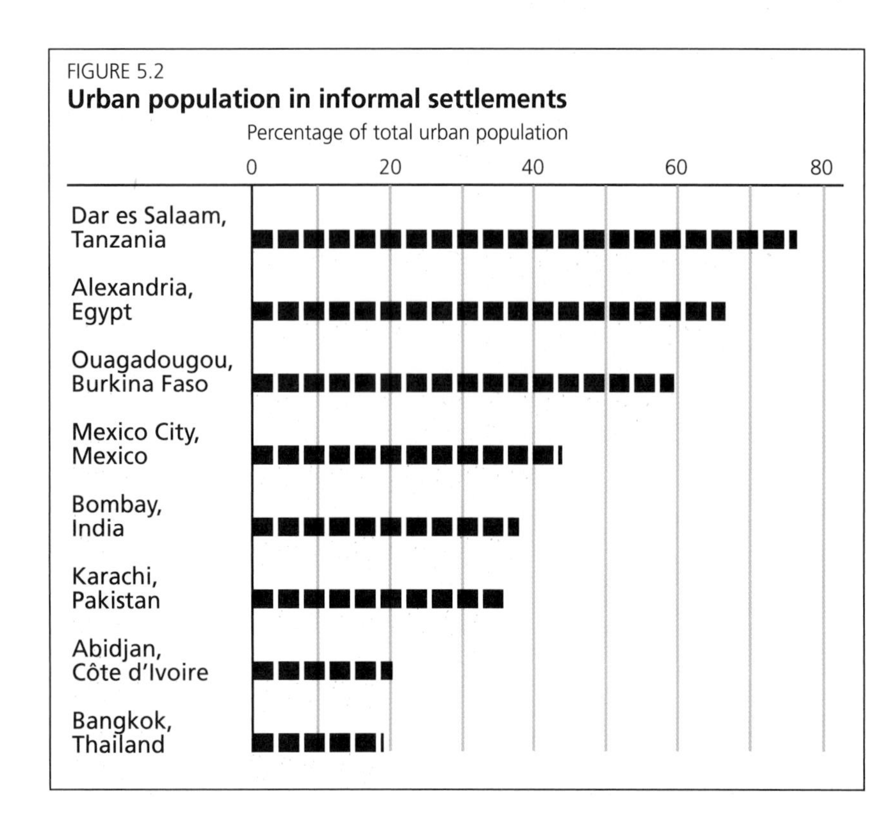

Failed attempts to turn the tide

Governments in most developing countries recognise the difficulty (or impossibility) of reversing urbanisation, or even slowing urban growth appreciably.

Given the inevitability of urbanisation, planners now are trying to develop rural areas and market towns and to manage cities more effectively to facilitate economic growth.

Since the 1950s several policies and programmes have tried to influence where people live and discourage their migration to cities. These programmes have failed because the pull from the urban areas and the push from the rural areas have accelerated urbanisation. In addition, governments implicitly favoured urban development through their preferential treatment of cities for industrial development, for pricing policies, for infrastructure investments, for social services and for food and other subsidies.

- Integrated rural development was intended to raise agricultural productivity and persuade people to remain on the farm. But such projects were overly complex and lacked the necessary manpower for effective implementation, especially in Africa. The main deficiency of these projects was that they did not reach enough rural poor to stop their migration.

• Colonisation projects, opening new lands for settlement, usually in lieu of land reform, have benefitted only a small proportion of the rural poor, and have not restrained rural-urban migration.

• Another measure to restrict urban growth has been the setting of minimum rural wages to reduce rural-urban wage differences, a major incentive to migration. But where this measure has been implemented, it supplanted permanent labourers with seasonal workers and accelerated farm mechanisation. The minimum rural wage thus led to rural unemployment and declining income, probably stimulating rural-urban migration.

• Urban policies to discourage migration have usually focussed on the destruction of new squatter settlements, while tolerating older squatter settlements, legalising them and sometimes providing them with services. A variation on this approach in Africa has been the periodic expulsion of unemployed migrants, as in Congo, Niger, Tanzania and Zaire. These and similar attempts to expel pavement dwellers and street vendors are usually directed at recent arrivals. Highly unpopular, the attempts have had little permanent effect. Indonesian officials tried in the early 1970s to regulate migration to Jakarta by issuing temporary permits requiring new migrants to find housing and employment in six months or face deportation. Largely ineffective, the controls were soon discontinued.

• Governments in developing countries also tried in the 1960s and 1970s to create growth poles and to stimulate development along major highways connecting the poles. Intended to slow the growth of the largest metropolitan areas and balance the distribution of the urban population, the policies for growth poles sought to stimulate the development by investing in heavy manufacturing or industrial activities. Often, however, the industrial growth poles did not have enough physical infrastructure, services or utilities to support industry. The transport costs to major metropolitan areas and ports were high, and industries were not linked effectively to local markets for labour and raw materials. The industrial growth poles became small economic enclaves that stimulated little regional growth or actually drained resources from the regional economies. Moreover, much of the development along the corridors to the growth poles was close to large metropolitan areas, expanding these areas further and increasing their congestion.

• There have also been some attempts to depopulate urban areas, all coercive and all abandoned. China's rustication programme sent many millions of people from cities to rural areas between 1961 and 1976. The programme demanded strong administrative control and was resented both by the rusticated and by those who had to receive them. Once the controls were removed or weakened, the process reversed itself naturally. The most drastic rustication effort was made by the Khmer Rouge armies upon their entry to Phnom Penh in April 1975. Perhaps as many as three out of four million people were forcibly sent to the countryside, but most eventually returned.

Thus it is that migration continues regardless of official attitudes, for the migrants see benefits in moving, both for themselves and for the rural households they leave behind. As long as differences exist between rural and urban areas, people will move to try to take advantage of better schools and social services, higher income opportunities, cultural amenities, new modes of living, technological innovations and links to the world.

Governments are discovering the impossibility of reversing urbanisation, or even of slowing it down

In many cases urbanisation contributes to rural development. Among the leading benefits is the increased economic security of households and extended families that draw on rural as well as urban incomes. Many cities, particularly smaller ones, provide casual or temporary employment for rural labour when it is not required on the farm, and many migrants send back a considerable share of their incomes as remittances to support their families in the village. In Kenya urban wage employment is the main source of nonagricultural cash income to smallholder households, a direct

Rapid urbanisation is neither a crisis nor a tragedy. It is a challenge for the future

response to cash and credit constraints on smallholder agriculture. Income from regular wage employment (predominantly urban) and remittances from relatives (almost entirely urban) make up more than three-quarters of Kenya's nonagricultural cash income.

Urban areas also absorb the excess rural labour from natural population growth and mechanised agriculture. The stronger, healthier and better educated villagers tend to migrate, and the village often provides initial support while they find employment. This cushion against failure imposes many of the costs of urbanisation on migrants' families, but the migrants repay them by sending back cash, new skills and innovations.

Managing the cities — four issues for the 1990s

Rapid urbanisation is neither a crisis nor a tragedy. It is a challenge for the future. The process of urbanisation has created a host of new opportunities intermeshed with new and ill-understood problems. Too often, conventional approaches to these problems have foundered on the failure of bureaucracies to understand the needs, motives and perseverance of urban immigrants. Decrees to restrict land use, enforce building codes, demolish slums and implement public housing schemes have often proved costly and irrelevant.

Cities are the prime place for creating wealth. They provide the basic infrastructure for economic growth and social transformation, and they continue to attract people precisely because they offer opportunities to share in that wealth and growth. The focus on today's cities must move decisively towards better urban management, with past failures giving way to more appropriate policies and practices. Urban management must do more to mobilise urban wealth for the benefit of the urban community as a whole, to maintain and develop the infrastructure and service networks necessary for urban activities and to care for the growing numbers of urban poor. The challenge for urban planners and policymakers in developing countries during the 1990s is to identify and implement innovative programmes that deal with four critical issues:

- Decentralising power and resources from central government to municipalities.
- Mobilising municipal revenue through local sources with the active participation of private sector and community organisations.
- Emphasising "enabling" strategies for shelter and infrastructure, with special assistance targetted to weaker groups.
- Improving the quality of the urban environment, especially for the vast majority of the urban poor in slums and squatter settlements.

Decentralisation

City management is low on the list of priorities for most governments in developing countries. Only a few cities can elect their own administrations and have reliable access to resources beyond the control of central government. There is no inherent reason, however, for central authorities to transfer powers unless they are under pressure to provide services and cannot deliver. There is also a risk that they will transfer powers only to pass the blame — or that they will transfer responsibility without the funds to enable municipalities to follow through.

Cities cannot hope to manage themselves properly until they have enough manpower in place. The shortages of trained professionals to plan, build, manage and maintain complex service systems point to the need for greater administrative and revenue-raising freedom. The growth of the modern civil service is generally supported by a growing, better organised, and more powerful urban citizenry that has developed its own civic value systems — emphasising clean government, planning, adherence to laws and regulations, promotion based on performance, public order, welfare, property taxation and public accountability. Urban dwellers in Africa, for example, are demanding better management of their cities, and some governments have begun to respond.

In most developing countries, however, political power is still highly centralised. Few cities can govern themselves by electing mayors and city councils. A tradition of responsible local government and a cadre of competent local officials have been slow to develop, and even where they exist they tend to function as extensions of the central government.

Realistic policies towards urban growth try to redirect it from primary cities to smaller cities that are more accessible to the rural areas, directing investments in public infrastructure and services away from the capital to support markets and small industries with strong rural links. An important element in this is financial decentralisation, which allows the smaller cities to control their development budgets and raise local taxes. Colombia's decentralisation law of 1986 is typical of the efforts in many developing countries.

Several countries have national plans and policies for strengthening small and intermediate cities. Ecuador's new national development plan focusses on 16 intermediate cities as centres for agricultural processing industries. Thailand has five regional cities for development and has decentralised its university system, increased investments in infrastructure and slum upgrading and strengthened local planning and taxation capacities. Kenya is also planning to invest in the infrastructure of its provincial cities. Just how effective intermediate cities will be in absorbing migrants — or in simply being a stopping point — remains to be seen.

Small and intermediate cities, with forward and backward linkages to the rural countryside, clearly promote both rural and urban development, particularly when they are in enterprising and economically active rural regions. The Upper Valley of Rio Negro and Neuquen in Argentina shows how relationships can develop between town and countryside, providing nonfarm employment to surplus rural populations and shortening the distance of migration.

Generating municipal revenues

If cities are to manage themselves, solve their own problems, and pay their own way, where is the money going to come from?

Municipal finance systems in six major Asian cities — Bangkok, Calcutta, Jakarta, Karachi, Osaka and Seoul — exhibit a high degree of fiscal self-sufficiency, with between 70% and 100% of local government budgets funded from local revenues. But all six cities had serious service deficiencies or deteriorating physical environments, and all six needed additional resources.

A simple increase in revenue transfers from the central government was not the answer, however. Each of the city governments could have generated the needed funds if allowed to exploit the local revenue bases more effectively — either by relaxing restrictions on tax rates, surcharges and assessment criteria or by gaining control over levies and collections from other levels of government that had less incentive to perform those functions effectively. The seeming fiscal independence in each of the six cities is thus illusory.

Most cities can generate the resources for their own improvement

The fundamental principle for improving municipal finance systems is for cities to pay their own way. This must be accomplished gradually by removing central government contributions and grants to cities and replacing them with revenues from local sources.

Urban residents are usually willing to pay for services, if they can see a direct benefit. The benefits are most obvious for water and electricity, but they are also clear for roads, drains, sewers and police and fire protection — if the services are close enough to raise property values.

Many countries use property taxes as a revenue base for municipal infrastructure and operations. Property owners gain from the increased value of their land and housing, and tax-supported services return part of the gain to the community creating it.

The urban property tax system in Brazil is a good framework for property tax administration (box 5.3). Although the logic is clear enough, municipal governments have generally been slow or ineffective in putting it into practice. Property owners, particularly those holding vacant land, have resisted the tax, arguing that because their land does not generate income, they have no means to pay. One way to overcome this difficulty is to defer taxes on vacant land, with interest, until the land is sold — in effect, lending the tax to landlords and collecting later.

Taxes on urban wealth, and particularly on the gains from rising property values, can more than offset the cost of adequate services for urban areas. But such taxes require great leaps of faith by those paying the higher taxes — and great changes in the behaviour of those administering the revenues.

Administrative capacity is the chief impediment to mobilising the resources for urban services and infrastructure. That capacity is blunted by severe shortages of trained manpower, particularly trained accountants and financial managers, by low morale because of low wages and limited career opportunities and by ineffective monitoring and evaluation systems.

User charges also offer promise. For example, Mexico City plans to augment its water supply in the 1990s with water from Tecolutla, some 200 kilometres away. Because Tecolutla is 2,000 metres lower than the capital, six 1,000-megawatt power plants will be needed to deliver the water. The construction of these plants will cost at least $6 billion, roughly half the annual interest repayments on Mexico's external debt. But if the cost is distributed among the urban population over 10 years, the additional water charges would be less than $0.10 a person a day.

BOX 5.3

Urban property taxes in Brazil

Brazil's experience with urban property tax yields several lessons on both the practice and process of implementing administrative reforms.

- Property tax procedures must continually identify and incorporate new construction onto the tax rolls. In Brazil the municipalities use ongoing field surveys to update property data, augmented by cross-referencing with data from other agencies to flag changes in the tax base.
- Converting visible property characteristics into an estimate of market value requires sophisticated knowledge of property markets. To solve this problem, Brazil uses a highly simplified form of mass appraisal, based on a few readily observable, measurable characteristics of each property. No skills, other than the ability to measure and write, are required of valuers in the field.
- Accurate data on market prices are essential for making defensible valuations. Brazil's municipalities use construction cost data from industry sources.
- Collection depends on identifying the person liable for paying the tax. Because records of ownership or tenancy may not be accessible to the taxing authority, Brazil's municipalities extend liability to any person in beneficial occupation of the property. This broad definition relieves the municipalities of any legal obligation to prove legal ownership before imposing the tax.

One of the ironies of life in today's third world cities is that the poor, although least able to pay, may be the main supporters of higher charges, as long as they get the services. In general, they pay considerably more for piped water than the rich. Lima's poor pay 18 times more for every gallon of their water than wealthier groups do. It would thus be more equitable to levy user charges for urban services but ensure that the services do reach the poor.

Enabling strategies for shelter and infrastructure

The limited financial and human resources of municipalities and central governments make it particularly important to use the energies of all actors on the urban scene. The best way to release these energies is for governments to shift from directly providing services to enabling others to provide them — be they formal and informal producers, community-based and nongovernmental organisations or the urban residents themselves. Enabling strategies can yield the highest returns in the provision of shelter and urban infrastructure.

The Global Shelter Strategy for the Year 2000, endorsed by the United Nations General Assembly, supports an enabling approach for urban dwellers to develop their own shelter. It recommends that governments direct their attention and resources towards agencies that can deliver building materials, infrastructure and finance for home construction — avoiding showcase housing projects, such as high-rise flats and expensive sites-and-services projects, which benefit relatively few people at great expense to others.

Materials, infrastructure and finance are much more valuable to home owners than are promises for government housing units. Governments should thus focus on increasing the supply of urban land, adjusting rules and regulations to prevailing needs, practices and conditions in the informal sector, improving infrastructure in existing settlements, guaranteeing land tenure and adjudicating disputes when people are threatened with eviction (box 5.4).

Public services — for roads, walkways,

water, sanitation, electricity and drainage — are the most effective way of expanding urban housing supply, dampening land price increases and stimulating private investment in shelter (box 5.5). In many developing countries, the central government builds the main line and leaves outer networks to others. In Bangkok, an informal road-building programme has greatly extended major intercity highways and brought a steady increase in housing opportunities. Almost a quarter of the land parcels in one fringe district were accessible by roads, thanks to the ingenuity and negotiating powers of local land brokers, developers and village headmen.

Cost recovery — and the decisions about who will bear the burdens of efficient urban expansion — can be approached in various ways. One is the land pooling and readjustment system, developed in the Federal Republic of Germany and commonly practised in Japan and the Republic of Korea. Such systems demand a high degree of trust and organisation, however, and they have not been easy to copy.

The private and informal sectors can do much to develop urban services, particularly housing, but a balance must be struck between privatisation and overregulation. The human skills and organisational resources available through voluntary organisations and the private sector must complement those in municipal and central governments — as in the growth of informal public transport in Nairobi (box 5.6) and the development of affordable housing projects in Bangkok.

Nongovernmental organisations and community-based organisations can provide much valuable organisational support and expertise to municipal governments. In many large cities, hundreds of these organisations are in direct contact with the urban poor. They understand and can articulate the needs of the poor. And they can be critical in the interface between large government bureaucracies and nascent communities because of the greater trust that donors and the people have in their administration of funds.

Self-help construction and maintenance of services by communities can be cheaper and better than similar services from cumbersome municipal authorities. In Khartoum, Sudan, a strong tradition of self-help bolsters local projects to provide primary schools, health centres, water supply networks and roads to more than 90 spontaneous housing areas scattered on the fringes of the city. Binding these communities are the active participation of 367 local neighbourhood councils, the cultural homogeneity of each community and the city's unwillingess to offer support unless people organise on their own. These self-help areas have been found to be remarkably clean and well planned.

BOX 5.4

Land sharing — not eviction — in Bangkok

With much of the prime land in the centre of Bangkok occupied by the poor, more than 5,000 families were evicted in 1985 and 1986 to make way for road construction, commercial development and public buildings.

The threat of eviction has strengthened communities, and the best-organised ones have resisted outright eviction by negotiating with their landlords to share land. The community agrees to evacuate the better part of a site earmarked for commercial development in return for the right to buy or lease back the balance to rehouse original residents.

Five land-sharing agreements were signed in Bangkok before 1985, and three since. The largest is in Klong Toey, where the Port Authority released a large plot of land to rehouse 1,300 families who agreed to make way for a new container port.

Because land eviction almost always is highly sensitive, land-sharing arrangements — by providing a way for all parties to compromise — have considerable political support.

BOX 5.5

Upgrading Jakarta's kampungs

Most urban Indonesians live in kampungs, heterogeneous communities of poor, middle-income and sometimes even rich families. Infrastructure is often poor, and densities are sometimes more than 21,500 people per square kilometre. Many dwellings are temporary, with a large percentage of them replaced every year.

The municipal administration of Jakarta introduced a Kampung Improvement Programme in 1969 to upgrade physical infrastructure. It has managed to improve living conditions in about 500 kampungs with a total population of 3.8 million.

The programme's first five-year cycle had a strong top-down character, focussing heavily on such public works as roads, footpaths, drainage, and water supply. Loans from the World Bank enabled the programme's second and third five-year cycles to be broadened and extended to about 200 other cities.

Some of the new activities in selected kampungs include garbage collection and disposal, the construction of sanitary facilities, horticultural and health training and vocational and nonformal education. The programme does not cover housing.

Public support for the programme is strong, and there has been a substantial increase in private household investment because of the improvements in infrastructure. The programme has also increased property values and rents.

Improving the urban environment

While cities contribute to economic development in developing countries, rapid population growth and uncontrolled industrial development degrade the urban environment, straining the natural resource base and undermining sustainable and equitable development.

The wastes of urban communities outstrip the capacity of cities to collect and dispose them safely and effectively. The textile, brewing, chemical and pulp and paper industries in many cities discharge wastewater into rivers and open drains, threatening the public health and reducing the quality of urban life. Inappropriate land development and improperly disposed toxic wastes damage groundwater, wetlands and other sensitive ecosystems. In Bangkok and Shanghai, which are near river estuaries, excessive pumping from groundwater wells is allowing salt water to intrude the aquifers, causing the ground to subside and increasing the problems of flooding and drainage.

Air pollution due to emissions from fossil fuels used for transport, industry and home heating is also getting worse. Air quality standards in most large cities of developing countries are far below internationally accepted standards for public health, particularly in industrial cities that depend on coal and wood for fuel. For example, more than 50% of Calcutta's residents suffer from respiratory diseases related to air pollution.

The deteriorating urban environment is severely affecting the urban poor in slums and squatter settlements, where residents suffer disproportionately from gastroenteric and respiratory disease because of their high density and the inadequacy of public health and sanitation services. They often live in the most polluted fringe areas of major cities, close to unsanitary landfills, waste disposal dumps and low-lying, poorly drained areas.

Nongovernmental organisations and international agencies, in cooperation with the governments in developing countries, have been identifying innovative approaches to improve the quality of the urban environment. The Orangi pilot project in Karachi is building a low-cost sewerage system with the participation of residents in one of the largest squatter settlements (box 5.7). A pilot project in Peru, financed by the World Bank and the United Nations Development Programme, shows that there are economical ways of treating wastewater to make it pure enough for irrigating vege-

BOX 5.6

Informal public transport in Africa

In Kinshasa they are the fula-fula (covered 10-ton trucks) and kimalu-malu (station wagons whose name means "those that go very fast"). In Dakar they are the car rapides. In Dar es Salaam, they are the dala dalas, named for the silver coin they charge as a fare. In Nairobi they are the matatus, minibusses that got their name from the 30-cent flat fee they used to charge.

The ridership of these private, small-scale operators has grown dramatically in the 1980s, and in many African cities more than 50% of all trips by public transport are taken in their vehicles.

The small-scale transport operators are also major employers. In Dakar 523 car rapides gave employment to 3,420 people in 1975 and supported 22,230 people. In Kenya the small-scale public transport sector employed 17,000 persons in 1986 and generated fare revenues of $140 million.

Nairobi's monopoly bus company, KBS, carried only 42% of the passengers in the early 1980s, the matatus most of the rest.

About 15% of the matatus belonged to individuals who owned only one vehicle, the rest to businesses with larger fleets. These fleets created 50% more jobs per passenger than the public bus service, and most of these jobs involved the poor.

The matatus were fully legalised in 1984, increasing private investment and improving service. Problems remain, however, notably the disregard for passenger safety and the poor vehicle maintenance. Self-regulation of informal public transport could do much to make it safer for, and more attractive to, the riders — and more profitable for the operators.

BOX 5.7

Community-based sanitation in Karachi

Residents of Karachi's low-income Orangi community are getting sewerage connections with the help of an enterprising nongovernmental organisation.

A sprawling, unauthorised settlement, Orangi is home to 700,000 people, 10% of the capital's resident population.

Plots of land in Orangi line unpaved lanes with no provision for public services. The Karachi Metropolitan Corporation (KMC) and the Karachi Development Authority (KDA) are responsible for providing trunk services, including major roads, drains, sewerage and water pipes, but residents are expected to pay for connections to the mains. In the case of sewerage, the official cost of a household connection is 100 rupees (roughly equal to $5) for every square yard on which the house is situated — a charge considered too high by families that have spent only Rs. 20,000 to construct their house.

In 1980 a nongovernmental organisation started up in the community to build a sewerage network that would be cheaper than state utilities. The NGO organised the community into lanes of 15 to 20 houses and looked for low-cost technologies for pipes, septic tanks and manholes. Each lane was in charge of its own accounts and took responsibility for the purchase of the necessary materials.

It took three years to develop technical solutions for servicing individual sites, but the cost of construction was brought down to roughly a twentieth of the official rate. Costs of pipe were slashed from 100 rupees per running foot to 13, those for septic tanks from 2,000 rupees to 160 and those for manholes from 500 rupees to 70. By the end of 1985, more than half of Orangi's 10,000 lanes had functioning sewerage systems — showing the effectiveness of NGOs in organising community self-help for cost effective services.

tables and raising fish. And by marshalling support for garbage and refuse collection, the city of Shanghai has generated revenues and employment opportunities through waste recycling (box 5.8).

To reverse urban environmental deterioration in the 1990s, the governments of developing countries must:

- Improve municipal waste collection coverage and efficiency.
- Adopt environmentally sound municipal waste treatment and disposal practices.
- Coordinate pollution control actions across levels of government and urban subsectors.
- Incorporate environmental planning and management techniques into citywide strategic planning and implementation.
- Facilitate the participation of the private sector in mobilising resources for environmental improvement.

Rapid urbanisation is transforming the developing countries, creating ever new problems but also offering ever new opportunities. To solve the growing problems of cities — and to unleash the many possibilities for human development — is going to depend heavily on better urban management, considerably better.

BOX 5.8

Recycling urban waste in Shanghai

Industrial and household wastes mean money and jobs in Shanghai.

Since 1957 the city has developed a network of 502 waste collection stations with 1,500 purchasing agents on the rural periphery, who are paid a commission to acquire materials for recycling.

The Shanghai Environmental Sanitation Administration collects from a population of about 12 million people over an area of 6,000 square kilometres. It has 26 recycling centres to process materials from industry, businesses and households, and it runs a network of sales outlets for reclaimed products.

Recovered materials include ferrous and nonferrous metals, rubber, plastics, paper, rags, cotton, chemical fibre, animal bones, human hair, bottles, broken glass, old machine parts, chemical residues and used oils. There are subsidiaries for refining copper, recovering precious metals and producing new oils from old.

The administration gives full-time employment to 29,000 people and provides many more with part-time work. It also employs 3,600 people to advise factories on the installation of systems for the sorting and collection of wastes.

ANNEX TABLE 1

Child survival and immunisation targets for the year 2000

	Under-five mortality target of 70 per 1,000 live births by 2000 (or a reduction of 50% if the rate in 1980 was less than 140)					Immunisation target of 100%				
	Under-five mortality rate (per 1,000 live births)		Average annual reduction rate (%)	Annual reduction rate needed to reach target	Year reaching target using past reduction rate	Percentage of one year olds immunised		Average annual growth rate (%)	Annual growth rate needed to reach target	Year reaching target using past reduction rate
	1960	1988	1960-88	by 2000		1981	1987-88	1981-88	by 2000	
Low human development										
1 Niger	320	228	1.2	9.8	2086	15	24	7.60	12.19	2008
2 Mali	370	292	0.8	11.9	2100 +	19	31	7.69	9.89	2004
3 Burkina Faso	362	233	1.6	10.0	2065	11	46	24.85	6.50	1992
4 Sierra Leone	386	266	1.3	11.1	2089	23	40	9.17	7.55	1998
5 Chad	326	223	1.3	9.7	2074	..	21	..	13.41	..
6 Guinea	346	248	1.2	10.5	2095	10	23	14.19	12.67	1999
7 Somalia	294	221	1.0	9.6	2100 +	3	28	44.82	10.80	1991
8 Mauritania	320	220	1.3	9.5	2074	35	45	4.17	6.60	2008
9 Afghanistan	380	300	0.8	12.1	2100 +	5	27	29.62	11.04	1993
10 Benin	310	185	1.8	8.1	2041	..	35	..	8.76	..
11 Burundi	258	188	1.1	8.2	2076	35	54	6.94	5.09	1997
12 Bhutan	297	197	1.5	8.6	2059	20	67	20.21	3.26	1990
13 Mozambique	330	298	0.4	12.1	2100 +	..	42	..	7.14	..
14 Malawi	364	262	1.2	11.0	2100 +	71	83	2.28	1.55	1997
15 Sudan	293	181	1.7	7.9	2044	2	58	75.24	4.53	1989
16 Central African Rep.	308	223	1.1	9.7	2089	17	33	11.12	9.34	1999
17 Nepal	297	197	1.5	8.6	2059	13	71	30.31	2.75	1989
18 Senegal	313	136	2.9	5.5	2011	..	57	..	4.60	..
19 Ethiopia	294	259	0.5	10.9	2100 +	8	18	14.42	14.70	2001
20 Zaire	251	138	2.1	5.7	2020	23	46	11.16	6.36	1995
21 Rwanda	248	206	0.7	9.0	2100 +	31	82	16.00	1.60	1989
22 Angola	346	292	0.6	11.9	2100 +	..	28	..	10.64	..
23 Bangladesh	262	188	1.2	8.2	2072	1	18	55.66	14.83	1992
24 Nigeria	318	174	2.1	7.6	2031	32	62	10.84	3.97	1993
25 Yemen Arab Rep.	378	190	2.4	8.3	2029	26	32	2.97	9.61	2027
26 Liberia	258	147	2.0	6.2	2025	63	43	-5.56	6.94	..
27 Togo	305	153	2.4	6.5	2020	27	73	16.31	2.58	1990
28 Uganda	224	169	1.0	7.4	2076	14	52	21.95	5.41	1991
29 Haiti	294	171	1.9	7.4	2035	26	50	10.89	5.66	1995
30 Ghana	224	146	1.5	6.1	2037	34	42	3.28	7.14	2015
31 Yemen, PDR	378	197	2.3	8.6	2033	6	37	31.47	8.28	1992
32 Côte d'Ivoire	264	142	2.2	5.9	2020	44	37	-2.66	8.40	..
33 Congo	241	114	2.6	4.6	2009	56	76	4.69	2.25	1994
34 Namibia	262	176	1.4	7.7	2053	..	..	..	..	..
35 Tanzania, United Rep.	248	176	1.2	7.7	2064	65	86	4.34	1.21	1992
36 Pakistan	277	166	1.8	7.2	2036	5	65	49.56	3.51	1989
37 India	282	149	2.3	6.3	2022	17	63	22.77	3.73	1990
38 Madagascar	364	184	2.4	8.1	2028	33	44	4.68	6.84	2006
39 Papua New Guinea	247	81	3.9	3.1	1998	49	55	1.97	4.86	2018
40 Kampuchea, Dem.	218	199	0.3	8.7	2100 +	..	47	..	6.32	..
41 Cameroon	275	153	2.1	6.5	2026	9	52	32.23	5.33	1990
42 Kenya	208	113	2.2	4.4	2013	..	74	..	2.44	..
43 Zambia	228	127	2.1	5.0	2017	54	84	7.19	1.40	1991
44 Morocco	265	119	2.8	4.4	2007	44	65	6.06	3.57	1995
Medium human development										
45 Egypt	300	125	3.1	4.8	2007	76	85	1.75	1.36	1998
46 Lao PDR	232	159	1.3	6.8	2049	6	20	19.60	13.74	1997
47 Gabon	288	169	1.9	7.4	2035	..	76	..	2.25	..
48 Oman	378	64	6.1	-0.8	1988	18	90	27.71	0.89	1988
49 Bolivia	282	172	1.8	7.5	2039	19	38	11.25	8.16	1997

		Under-five mortality target of 70 per 1,000 live births by 2000 (or a reduction of 50% if the rate in 1980 was less than 140)					Immunisation target of 100%				
		Under-five mortality rate (per 1,000)		Average annual reduction rate	Annual reduction rate needed to reach target	Year reaching target using past reduction	Percentage of one year olds immunised		Average annual growth rate (%)	Annual growth rate needed to reach target	Year reaching target using past reduction
		1960	1988	1960-88	by 2000	rate	1981	1987-8	1981-88	by 2000	rate
50	Myanmar	229	95	3.1	4.0	2004	10	24	14.23	12.19	1999
51	Honduras	232	107	2.7	3.5	2004	40	76	10.54	2.19	1991
52	Zimbabwe	182	113	1.7	4.5	2020	49	81	7.85	1.75	1991
53	Lesotho	208	136	1.5	5.5	2032	60	81	4.68	1.73	1993
54	Indonesia	235	119	2.4	4.4	2010	55	71	3.89	2.84	1997
55	Guatemala	230	99	3.0	3.5	2002	30	49	7.53	5.96	1998
56	Viet Nam	233	88	3.4	3.5	2000	..	58	..	4.45	..
57	Algeria	270	107	3.3	3.5	2001	35	71	11.56	2.81	1991
58	Botswana	174	92	2.3	4.3	2011	71	90	3.77	0.85	1991
59	El Salvador	206	84	3.2	3.5	2002	43	63	6.08	3.80	1996
60	Tunisia	255	83	3.9	3.1	1998	51	88	8.74	1.07	1990
61	Iran, Islamic Rep.	254	90	3.6	2.7	1997	33	81	15.03	1.73	1990
62	Syrian Arab Rep.	218	64	4.3	3.1	1997	20	63	19.84	3.73	1991
63	Dominican Rep.	200	81	3.2	3.9	2003	30	70	13.99	2.87	1991
64	Saudi Arabia	292	98	3.8	3.3	1999	42	88	12.21	1.05	1989
65	Philippines	135	73	2.2	4.4	2013	52	82	7.31	1.58	1991
66	China	202	43	5.4	3.6	1996	..	96	..	0.33	..
67	Libyan Arab Jam.	268	119	2.9	4.4	2007	56	62	1.72	3.90	2016
68	South Africa	192	95	2.5	3.8	2007	..	..	..	..	..
69	Lebanon	92	51	2.1	4.2	2012	..	88	..	1.06	..
70	Mongolia	158	59	3.5	3.5	2001	84	67	-3.42	3.29	..
71	Nicaragua	210	95	2.8	3.0	2001	40	70	8.87	2.95	1992
72	Turkey	258	93	3.6	2.7	1998	57	71	3.45	2.81	1998
73	Jordan	218	57	4.7	3.0	1996	..	71	..	2.75	..
74	Peru	233	123	2.3	4.7	2013	31	66	12.40	3.41	1992
75	Ecuador	183	87	2.6	4.0	2007	40	62	7.25	3.86	1995
76	Iraq	222	94	3.0	4.5	2006	35	84	14.72	1.38	1989
77	United Arab Emirates	239	32	6.9	3.1	1994	38	73	10.79	2.55	1991
78	Thailand	149	49	3.9	3.1	1998	48	79	7.90	1.88	1991
79	Paraguay	134	62	2.7	4.8	2009	28	65	13.70	3.57	1991
80	Brazil	160	85	2.2	4.1	2011	70	68	-0.61	3.19	..
81	Mauritius	104	29	4.5	2.7	1996	..	84	..	1.43	..
82	Korea, Dem. Rep.	120	33	4.5	3.4	1998	47	59	3.73	4.31	2002
83	Sri Lanka	113	43	3.4	3.3	2000	51	79	7.12	1.88	1991
High human development											
85	Malaysia	106	32	4.2	3.5	1998	69	74	1.05	2.49	2017
86	Colombia	148	68	2.7	4.6	2009	31	85	16.70	1.28	1989
87	Jamaica	88	22	4.8	3.2	1997	38	82	12.61	1.58	1990
88	Kuwait	128	22	6.1	2.2	1993	65	51	-3.74	5.53	..
89	Venezuela	114	44	3.3	4.7	2005	62	62	-0.19	3.97	..
91	Mexico	140	68	2.5	4.0	2008	50	74	6.27	2.41	1993
92	Cuba	87	18	5.5	2.1	1994	74	93	3.59	0.60	1990
93	Panama	105	34	3.9	3.6	2000	57	79	4.98	1.96	1993
94	Trinidad and Tobago	67	23	3.7	3.6	2001	54	78	5.97	2.01	1992
96	Singapore	50	12	5.0	3.4	1998	79	95	2.97	0.39	1990
97	Korea, Rep.	120	33	4.5	3.4	1998	43	89	11.99	0.96	1989
99	Argentina	75	37	2.5	4.0	2007	55	68	3.38	3.10	2000
102	Uruguay	56	31	2.1	2.9	2006	72	84	2.42	1.45	1996
103	Costa Rica	121	22	5.9	2.7	1994	80	89	1.70	0.91	1995
107	Chile	142	26	5.9	1.4	1992	97	96	-0.04	0.31	..
108	Hong Kong	65	10	6.5	3.0	1994	92	91	-0.22	0.76	..

ANNEX TABLE 2
Child nutrition targets for the year 2000

		Eliminate severe child malnutrition and reduce moderate child malnutrition by half			
		Child malnutrition (percentage of under-fives underweight)			Annual reduction rate needed to reach target by 2000 (%)
	Reference year	Moderate	Severe	Total	
Low human development					
1 Niger	1985	34.4	15.0	49.4	6.8
2 Mali	1987	21.6	9.4	31.0	7.8
4 Sierra Leone	1978	20.8	2.4	23.2	3.6
8 Mauritania	1981	23.1	7.9	31.0	5.1
11 Burundi	1987	27.9	10.4	38.3	7.5
14 Malawi	1981	17.9	6.0	23.9	5.1
17 Nepal	1975	62.5	7.1	69.6	3.2
18 Senegal	1986	16.1	5.5	21.6	6.8
19 Ethiopia	1982	27.8	10.3	38.1	5.5
20 Zaire	1975	19.8	8.6	28.4	4.1
21 Rwanda	1976	21.4	6.4	27.8	3.9
23 Bangladesh	1986	51.2	9.2	60.4	5.9
25 Yemen Arab Rep.	1979	54.9	6.3	61.2	3.7
26 Liberia	1976	16.0	4.3	20.3	3.8
27 Togo	1977	17.6	7.7	25.3	4.5
29 Haiti	1978	33.0	4.4	37.4	3.6
30 Ghana	1988	21.4	5.7	27.1	7.4
31 Yemen, PDR	1983	23.2	2.6	25.8	4.6
32 Côte d'Ivoire	1986	10.2	2.2	12.4	6.1
33 Congo	1987	16.4	7.2	23.6	7.8
35 Tanzania	1988	42.0	6.0	48.0	6.7
36 Pakistan	1987	38.6	12.9	51.5	7.3
37 India	1982	34.8	6.1	40.9	4.6
38 Madagascar	1984	24.5	8.3	32.8	6.0
39 Papua New Guinea	1984	33.8	0.9	34.7	4.4
41 Cameroon	1978	14.2	3.1	17.3	4.0
44 Morocco	1987	12.1	3.6	15.7	7.1
Medium human development					
45 Egypt	1978	15.1	1.5	16.6	3.5
46 Lao PDR	1984	27.8	8.9	36.7	5.9
49 Bolivia	1981	13.5	1.0	14.5	3.9
50 Myanmar	1985	31.7	6.3	38.0	5.7
51 Honduras	1987	16.6	4.0	20.6	6.8
52 Zimbabwe	1988	10.2	1.3	11.5	6.6
54 Indonesia	1987	50.0	1.3	51.3	5.4
55 Guatemala	1987	25.1	8.4	33.5	7.3
56 Viet Nam	1986	39.0	12.5	51.5	6.7
58 Botswana	1987	13.3	1.7	15.0	6.1
60 Tunisia	1975	16.4	4.8	21.2	3.7
61 Iran, Islamic Rep.	1980	32.3	10.8	43.1	4.8
63 Dominican Rep.	1986	9.8	2.7	12.5	6.5
65 Philippines	1982	29.6	3.0	32.6	4.3
71 Nicaragua	1982	9.6	0.9	10.5	4.3
74 Peru	1984	11.1	2.3	13.4	5.4
75 Ecuador	1987	15.8	0.7	16.5	5.5
78 Thailand	1987	21.5	4.3	25.8	6.5
80 Brazil	1986	9.9	2.8	12.7	6.5
81 Mauritius	1985	17.9	6.0	23.9	6.4
83 Sri Lanka	1987	29.5	8.6	38.1	7.0
High human development					
86 Colombia	1986	9.9	2.0	11.9	6.1
87 Jamaica	1978	7.5	1.8	9.3	4.0
88 Kuwait	1984	5.0	1.4	6.4	5.7
89 Venezuela	1982	8.5	1.7	10.2	4.8
93 Panama	1980	12.3	3.4	15.7	4.6
94 Trinidad and Tobago	1987	6.5	0.4	6.9	5.6
96 Singapore	1972	11.1	3.1	14.2	3.3
103 Costa Rica	1982	4.8	1.2	6.0	4.9
107 Chile	1986	2.0	0.5	2.5	6.5

ANNEX TABLE 3

Primary enrolment targets for the year 2000

		Net primary school enrolment target of 100%				
		Net primary school enrolment ratio		Average annual growth rate (%)	Annual growth rate needed to reach target by 2000	Year reaching target using past growth rate
		1980	1986-88	1980-87		
Low human development						
2	Mali	20	19	-1.11	13.86	..
3	Burkina Faso	15	27	8.76	10.60	2003
5	Chad	..	38	..	7.84	..
6	Guinea	..	23	..	11.97	..
7	Somalia	20	15	-4.49	16.01	..
10	Benin	..	50	..	5.48	..
11	Burundi	21	42	10.22	7.00	1997
13	Mozambique	36	45	3.24	6.33	2013
14	Malawi	43	49	1.73	5.72	2030
16	Central African Rep.	57	49	-2.14	5.64	..
17	Nepal	..	56	..	4.63	..
18	Senegal	37	50	4.40	5.48	2004
19	Ethiopia	..	27	..	10.60	..
21	Rwanda	59	64	1.17	3.49	2026
23	Bangladesh	..	56	..	4.63	..
27	Togo	..	73	..	2.45	..
28	Uganda	..	41	..	7.20	..
29	Haiti	37	44	2.34	6.61	2024
35	Tanzania, United Rep.	68	51	-4.16	5.40	..
44	Morocco	62	57	-1.19	4.42	..
Medium human development						
48	Oman	50	80	6.94	1.73	1991
49	Bolivia	77	83	1.08	1.44	2005
52	Zimbabwe	..	100	..	-	..
54	Indonesia	88	98	1.55	0.16	1989
57	Algeria	81	89	1.35	0.90	1996
58	Botswana	76	89	2.28	0.90	1993
59	El Salvador	..	62	..	3.81	..
60	Tunisia	83	95	1.87	0.44	1991
61	Iran, Islamic Rep.	..	94	..	0.52	..
62	Syrian Arab Rep.	91	97	0.92	0.23	1991
63	Dominican Rep.	..	79	..	1.83	..
64	Saudi Arabia	50	56	1.63	4.56	2023
65	Philippines	94	94	0.00	0.48	..
66	China	..	95	..	0.40	..
71	Nicaragua	74	77	0.48	2.08	2044
73	Jordan	93	88	-0.79	0.99	..
76	Iraq	100	87	-2.05	1.12	..
77	United Arab Emirates	73	89	2.79	0.94	1992
79	Paraguay	87	85	-0.33	1.26	..
81	Mauritius	..	94	..	0.48	..
83	Sri Lanka	..	100	..	..	..
High human development						
86	Colombia	..	73	..	2.45	..
88	Kuwait	84	79	-0.87	1.83	..
92	Cuba	98	95	-0.52	0.44	..
93	Panama	..	90	..	0.86	..
94	Trinidad and Tobago	88	88	-0.08	1.03	..
96	Singapore	99	100	0.14	..	..
97	Korea, Rep.	99	100	0.07	..	..
103	Costa Rica	89	85	-0.65	1.26	..
108	Hong Kong	95	95	0.00	0.40	..

ANNEX TABLE 4

Literacy targets for the year 2000

		Halving the estimated 1990 illiteracy rate					Female illiteracy no higher than male				
		Adult illiteracy rate (%)		Average annual reduction rate (%) 1970-85	Annual reduction rate needed to reach target by 2000	Year reaching target using past reduction rate	Female illiteracy rate (%)		Average annual reduction rate (%) 1970-85	Annual reduction rate needed to reach target by 2000	Year reaching target using past reduction rate
		1970	1985				1970	1985			
Low human development											
1	Niger	96	86	0.73	4.75	2080	98	91	0.49	5.11	2100 +
2	Mali	93	83	0.73	4.75	2080	96	89	0.50	5.21	2100 +
3	Burkina Faso	92	86	0.42	4.65	2100 +	97	94	0.21	5.18	2100 +
4	Sierra Leone	87	70	1.41	4.97	2034	92	79	1.01	5.70	2072
5	Chad	89	74	1.20	4.90	2043	98	89	0.64	6.04	2100 +
6	Guinea	86	71	1.24	4.91	2041	93	83	0.76	5.86	2100 +
7	Somalia	97	88	0.67	4.73	2089	99	94	0.34	5.17	2100 +
9	Afghanistan	93	76	1.30	4.93	2038	98	92	0.42	6.14	2100 +
10	Benin	85	73	0.94	4.82	2059	92	84	0.60	5.67	2100 +
11	Burundi	81	65	1.38	4.96	2035	90	73	1.39	5.66	2048
13	Mozambique	79	61	1.64	5.04	2027	86	78	0.65	6.56	2100 +
14	Malawi	70	58	1.21	4.90	2042	82	69	1.14	5.96	2066
15	Sudan	83	77	0.54	4.69	2100 +	94	85	0.67	5.35	2100 +
16	Central African Rep.	84	59	2.37	5.27	2014	94	71	1.85	6.47	2039
17	Nepal	87	74	1.07	4.86	2050	97	88	0.65	5.95	2100 +
18	Senegal	89	72	1.37	4.96	2036	95	81	1.06	5.71	2068
20	Zaire	59	38	2.84	5.43	2010	78	55	2.30	7.73	2037
21	Rwanda	68	53	1.67	5.05	2027	79	67	1.09	6.54	2078
22	Angola	89	59	2.67	5.37	2011	93	67	2.16	6.17	2029
23	Bangladesh	76	67	0.84	4.78	2068	88	78	0.80	5.74	2096
24	Nigeria	76	57	1.81	5.10	2023	86	69	1.46	6.25	2051
25	Yemen Arab Rep.	95	75	1.56	5.02	2029	99	93	0.42	6.37	2100 +
26	Liberia	83	65	1.56	5.02	2030	92	77	1.18	6.07	2065
27	Togo	83	59	2.21	5.22	2017	93	72	1.69	6.44	2044
28	Uganda	59	42	2.18	5.21	2017	70	55	1.59	6.85	2152
29	Haiti	79	62	1.51	5.00	2031	83	65	1.62	5.25	2035
30	Ghana	70	46	2.65	5.37	2011	82	57	2.40	6.65	2028
31	Yemen, PDR	80	58	2.14	5.20	2017	91	75	1.28	6.83	2068
32	Côte d'Ivoire	82	58	2.26	5.24	2016	90	69	1.76	6.32	2041
33	Congo	66	37	3.74	5.72	2004	81	45	3.84	6.94	2013
35	Tanzania, United Rep.	67	..	..	..	..	82	12	12.03	..	..
36	Pakistan	80	70	0.84	4.79	2067	89	81	0.63	5.71	2100 +
37	India	67	57	0.97	4.82	2057	80	71	0.79	6.16	2100 +
38	Madagascar	51	32	3.01	5.48	2008	57	38	2.67	6.57	2023
39	Papua New Guinea	69	55	1.41	4.97	2034	76	65	1.04	5.97	2074
40	Kampuchea, Dem.	..	25	..	..	..	77	35	5.12	..	..
41	Cameroon	67	..	..	..	..	81	*51*	3.04	..	..
42	Kenya	69	40	3.52	5.65	2005	81	51	3.04	7.17	2022
43	Zambia	49	24	4.48	5.96	2001	63	33	4.22	7.84	2014
44	Morocco	78	66	1.11	4.87	2048	90	78	0.95	5.92	2082
Medium human development											
45	Egypt	65	55	1.11	4.87	2048	80	70	0.89	6.39	2097
46	Lao PDR	68	16	9.14	7.52	1993	72	24	7.06	9.97	2007
47	Gabon	68	38	3.70	5.71	2004	78	47	3.32	6.97	2018
49	Bolivia	43	25	3.46	5.63	2005	54	35	2.85	7.63	2027
50	Myanmar	29	*21*	2.13	5.20	2018	43	..	..	..	..
51	Honduras	48	41	1.05	4.85	2051	50	42	1.16	5.08	2053
52	Zimbabwe	45	26	3.61	5.68	2004	53	33	3.11	7.18	2021
53	Lesotho	39	27	2.24	5.23	2016	26	16	3.18	1.77	1994
54	Indonesia	46	26	3.74	5.72	2004	58	35	3.31	7.58	2021
55	Guatemala	56	45	1.43	4.97	2033	63	53	1.15	5.99	2066
57	Algeria	75	50	2.67	5.37	2011	89	63	2.28	6.82	2032
58	Botswana	60	29	4.70	6.04	2000	56	31	3.87	6.47	2011
59	El Salvador	43	28	2.83	5.43	2010	47	31	2.74	6.08	2019
60	Tunisia	70	45	2.86	5.43	2009	83	59	2.25	7.13	2034
61	Iran, Islamic Rep.	72	49	2.49	5.31	2013	83	61	2.03	6.69	2036

		Halving the estimated 1990 illiteracy rate					Female illiteracy no higher than male				
		Adult illiteracy rate (%)		Average annual reduction rate (%) 1970-85	Annual reduction rate needed to reach target by 2000	Year reaching target using past reduction rate	Female illiteracy rate (%)		Average annual reduction rate (%) 1970-85	Annual reduction rate needed to reach target by 2000	Year reaching target using past reduction rate
		1970	1985				1970	1985			
62	Syrian Arab Rep.	60	40	2.67	5.37	2011	80	57	2.23	7.58	2038
63	Dominican Rep.	33	22	2.67	5.37	2011	35	23	2.76	5.65	2017
64	Saudi Arabia	92	..	..	..	..	98	69	2.31	..	..
65	Philippines	18	14	1.48	4.99	2032	19	15	1.56	5.42	2039
67	Libyan Arab Jam.	64	34	4.08	5.83	2002	87	50	3.63	8.22	2020
69	Lebanon	32	22	2.29	5.25	2015	42	31	2.00	7.33	2042
70	Mongolia	20	..	..	..	..	26	*13*	4.52	..	..
71	Nicaragua	43	12	8.09	7.16	1994	43	..	..	..	..
72	Turkey	49	26	3.99	5.80	2003	66	38	3.61	8.08	2020
73	Jordan	54	25	4.86	6.09	1999	71	37	4.25	8.43	2016
74	Peru	30	15	4.41	5.94	2001	40	22	3.91	8.31	2018
75	Ecuador	29	17	3.39	5.61	2006	32	20	3.08	6.62	2018
76	Iraq	66	11	11.26	8.24	1991	82	13	11.55	9.26	1997
78	Thailand	21	9	5.48	6.29	1998	28	12	5.49	8.07	2008
79	Paraguay	20	12	3.33	5.59	2006	25	15	3.35	6.96	2017
80	Brazil	34	22	2.72	5.39	2011	37	24	2.84	5.80	2017
81	Mauritius	32	17	4.16	5.86	2002	41	23	3.78	7.76	2017
83	Sri Lanka	23	13	3.72	5.71	2004	31	17	3.93	7.37	2014
High human development											
85	Malaysia	41	26	2.91	5.45	2009	52	34	2.79	7.13	2025
86	Colombia	23	..	..	..	..	24	*12*	4.52	..	..
88	Kuwait	47	30	2.88	5.44	2009	58	37	2.95	6.75	2021
89	Venezuela	25	13	4.27	5.89	2001	29	15	4.30	6.79	2009
91	Mexico	27	10	6.29	6.56	1996	31	12	6.13	7.69	2004
92	Cuba	14	4	7.79	7.06	1994	13	4	7.56	7.06	1999
93	Panama	19	11	3.58	5.67	2005	19	12	3.02	6.21	2017
94	Trinidad and Tobago	8	4	4.52	5.98	2000	11	5	5.12	7.37	2007
96	Singapore	27	14	4.11	5.84	2002	45	21	4.95	8.30	2011
97	Korea, Rep.	13	..	..	..	..	19	*9*	4.86	..	..
99	Argentina	7	4	2.91	5.45	2009	8	4	4.52	4.71	2001
102	Uruguay	7	5	2.22	5.23	2016	7	..	..	..	..
103	Costa Rica	13	7	4.26	5.89	2001	13	7	4.04	6.35	2009
107	Chile	11	2	10.74	8.07	1992	12	3	8.83	10.52	2004
108	Hong Kong	23	12	4.13	5.85	2002	36	19	4.17	8.58	2017

ANNEX TABLE 5

Safe water targets for the year 2000

	100% access to safe water				
	Percentage of people with access		Average annual growth rate (%) 1975-86	Annual growth rate needed to reach target by 2000	Year reaching target using past growth rate
	1975	1985-87			
Low human development					
1 Niger	27	47	5.17	5.54	2001
2 Mali	..	17	..	13.49	..
3 Burkina Faso	25	67	9.38	2.90	1991
4 Sierra Leone	..	25	..	10.41	..
5 Chad	26	..	..	..	..
6 Guinea	14	19	2.82	12.59	2046
7 Somalia	38	34	-1.01	8.01	..
9 Afghanistan	9	21	8.01	11.79	2007
10 Benin	34	52	3.94	4.78	2003
11 Burundi	..	26	..	10.10	..
13 Mozambique	..	16	..	13.99	..
14 Malawi	..	56	..	4.23	..
15 Sudan	..	21	..	11.79	..
17 Nepal	8	29	12.42	9.24	1997
18 Senegal	..	53	..	4.64	..
19 Ethiopia	8	16	6.50	13.99	2016
20 Zaire	19	33	5.15	8.24	2009
21 Rwanda	68	50	-2.76	5.08	..
22 Angola	..	30	..	8.98	..
23 Bangladesh	56	46	-1.77	5.70	..
24 Nigeria	..	46	..	5.70	..
25 Yemen Arab Rep.	..	42	..	6.39	..
26 Liberia	..	55	..	4.36	..
27 Togo	16	55	11.88	4.36	1992
28 Uganda	35	20	-4.96	12.18	..
29 Haiti	12	38	11.05	7.16	1996
30 Ghana	35	56	4.37	4.23	2000
31 Yemen, PDR	..	54	..	4.50	..
32 Côte d'Ivoire	..	19	..	12.59	..
33 Congo	38	21	-5.25	11.79	..
35 Tanzania, United Rep.	39	56	3.34	4.23	2004
36 Pakistan	25	44	5.27	6.04	2002
37 India	31	57	5.69	4.10	1997
38 Madagascar	25	32	2.27	8.48	2037
39 Papua New Guinea	20	27	2.77	9.80	2034
40 Kampuchea, Dem.	..	3	..	28.46	..
41 Cameroon	..	33	..	8.24	..
42 Kenya	17	30	5.30	8.98	2010
43 Zambia	42	59	3.14	3.84	2004
44 Morocco	..	60	..	3.72	..
Medium human development					
45 Egypt	..	73	..	2.27	..
46 Lao PDR	..	21	..	11.79	..
47 Gabon	..	92	..	0.60	..
48 Oman	..	53	..	4.64	..
49 Bolivia	34	44	2.37	6.04	2022
50 Myanmar	17	27	4.30	9.80	2018
51 Honduras	41	50	1.82	5.08	2025
53 Lesotho	17	36	7.06	7.57	2001
54 Indonesia	11	38	11.93	7.16	1995
55 Guatemala	39	38	-0.24	7.16	..
56 Viet Nam	..	46	..	5.70	..
57 Algeria	77	68	-1.12	2.79	..
58 Botswana	..	54	..	4.50	..
59 El Salvador	53	52	-0.17	4.78	..
60 Tunisia	..	68	..	2.79	..
61 Iran, Islamic Rep.	51	76	3.69	1.98	1994
62 Syrian Arab Rep.	..	76	..	1.98	..
63 Dominican Rep.	55	63	1.24	3.36	2024
64 Saudi Arabia	64	97	3.85	0.22	1987
65 Philippines	40	52	2.41	4.78	2014

		100% access to safe water				
		Percentage of people with access		Average annual growth rate (%) 1975-86	Annual growth rate needed to reach target by 2000	Year reaching target using past growth rate
		1975	1985-87			
67	Libyan Arab Jamahiriya	87	97	0.99	0.22	1990
69	Lebanon	..	93	..	0.52	..
71	Nicaragua	46	49	0.58	5.23	2111
72	Turkey	68	78	1.26	1.79	2006
73	Jordan	..	96	..	0.29	..
74	Peru	47	55	1.44	4.36	2028
75	Ecuador	36	58	4.43	3.97	1999
76	Iraq	66	87	2.54	1.00	1992
78	Thailand	25	64	8.92	3.24	1992
79	Paraguay	13	29	7.57	9.24	2003
80	Brazil	..	78	..	1.79	..
81	Mauritius	..	100	..	..	..
83	Sri Lanka	19	40	7.00	6.76	2000
High human development						
85	Malaysia	34	84	8.57	1.25	1989
86	Colombia	64	92	3.35	0.60	1989
87	Jamaica	86	96	1.01	0.29	1991
88	Kuwait	89	..	..	..	..
89	Venezuela	..	90	..	0.76	..
91	Mexico	62	77	1.99	1.88	2000
93	Panama	77	83	0.68	1.34	2014
94	Trinidad and Tobago	93	98	0.48	0.14	1991
96	Singapore	..	100	..	..	..
97	Korea, Rep.	66	77	1.41	1.88	2005
99	Argentina	66	56	-1.48	4.23	..
102	Uruguay	98	85	-1.29	1.17	..
103	Costa Rica	72	91	2.15	0.68	1991
107	Chile	70	94	2.72	0.44	1989

Technical notes

1. Statistical measures of development

The early leaders of quantification in economics kept their main focus on people, a focus that in recent years has been blurred. Although development has been a constant concern of government policymakers, economists and other social scientists — and has touched the lives of more people than ever before — there has been little agreement on what constitutes development, how it is best measured and how it is best achieved. One reason for this lack of agreement is that dissatisfaction with the pace and character of economic and social change has instilled a desire to redefine the aims and measures of development.

While the pioneers of measurement of national output and income stressed the importance of social concerns, economic growth became the main focus after the Second World War. Growth in the capital stock was seen as the means of achieving development, and the growth rate of per capita GDP became the sole measure of development.

Income was first developed as a way of measuring welfare and well-being by Pigou, who described economic welfare as the measurable part of human welfare — the part that could be brought into a relationship with "the measuring rod of money". As a measure of well-being, income pertains to individuals or to households. It was seen as a forward-looking measure of benefits yet to come rather than as a record of what had already transpired.

But production and distribution processes constrain the income of an individual or household. Thus, income is also a record of economic activity, of the production of goods and services already achieved. This backward-looking, recording aspect came to the fore during the Second World War. Income at the national level — GDP or GNP, as it came to be called — became a measure of activity of the total mass of quantity of goods and services produced, weighted by their respective prices, rather than a measure of individual well-being.

As GNP became the goal of development in the 1950s and 1960s, the question of promoting individual well-being receded. It was assumed that well-being would follow automatically from economic growth. A tenuous link between income and well-being was made through the notion of income per capita, which compounded the shift of emphasis from welfare to production by its insensitivity to distribution. In time, distribution was altogether forgotten, and the argument of "trickle down" was made to defend such neglect. Thus, income moved from an admittedly partial monetary measure of well-being to centre stage as a measure of production and as the sole measure of welfare in its per capita form.

By the 1960s, it was clear from many developing countries that income growth had not tackled the problem of mass poverty. Income distribution and equity came to the forefront as an additional objective of development. The focus of development was turned towards the alleviation of poverty, a change that led to a re-examination of the concept of income and its adequacy as a measure of development.

Against this central dominance of income, several voices were raised. In a pioneering effort at UNRISD, McGranahan and associates examined several development indicators — some relating to mortality and morbidity, others to such social factors as urbanisation and still others to economic factors. These indicators were correlated with each other and used jointly to describe socioeconomic development. Each indicator was related to per capita GDP in a series of regressions that allowed the identification of a threshold level of development. Below this threshold a country was underdeveloped and above it, developed.

There remained the problem of combining these various indicators into a single measure of development, in analogy with income. Income is a price-weighted sum of quantities of different goods and services exchanged in the marketplace. Prices are by no means ideal weights. They may overvalue or undervalue goods and services for which the market is imperfect, and still worse, they totally ignore those for which the market does not exist. But prices are in some sense "natural" weights, since they are part of people's everyday experience. A price conveys the relative importance of one good compared with another in terms of income.

Any synthetic index combines diverse indicators. Weighting can be equal or determined by such data-driven statistical techniques as factor analysis. Weights have a statistical interpretation, but they cannot be explained either by daily experience or by the relative importance of the indicators. By contrast, income provides an indication about the tradeoff a consumer or producer is willing to make among different goods.

Another concern of measuring development is deciding which indicators to include and which to leave out. The income measure includes all goods and services that are produced and marketed, among them harmful goods that pollute the atmosphere or injure health. In this sense, income is comprehensive, a quality that alternative indexes lack. The more comprehensive they seem to be, the more indicators they include, and the less they are transparent and relevant to daily experience.

In response to such considerations, M.D. Morris put forward the Physical Quality of Life Index (PQLI). He saw the UNRISD effort as measuring development as an *activity*. He wished to focus on development as *achieved well-being* and chose three indicators — infant mortality, life expectancy at age one and literacy, combining them in a simple unweighted index to give the PQLI. There obviously is considerable overlap between the first two indicators, particularly for developing countries, as they both relate to longevity and are connected by a precise relationship.

The perception of development has since shifted — first, from *economic* development to *socioeconomic* development, with a new emphasis on poverty. Now the shift is to human development. It emphasises the development of human choices and returns to the centrality of people. It is reflected in measuring development not as the expansion of commodities and wealth but as the widening of human choices. The outcome is the human development index (HDI) used in this Report.

2. Statistical measures of poverty

The measurement of poverty has a history of more than a hundred years. Pioneering work by Booth (1889-92) and Rowntree (1901) tried to measure the extent of urban poverty in London and York. Less known, but perhaps more ambitious, was the attempt to measure national poverty by the Indian politician and economist Dadabhai Naoroji at the beginning of this century.

The primary task of these studies was to define a poverty standard, or poverty line, to separate the "poor" from the "nonpoor". Subsequent debates have continued this focus, and the task of providing the investigator with a standard to distinguish poverty has remained central in poverty studies.

Poverty measures vary according to the variables deemed important: commodities and characteristics of commodities, needs and requirements, or income and expenditure. Typically a poverty measure starts from a notion of (basic) needs, such as nourishment, and translates those needs into commodity bundles (foodstuffs) directly or indirectly through characteristics of commodities (calories and protein). It then multiplies the quantities by appropriate prices to arrive at an expenditure-income level.

A central issue in all debates on poverty is whether poverty should be defined in absolute or in relative terms. It is normally assumed that the two definitions are exclusive because of a lack of clarity about the units of poverty. An absolute measure will typically reflect basic biological and physiological needs. A relative measure will focus more on a notion of requirements that vary depending on circumstances — such as a country's level of development or the disparities between rich and poor or other social and ethnic groups.

Absolute poverty

Poverty is defined in absolute terms if the content of a poverty standard (whether defined by commodities or characteristics) is taken to be fixed across time and space. A historical notion of subsistence — reflecting a very minimal list of basic needs — is at the base of this notion, where the defining variables are commodities or their characteristics. The argument is often made that there is no poverty because, compared with the late nineteenth century, the poor are much better off today. It is also argued that there are no poor in developed countries since, compared with the "really poor" in developing countries, the poor in developed countries are almost affluent.

A common approach in delineating the poverty line is to specify a minimum calorie intake. This calorie level is then converted into foodstuffs adequate to meet the level, given typical consumption patterns in a society. The cost of this amount of food is then determined to yield a poverty level. It has often been the practice, though much criticised, to take a constant calorie intake for everyone. In the poverty level prescribed by the Indian Planning Commission in the early 1960s, 2,250 calories per day per person in rural areas was specified as the minimum level. A similar figure of 2,100 calories has been mentioned for Pakistan in a poverty study for 1963-64, and 2,122 calories for Bangladesh.

While it is impossible to specify a separate level for each individual, it is possible to specify a required calorie level as a function of age, gender, type of activity and health status. This approach was adopted in Altimir's pioneering study of poverty in Latin America. This specificity would mean that even for an individual the required calorie level would alter over time.

The conversion of the calorie intake into a commodity basket must be culture-specific, no matter how absolute the standard. The specification of typical foodstuffs requires a survey of prevailing consumption practices. In ECLAC's studies of poverty in Latin America, the commodity basket required to meet the calorie intake (calibrated by age, gender, activity and health status) was obtained from a sample survey of nonpoor households. Frequently bought foodstuffs were isolated, and a minimum cost list was chosen from them. This procedure frees the method from exclusive reliance on poor families' consumption patterns, which might reflect the restricted choice of poor households.

In pricing the consumption basket to arrive at a level of expenditure for the poverty level, nonfood items necessary for subsistence need to be considered. This problem is frequently tackled by multiplying the money required to buy the food basket by a coefficient known as the Engel coefficient, the reciprocal of the ratio of food expenditure to total expenditure. The choice of the ratio is not straightforward. By Engel's law, the food ratio will be higher for the poor and lower for the rich. By implication, the multiplier is higher if the ratio chosen is that of nonpoor households.

Even in absolute concepts of poverty, there are relative levels. A distinction is made even in absolute poverty calculations between indigence and poverty. If the income is less than the required food expenditure, the household is termed indigent (primary poverty). This is the practice in the ECLAC poverty studies. If the income is below the multiple of food expenditure as given by the Engel coefficient, the family is termed poor (secondary poverty).

Once the poverty line has been established, it must be adjusted for changes in time. A crude method is to index the poverty line according to some overall consumer price index. A better method is to treat the price index of food separately from other items. This would account for the different inflation rates of food and other items. A further refinement would be to allow substitution of items that enter the basic basket and recompute the food expenditure. The poverty line would also be less arbitrary if new trends in consumer expenditure could be captured by a recalculation of the Engel coefficient.

The absolutist approach, though popular, is not free of conceptual problems. But its narrow economic and physiological basis, its seeming objectivity and its ease of computation make it the most frequently used approach. Poverty line calculations in Latin America and South Asia are based on this method. So are those in the United States. The U.S. poverty standard is based on nutritional guidelines laid down in 1955 and not revised since. A range of critics accept the absolute approach but criticise details of method, such as the calculation of the poverty line and the evaluation of the actual resources of a household.

The derivation of the food basket has been a matter for debate. If we look at what the poor actually consume rather than what they *could* consume if they had the resources, we would arrive at a distorted consumption pattern. The food basket can and should be derived from the consumption pattern of nonpoor households. A minimum cost basket can then be derived from this larger basket. As the poverty line is recalculated over time, there should be allowance for substitution between foodstuffs as relative prices change, requiring an econometric specification of the expenditure pattern to allow for accurate estimation of income and substitution effects.

Much of the criticism of the poverty line relates to the assumption of a common, constant calorie intake unrelated to an individual's personal characteristics. This is not, however, a necessary part of an absolute approach, as the Altimir approach has demonstrated for Latin America. Another problem in calculating the poverty line is aggregating members of a household. There is growing evidence of intrahousehold inequalities in consumption. The consumption and nutritional level of children are often better indicators of poverty than any other variable, and they merit further enquiry.

Having defined a poverty line, the problem is to measure the resources of a household before labelling it poor or not. This is the tricky problem of defining and measuring income, which raises several questions. Should it be actual or permanent income? How should nonmarket transactions be imputed? How should assets be taken into account? And so on. On the criterion of actual income, one can frequently have households going in and out of poverty as defined by the poverty line. This requires distinguishing the "always poor" from the "frequently poor".

A different approach of an absolute measure is the Dissatisfaction of Basic Needs. Here a number of indicators of basic needs are identified. In the Latin American studies, for instance, there are questions about the quality of housing, access to primary schools, the dependency ratio, and the level of education of the head of household. If the answer to any one of these questions indicates inadequate levels, the household is declared poor. The dissatisfactions in various dimensions are not weighted and aggregated on a single scale, and different basic needs can be emphasised. The method is less sensitive to price fluctuations, but it does not allow for substitution between different needs.

Relative poverty

The relativist approach defines requirements not merely for existence but for leading a full life as members of a social community. The living standard can be defined by conducting a survey of actual consumption practices or by surveying a sample of households for what they consider adequate consumption practices. A third method is the Leyden method, which asks respondents what their income is and whether they consider it to be too low, adequate or more than adequate. The relativist approach is thus sociocultural rather than narrowly physiological. Since it goes beyond commodities as well as characteristics and consumption practices, it demands more data.

A shortcut through the problem of measurement of relative poverty is to consider poverty a type of inequality. One method of defining poverty is by taking the poverty threshold as some function of median income. By definition, such a measure does not take into account needs of different households or the broader issues posed by the social approach of relativism. But it is an objective economic measure relying on income rather than commodities or characteristics. It is a positive rather than normative measure, the only judgement being in determining the fraction of median income that is to be the cutoff point.

A similar but more limited approach is to define poverty as a function of average earning, which implies that wage or salary employment is the predominant way of earning income. This approach is clearly inapplicable to developing countries, where agriculture and the informal sector provide substantial employment.

The usefulness of measuring poverty

The study of poverty goes well beyond measuring poverty, which is only one step in the process. Measurements should be useful for several purposes. If the purpose is to record levels of well-being, the measurement of income and the analysis of what income can buy will be relevant and informative. As this Report has shown, however, being poor means different things in different countries. If the government provides a social safety net, it is easier for the poor to get by — at least for some time. But if policies for the poor are lacking, it may be harder for the poor to get by. Measurements of poverty thus have to be interpreted in their context.

If the purpose of the study of poverty is to get at the root causes of the problem, the foregoing poverty measures may have to be expanded. One would have to ask first: What makes people poor? They may lack access to assets and to employment or learning opportunities, live in households with a high dependency ratio, belong to ill-served minorities and so on. Measurements of poverty would focus on the key variables of people's deprivation. In the terminology of this Report, action-oriented poverty measures would focus on the access, or lack of access, that people have to various options for human development. This would then allow saying *how* poor people are and *why* they are poor — and *where* corrective policy interventions should break the process of poverty.

3. A mathematical formulation of the human development index

The human development index (HDI) is constructed in three steps. The *first* step is to define a measure of deprivation that a country suffers in each of the three basic variables — life expectancy (X_1), literacy (X_2), and (the log of) real GDP per capita (X_3). A maximum and a minimum value is determined for each of the three variables given the actual values. The deprivation measure then places a country in the range of zero to one as defined by the difference between the maximum and the minimum. Thus I_{ij} is the deprivation indicator for the jth country with respect to the ith variable and it is defined as:

$$I_{ij} = \frac{(\max_j X_{ij} - X_{ij})}{(\max_j X_{ij} - \min_j X_{ij})} \quad (1)$$

The *second* step is to define an average deprivation indicator (I_j). This is done by taking a simple average of the three indicators:

$$I_j = \sum_{i=1}^{3} I_{ij} \quad (2)$$

The *third* step is to measure the human development index (HDI) as one minus the average deprivation index:

$$(HDI)_j = (1 - I_j) \quad (3)$$

To illustrate, the application of this formula to Kenya is as follows:

Maximum life expectancy	=	78.4
Minimum life expectancy	=	41.8
Maximum adult literacy rate	=	100.0
Minimum adult literacy rate	=	12.3
Maximum real GDP per capita (log)	=	3.68
Minimum real GDP per capita (log)	=	2.34
Kenya life expectancy	=	59.4
Kenya adult literacy rate	=	60.0
Kenya real GDP per capita (log)	=	2.90

Kenya's life expectancy deprivation (1)
= (78.4 - 59.4) / (78.4 - 41.8) = 0.519

Kenya's literacy deprivation
= (100.0 - 60.0) / (100.0 - 12.3) = 0.456

Kenya's GDP deprivation
= (3.68 - 2.90) / (3.68 - 2.34) = 0.582

Kenya's average deprivation (2)
= (0.519 + 0.456 + 0.582) / 3 = 0.519

Kenya's Human Development Index (HDI) (3)
= 1 - 0.519 = 0.481

4. A female and male human development index

It would be desirable to present separate HDIs for females and males because of the considerable gender inequality that persists. The narrowing of gender disparities should therefore be carefully monitored, and that requires relevant information.

Data limitations pose several problems, however.

- Income, expressed as the log of the real (purchasing-power-parity adjusted) gross domestic product (GDP) per capita, does not differentiate between males and females. In reality, however, we know that the per capita income of females is far less than that of males in all countries.
- For adult literacy, the great majority of countries with gender-specific estimates show female literacy rates significantly below those for males, a disparity that steadily narrows in moving up the HDI scale. There are, however, no reliable comparable gender-specific estimates for many countries.
- There is thus only one indicator for which gender-specific estimates are fully available — life expectancy.

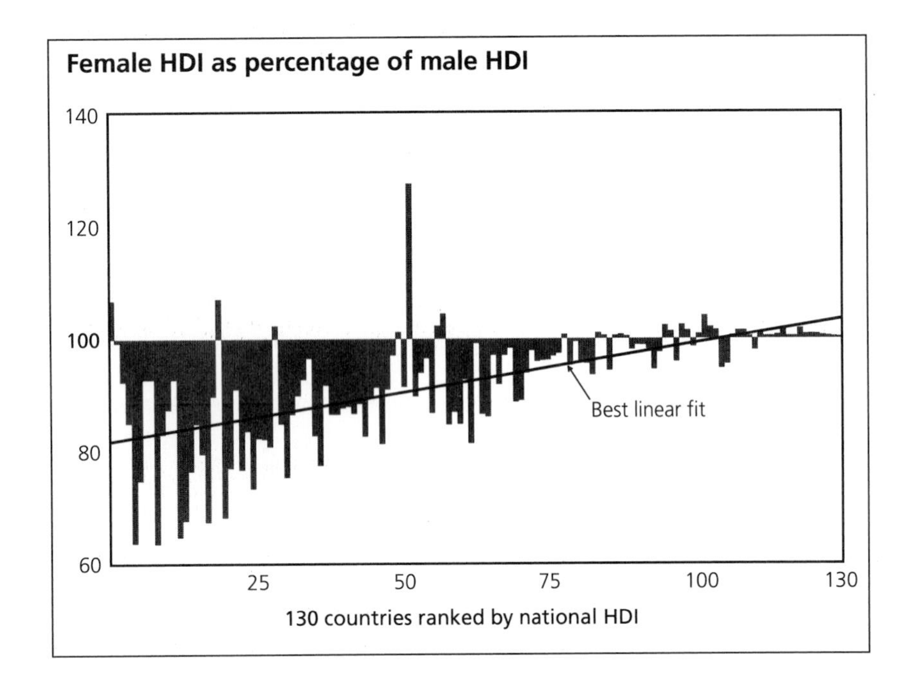

Despite these constraints, it is interesting to compare the two gender-specific HDIs constructed on the basis of existing and estimated data (see the figure and table). The intercountry differences lead to two conclusions.

First, as countries move up the HDI scale, there is a clear overall tendency for the female index to approach and finally overtake the male index. This is primarily the effect of the lower female adult literacy levels dampened down by the effect of higher female life expectancy levels.

Second, among countries with very similar HDIs, there is enormous variation in the female-male disparity, particularly among countries belonging to the low and medium HDI groups.

For example, Tanzania, Pakistan and India are next to each other in their low HDI rank (35 to 37), yet their female-male disparities are very different. The female HDI as a percentage of the male HDI ranges from 96 in Tanzania to 83 in Pakistan to 77 in India. Similarly, the Philippines, China and Saudi Arabia are next to each other in their medium HDI rank (64 to 66), yet their female-male HDI values range from 99% in the Philippines to 87% in China and to 82% in Saudi Arabia. Perhaps most interesting of all is the fact that the disparity range of 99 to 82 in the three adjacent *medium* HDI countries is not all that different from the disparity range of 96 to 77 in the three adjacent *low* HDI countries.

These comparisons show that national averages may conceal distressingly large gender disparities. More professional work needs to be done to bring out clearly the state of the human condition separately for men and women.

HDI: National, female and male

	Human development index	Male HDI	Female HDI
Low human development			
1 Niger	0.116	0.114	0.122
2 Mali	0.143	0.146	0.145
3 Burkina Faso	0.150	0.159	0.146
4 Sierra Leone	0.150	0.166	0.141
5 Chad	0.157	0.195	0.124
6 Guinea	0.162	0.189	0.142
7 Somalia	0.200	0.201	0.201
8 Mauritania	0.208	0.209	0.211
9 Afghanistan	0.212	0.265	0.171
10 Benin	0.224	0.247	0.205
11 Burundi	0.235	0.252	0.221
12 Bhutan	0.236	0.290	0.188
13 Mozambique	0.239	0.290	0.197
14 Malawi	0.250	0.286	0.219
15 Sudan	0.255	0.279	0.237
16 Central African Rep.	0.258	0.290	0.230
17 Nepal	0.273	0.327	0.220
18 Senegal	0.274	0.291	0.261
19 Ethiopia	0.282	0.275	0.296
20 Zaire	0.294	0.354	0.241
21 Rwanda	0.304	0.347	0.267
22 Angola	0.304	0.321	0.292
23 Bangladesh	0.318	0.361	0.277
24 Nigeria	0.322	0.354	0.295
25 Yemen Arab Rep.	0.328	0.380	0.280
26 Liberia	0.333	0.369	0.304
27 Togo	0.337	0.372	0.306
28 Uganda	0.354	0.395	0.320
29 Haiti	0.356	0.353	0.361
30 Ghana	0.360	0.391	0.333
31 Yemen, PDR	0.369	0.424	0.319
32 Côte d'Ivoire	0.393	0.425	0.368
33 Congo	0.395	0.418	0.376
34 Namibia	0.404	0.413	0.415
35 Tanzania, United Rep.	0.413	0.482	0.465
36 Pakistan	0.423	0.463	0.383
37 India	0.439	0.500	0.387
38 Madagascar	0.440	0.459	0.423
39 Papua New Guinea	0.471	0.509	0.441
40 Kampuchea, Dem.	0.471	0.502	0.435
41 Cameroon	0.474	0.491	0.430
42 Kenya	0.481	0.510	0.449
43 Zambia	0.481	0.517	0.450
44 Morocco	0.489	0.518	0.457
Medium human development			
45 Egypt	0.501	0.549	0.453
46 Lao PDR	0.506	0.535	0.479
47 Gabon	0.525	0.550	0.502
48 Oman	0.535	0.589	0.481
49 Bolivia	0.548	0.575	0.523
50 Myanmar	0.561	0.568	0.552
51 Honduras	0.563	0.560	0.566
52 Zimbabwe	0.576	0.598	0.553
53 Lesotho	0.580	0.505	0.648
54 Indonesia	0.591	0.625	0.559
55 Guatemala	0.592	0.609	0.573
56 Viet Nam	0.608	0.633	0.611
57 Algeria	0.609	0.652	0.567
58 Botswana	0.646	0.636	0.653
59 El Salvador	0.651	0.630	0.656
60 Tunisia	0.657	0.711	0.603
61 Iran, Islamic Rep.	0.660	0.702	0.610
62 Syrian Arab Rep.	0.691	0.748	0.635
63 Dominican Rep.	0.699	0.696	0.698
64 Saudi Arabia	0.702	0.757	0.621
65 Philippines	0.714	0.715	0.711
66 China	0.716	0.771	0.669
67 Libyan Arab Jam.	0.719	0.774	0.665
68 South Africa	0.731	0.741	0.721
69 Lebanon	0.735	0.766	0.704
70 Mongolia	0.737	0.757	0.738
71 Nicaragua	0.743	0.744	0.733
72 Turkey	0.751	0.798	0.709
73 Jordan	0.752	0.799	0.711
74 Peru	0.753	0.773	0.726
75 Ecuador	0.758	0.766	0.751
76 Iraq	0.759	0.772	0.743
77 United Arab Emirates	0.782	0.796	0.767
78 Thailand	0.783	0.795	0.771
79 Paraguay	0.784	0.799	0.777
80 Brazil	0.784	0.782	0.788
81 Mauritius	0.788	0.806	0.770
82 Korea, Dem. Rep.	0.789	0.801	0.798
83 Sri Lanka	0.789	0.807	0.775
84 Albania	0.790	0.809	0.776
High human development			
85 Malaysia	0.800	0.826	0.774
86 Colombia	0.801	0.775	0.783
87 Jamaica	0.824	0.824	0.826
88 Kuwait	0.839	0.861	0.817
89 Venezuela	0.861	0.859	0.864
90 Romania	0.863	0.862	0.867
91 Mexico	0.876	0.875	0.879
92 Cuba	0.877	0.886	0.872
93 Panama	0.883	0.887	0.878
94 Trinidad and Tobago	0.885	0.888	0.882
95 Portugal	0.899	0.907	0.893
96 Singapore	0.899	0.925	0.880
97 Korea, Rep.	0.903	0.900	0.884
98 Poland	0.910	0.900	0.925
99 Argentina	0.910	0.905	0.918
100 Yugoslavia	0.913	0.931	0.899
101 Hungary	0.915	0.905	0.927
102 Uruguay	0.916	0.906	0.919
103 Costa Rica	0.916	0.921	0.913
104 Bulgaria	0.918	0.915	0.923
105 USSR	0.920	0.901	0.938
106 Czechoslovakia	0.931	0.922	0.942
107 Chile	0.931	0.921	0.935
108 Hong Kong	0.936	0.963	0.917
109 Greece	0.949	0.972	0.931
110 German Dem. Rep.	0.953	0.951	0.956
111 Israel	0.957	0.973	0.943
112 USA	0.961	0.953	0.969
113 Austria	0.961	0.953	0.969
114 Ireland	0.961	0.961	0.962
115 Spain	0.965	0.973	0.960
116 Belgium	0.966	0.961	0.972
117 Italy	0.966	0.965	0.969
118 New Zealand	0.966	0.964	0.970
119 Germany, Fed. Rep.	0.967	0.963	0.972
120 Finland	0.967	0.957	0.978
121 United Kingdom	0.970	0.969	0.972
122 Denmark	0.971	0.971	0.974
123 France	0.974	0.963	0.986
124 Australia	0.978	0.974	0.984
125 Norway	0.983	0.979	0.989
126 Canada	0.983	0.978	0.990
127 Netherlands	0.984	0.980	0.990
128 Switzerland	0.986	0.983	0.991
129 Sweden	0.987	0.986	0.989
130 Japan	0.996	0.996	0.996

5. Deficiencies in social statistics

This first *Human Development Report* relies on readily available data. In many respects this has limited the scope and depth of its analysis. Not only are many relevant concerns not reflected adequately through existing statistics, but the data that are available often have an inadequate coverage of countries and are seriously lacking in timeliness.

Inadequate data

Many indicators and subject areas were omitted simply because there were too few countries with comparable, reliable data. The more important omissions make a formidable list: wages, unemployment and underemployment, public expenditures in the various sectors by provincial and local authorities, development assistance to individual countries by sector, capital flight, prices of the main staple foods or any satisfactory indicator of food access, access and use of social services by various income groups, the conditions of those living in urban slums, which is a rapidly growing problem, the internal allocations of health expenditures, the whole area of morbidity and health status, net secondary enrolment ratios, educational attainment (the stock of human development), educational achievement (the qualitative output of the education system), the brain drain, key rural-urban differentials such as income and age-specific mortality, health facilities, enrolment, dropout and literacy and key female-male differentials, such as income, age-specific mortality and health.

Incomplete country coverage

Turning to the indicators that are included in the tables, as many as a third of the some 120 indicators were not readily available in some comparable form in a third of the countries. This shows the crippling lack of key indicators of human development. Among the indicators absent in so many countries are some of the most important ones: access to health services or to safe water or to sanitation, total, rural and urban; underweight, wasted, and stunted children; breast-feeding; adult literacy, total, male and female; net primary enrolment ratios, total, urban and rural (a particularly shocking omission); scientists and technicians; educated unemployed; earnings per employee; GNP per capita and income share of the poorest 40%; ratio of the highest 20% to the lowest 20%; the Gini coefficient; urban and rural population below the poverty line; persons per habitable room; and deforestation. Just reading this list tells the story.

Lack of reliability and timeliness

In addition to availability and coverage are questions of reliability. Some indicators with limited coverage, such as the nutritional status of children, are very reliable. So are a number of economic indicators with comprehensive coverage, such as the balance of payments, debt, and trade. Other indicators such as literacy, access to health services, and maternal mortality are only very broad approximations: sometimes there is a lack of representative national data (access to health services), and sometimes there is difficulty in controlling the quality of the definition in practice (literacy).

As to timeliness, some indicators are fairly up-to-date because of their institutional origin (enrolment, from school records) or because they are processed very quickly from small-scale household surveys (assessments of health interventions or nutritional status). Other indicators — such as literacy, income distribution and poverty — are far less timely because they come from infrequent and complex household surveys (income and poverty), or have traditionally been obtained only from the decennial population censuses (literacy). It is sometimes necessary to go back to the beginning of the 1980s to cover a reasonable number of countries. There is a great need to ask fewer questions of fewer people more frequently.

Next steps

Which of all these statistical gaps and weaknesses should governments and international agencies address as priorities? In every country, no matter how poor, extensive statistical activities are taking place. There are the regular statistical operations allied to the administrative process, the large-scale (regular but infrequent) operations such as the population, housing and agricultural censuses, and a considerable number of surveys and case studies undertaken independently by various governmental and academic agencies. Programmes for improving statistics of human development should try to build on and rationalise existing activities, particularly the various surveys and case studies — being careful to avoid unnecessary duplication. They should also seek to link the improvement of the data situation to decisionmaking on development, especially the monitoring of overall trends in priority areas.

In light of the above discussion, it is difficult to establish clear-cut national priorities for improved data collection. One focus may be suggested, however. It is important that HDI calculations be improved and made more comparable among various countries. For this purpose, the following steps deserve priority:

- Better data collection and analysis should be organised for the three essential components of HDI: life expectancy, adult literacy and real income (purchasing-power-adjusted).
- Distribution of these three components by income groups should be investigated so that the HDI can be made distributionally sensitive.
- Distribution of these three components between males and females, between rural and urban

and among regional areas should also be investigated so that separate HDIs can be constructed that are sensitive to gender differences and geographical differences.

In addition, it is necessary to collect comprehensive information on social sector budgets, which are one of the most important instruments for improved human development. Data should cover all social sector expenditures, whether by government (central, provincial or local), by private sector or by NGOs. Detailed data should be collected for expenditure in environment and other social fields, with a break-up for each important subsector (such as primary, secondary and tertiary education; general and technical education; preventive and curative health care). Data should also be collected on all social subsidies, their coverage and their impact on various income groups. Many of these data are not easily available at present, except for a few selected countries. Meaningful analysis of meso interventions or concrete proposals for budget restructuring cannot be prepared unless such data are available for all countries on a comparable and continuous basis.

Bibliographic note

The sources for the text tables and figures, unless otherwise noted here, are the same as the sources for the human development indicators in the appendix. These sources are listed at the end of the appendix.

Chapter 1 draws from the following: Buhmann and others 1988, pp. 130-31; Griffin and Knight 1989; Haq 1988; Sen 1981a and 1985; United Nations 1986d and 1989d; UNDP 1988a; USAID 1989; and World Bank 1989. The references for box 1.4 are: Buhmann and others 1988; United Nations 1986d; USAID 1989; and World Bank 1989.

Chapter 2 draws from the following: African Development Bank, UNDP and World Bank 1990; Alexandratos 1988; Berg 1981 and 1987; Cohen 1989; ECLAC and Centro Latino Americano de Demografía 1988; FAO 1986 and various years; Fields 1989; Fordham Institute for Innovation in Social Policy 1989; IFAD 1989; ILO 1988 and various years; Patel 1989; Pinstrup-Andersen 1988; Sen 1981b; Serageldin 1989; UNCTAD 1988; UNDP 1988b and 1988c; UNICEF 1989b and 1989c; United Nations 1987 and 1988a; U.S. House of Representatives 1989; WHO 1989a and 1989b; World Bank 1986c; and Zuckerman 1988.

Additional sources for the sections include: Albanez and others 1989; Athreya 1984; Barcellos and others 1986; Berry 1980; Bramwell 1988; Brown and others 1989; Carlson and Wardlaw 1990; Caton 1990; Cernea 1985 and 1988; Chambers 1989; Chen, Huq and D'Souza 1981; Commonwealth Secretariat 1989; Cornia, Jolly and Stewart 1987; Cotic 1988; Das Gupta 1987; Davies and Saunders 1988; Drabek 1987; FAO 1988; Fei, Ranis and Kuo 1979; Findlay and Zvekic 1988; Ghai 1989; Ghai and de Alcantara 1989; ILO 1987 and various years; Jacobson 1988; Jamal and Weeks 1988; Jamison and Lau 1983; Leonard and contributors 1989; Mouly 1989; Nadelmann 1989; Newman 1989; Potter 1978; Preble 1989; Psacharopoulos 1980 and 1989; Rodgers 1989; Roussel 1986, p.933; Suarez-Berenguela 1987; Tibaijuka 1988; UNDRO and UNEP 1986; UNICEF 1988, 1989a and 1989b; United Nations 1985; 1986a; 1986c; 1989a; 1989e and 1989g; United Nations Economic Commission for Africa 1989; United States Census Bureau and Centre on Budget and Policy Priorities 1989; Uphoff 1986; van Ginneken 1976 and 1987; WHO 1988b; 1989c, 1989d and 1989e; World Bank 1983, 1986a and 1986b; World Commission on Environment and Development 1987; World Food Council 1987; and Zvekic and Mattie 1987.

References for the boxes are as follows: Box 2.1, Leonard and contributors 1989. Box 2.4, Davies and Saunders 1988. Box 2.5, Reid 1988.

References for the figures are as follows: Fig. 2.6, ILO 1987, p. 17. Fig. 2.17, Bramwell 1988.

References for the tables are as follows: Table 2.5, United Nations 1989e, p. 74. Table 2.6, Roussel 1986, p. 933.

Chapter 3 draws from the following: Fei, Ranis and Kuo 1979; Fields 1989; Halstead, Walsh and Warren 1985; IMF various years; and World Bank 1989.

References for the country case studies are: Adelman and Robinson 1978; Alailima 1984; Bannister 1987; Boyd 1988; Brundenius 1981, pp. 1083-96; Chen 1988; Davies and Saunders 1988; Departamento Nacional de Planeación and others 1989; Drèze and Sen forthcoming; Edirisinghe 1987; FEDESAROLLO 1989; Government of Colombia 1988; ILO 1972; Jamison and others 1984; Meerman 1979; Quinn and others 1988; Raczynski 1988; Rosero-Bixby 1985; Sahota 1980; Stewart 1985; Sul and Williamson 1988; UNICEF, Colombo 1988; World Bank 1987, 1988a and 1988b.

References for the boxes are as follows: Box 3.1, Stewart 1988. Box 3.2, Edirisinghe 1987, p.9; and UNICEF, Colombo 1988. Box 3.3, Jamison 1985; World Bank 1987; and Chen 1988.

Chapter 4 draws from the following: African Development Bank, UNDP and World Bank 1990; Alderman and Gertler 1989; Anderson 1987; Berg 1987; Chambers 1985; Demery and Addison 1987; Gertler and Glewwe 1989; Gertler and van der Gaag 1988; Jolly 1989; Kanbur 1988; Nelson and contributors 1989; Patel 1989; Pinstrup-Andersen 1988; Roth 1987; Stelcner, Arriagada and Moock 1987; United Nations 1989a, 1989b and 1989d; United Nations Centre on Transnational Corporations 1989; WHO 1987a and 1987b; and World Bank 1986b.

References for the boxes are as follows: Boxes 4.2 and 4.3, Ashe and Cosslett 1989. Box 4.6, Phua 1986, pp. 11-12. Box 4.7, Moon 1986, p. 20. Box 4.8, Sène 1986, pp. 4-6. Box 4.9 United Nations

1989a. Box 4.10, Haq 1984. Box 4.11, UNDP Development Study Programme and UNCTAD 1988; and UNCTAD 1988b. Box 4.12, WHO 1988a, p. 63; and Patel 1989.

Chapter 5 draws from the following: Asian Development Bank 1988; Cheema 1987; Hardoy and Satterthwaite 1986; Harpham, Vaughan and Rifkin 1985; Linn 1983; Rodgers 1989; Sivaramakrishnan and Green 1986; United Nations 1980 and 1986b; United Nations Centre for Human Settlements 1987, 1988 and 1989; and UNDP 1989.

References for the boxes are as follows: Box 5.1, World Commission on Environment and Development 1987, p. 235-36. Box 5.2, Pantumvanit and Liengcharernsit 1989, pp. 31-39; Sivaramakrishnan and Green 1986; Tiecouta 1989, pp. 176-202; and United Nations 1988b. Box 5.3, Dillinger 1989. Box 5.4, Angel and Pornchokchai 1989, p. 141. Box 5.5, Taylor 1987, pp. 47-51; and United Nations 1989c, pp. 28-29. Box 5.6, Republic of Kenya 1987, p. 170. Box 5.7, Khan 1983, pp. 12-18. Box 5.8, Gunnerson 1987.

References for the tables are as follows: Table 5.1, United Nations 1989f.

Technical note 1 draws from the following: Adelman and Taft-Morris 1973; Bardhan and Srinivasan 1988; Baster 1972; Chenery and others 1974; Haq 1976; McGranahan and Pizarro 1985; Morris 1979; and UNRISD 1972.

Technical note 2 draws from the following: Alamgir and Ahmed 1988; Altimir 1979; Bhalla and Vasistha 1988; Booth 1889 and 1891; Brannen and Wilson 1986; Burki 1988; Carlson 1987; Desai 1989; Harrington 1968; Kynch and Sen 1983; Naoroji 1901; Rowntree 1901; Townsend 1979; Watts 1968; and Wilson 1986.

References

Adelman, Irma, and Sherman Robinson. 1978. *Income Distribution Policy in Developing Countries: A Case Study of Korea*. New York: Oxford University Press.

Adelman, Irma, and Cynthia Taft-Morris. 1973. *Economic Growth and Social Equity in Developing Countries*. Stanford: Stanford University Press.

African Development Bank, UNDP, and World Bank. 1990. *The Social Dimensions of Adjustment in Africa: A Policy Agenda*. Washington, D.C.: World Bank.

Alailima, Patricia J. 1984. *Fiscal Incidence in Sri Lanka, 1973 and 1980*. WFP 2-32/WP 56. Geneva: ILO.

Alamgir, Mahiuddin, and Sadiq Ahmed. 1988. "Poverty and Income Distribution in Bangladesh." In Pranab K. Bardhan and T. N. Srinivasan, eds., *Rural Poverty in South Asia*. New York: Columbia University Press.

Albanez, Teresa, and others. 1989. *Economic Decline and Child Survival: The Plight of Latin America in the Eighties*. Innocenti Occasional Papers 1. Florence: UNICEF.

Alderman, Harold, and Paul Gertler. 1989. *The Substitutability of Public and Private Health Care for the Treatment of Children in Pakistan*. World Bank Living Standards Measurement Survey Working Paper 57. Washington, D.C.

Alexandratos, Nikos, ed. 1988. *World Agriculture: Toward 2000*. London: Bellhaven Press and Rome: FAO.

Altimir, Oscar. 1979. *La Dimension de la Pobreza en America Latina*. Santiago: ECLAC.

Anderson, Dennis. 1987. *The Public Revenue and Economic Policy Reform in African Countries: An Overview of Issues and Policies*. World Bank Discussion Paper 19. Washington, D.C.

Angel, Shlomo, and Sopon Pornchokchai. 1989. "Bangkok Slum Lands: Policy Implications of Recent Findings." *Cities* 6(2):136-46.

Ashe, Jeffrey, and Christopher E. Cosslett. 1989. *Credit for the Poor*. New York: UNDP.

Asian Development Bank. 1988. *Urban Policy Issues: Regional Services on Major National Urban Policy Issues*. Manila.

Athreya, Venkatesh B. 1984. "Valadamalaipuram: A Resurvey." Madras Institute of Development Studies Working Paper 50. Madras, India.

Bannister, Judith. 1987. *China's Changing Population*. Stanford: Stanford University Press.

Barcellos, T. M., and others. 1986. *Segregacao Urbana e Mortalidade em Porto Alegre*. Porto Alegre, Brazil: Fundacao de Economia e Estatistica.

Bardhan, Pranab K., and T. N. Srinivasan, ed. 1988. *Rural Poverty in South Asia*. New York: Columbia University Press.

Baster, Nancy, ed. 1972. *Measuring Development: The Role and Adequacy of Development Indicators*. London: Frank Cass.

Berg, Alan. 1981. *Malnourished People: A Policy View*. World Bank Poverty and Basic Needs Series. Washington, D.C.

_____. 1987. *Malnutrition: What Can be Done? Lessons from World Bank Experience*. Baltimore and London: Johns Hopkins University Press.

Berry, Albert. 1980. "Education, Income, Productivity and Urban Poverty." In *Education and Income*, World Bank Staff Working Paper 402. Washington, D.C.

Bhalla, Surjit S., and Prem S. Vasistha. 1988. "Income Distribution in India: A Reexamination." In Pranab K. Bardhan and T. N. Srinivasan, eds., *Rural Poverty in South Asia*. New York: Columbia University Press.

Booth, Charles. 1889 and 1891. *Labour and Life of the Peoples*. 2 vols. London: Macmillan.

Boyd, Derick. 1988. "The Impact of Adjustment Policies on Vulnerable Groups: The Case of Jamaica, 1973-1985." In Giovanni Andrea Cornia, Richard Jolly, and Frances Stewart, eds., *Adjustment with a Human Face*, vol. II. Oxford: Clarendon Press.

Bramwell, Anna, ed. 1988. *Refugees in the Age of Total War*. London: Unwin Hayman.

Brannen, Julia, and Gail Wilson. 1986. *Give and Take in Families*. London: Unwin Hayman.

Brown, Lester, C. Flavin, Sandra Postel, and others. 1989. *State of the World 1989*. New York: W. W. Norton.

Brundenius, Claes. 1981. "Growth and Equity: the Cuban Experience 1959-1980." *World Development* 9(11/12):1083-96.

Buhmann, Brigitte, Lee Rainwater, Guenther Schmaus, and Timothy M. Smeeding. 1988. "Equivalence Scales, Well-Being, Inequality, and Poverty: Sensitivity Estimates across Ten Countries Using the Luxembourg Income

Study (LIS) Database." *Review of Income and Wealth* 34(2):115-42.

Burki, Shahid J. 1988. "Poverty in Pakistan: Myth or Reality?" In Pranab K. Bardhan and T. N. Srinivasan, eds., *Rural Poverty in South Asia.* New York: Columbia University Press.

Carlson, Beverley A. 1987. *Core Indicators for the Interagency Food and Nutrition Surveillance Programme*. New York: UNICEF.

Carlson, Beverley A., and Tessa Wardlaw. 1990. "A Global, Regional and Country Assessment of Child Malnutrition." UNICEF Staff Working Paper 7. New York.

Caton, Carol L. M. 1990. *Homeless in America.* New York: Oxford University Press.

Cernea, Michael M., ed. 1985. *Putting People First: Sociological Variables in Rural Development.* New York: Oxford University Press.

_____. 1988. *Nongovernmental Organizations and Local Development.* World Bank Discussion Paper 40. Washington, D.C.

Chambers, Robert. 1985. "Shortcut Methods of Gathering Social Information for Rural Development Projects." In Michael Cernea, ed., *Putting People First: Sociological Variables in Rural Development*. New York: Oxford University Press.

_____. ed. 1989. "Editorial Introduction: Vulnerability, Coping and Policy." *IDS Bulletin* 20(2):1-7.

Cheema, G. Shabbir. 1987. *Urban Shelter and Services: Public Policies and Management Approaches*. New York: Praeger.

Chen, Lincoln C. 1988. "Health Policy Responses: An Approach Derived from the China and India Experiences." In David E. Bell and Michael Reich, eds., *Health, Nutrition, and Economic Crisis: Approaches to Policy in the Third World.* Dover, MA: Auburn House.

Chen, Lincoln C., Emdadul Huq, and Stan D'Souza. 1981. "Sex Bias in the Family Allocation of Food and Care in Rural Bangladesh." *Population and Development Review* 7(1):55-70.

Chenery, Hollis, Montele S. Ahluwalia, C. L. G. Bell, John H. Duloy, and Richard Jolly. 1974. *Redistribution with Growth*. London: Oxford University Press.

Cohen, C. Desmond. 1989. "Trends in Human Development in the United Kingdom." Brighton: University of Sussex School of Social Sciences.

Colombia, Government of. 1988. *Colombia Estadística,* vol. I. Bogotá: Departamento Administrativo Nacional de Estadística.

Commonwealth Secretariat. 1989. *Engendering Adjustment for the 1990s*. Report of a Commonwealth Expert Group on Women and Structural Adjustment. London: Commonwealth Secretariat.

Cornia, Giovanni Andrea, Richard Jolly, and Frances Stewart, eds. 1987. *Adjustment with a Human Face.* 2 vols. Oxford: Clarendon Press.

Cotic, Dusan. 1988. *Drugs and Punishment*. United Nations Social Defence Research Institute Publication 30. Rome.

Das Gupta, Monica. 1987. "Selective Discrimination against Female Children in the Rural Punjab." *Population and Development Review* 13(1):77-100.

Davies, Rob, and David Saunders. 1988. "Adjustment Policies and the Welfare of Children: Zimbabwe 1980-85." In Giovanni Andrea Cornia, Richard Jolly, and Frances Stewart, eds., *Adjustment with a Human Face,* vol. II. Oxford: Clarendon Press.

Demery, Lionel, and Tony Addison. 1987. *The Alleviation of Poverty under Structural Adjustment*. Washington, D.C.: World Bank.

Departamento Nacional de Planeación, Departamento Administrativo Nacional de Estadística, UNICEF and UNDP. 1989. *La Pobreza en Colombia, Tomo I.* Bogotá: Departamento Nacional de Planeación.

Desai, Meghnad. 1989. *Methodological Problems of Measurement of Poverty in Latin America*. London: Department of Economics, London School of Economics.

Dillinger, William. 1989. *Urban Property Taxation: Lessons from Brazil*. World Bank Policy, Planning and Research Working Paper 362. Washington, D.C.

Diouf, Mamadou. 1989. "Dakar: Urban Management and Municipal Administration." Toronto: Centre for Urban and Community Studies, University of Toronto.

Drabek, Anna G., ed. 1987. "Development Alternatives: The Challenge for NGOs." *World Development* 15(supplement):ix-xv.

Drèze, Jean, and Amartya K. Sen. Forthcoming. *Hunger and Public Action.* Oxford: Clarendon Press.

ECLAC and Centro Latino Americano de Demografía. 1988. *La Mortilidad en la Ninez en Centroamerica: Panama y Belize, 1970-85.* San Jose, Costa Rica: Centro Latino Americano de Demografía.

Edirisinghe, Neville. 1987. *The Food Stamp Scheme in Sri Lanka: Costs, Benefits, and Options for Modifications.* International Food Policy Research Institute Research Report 58. Washington, D.C.

FAO. 1986. *The Dynamics of Rural Poverty*. Rome: FAO.

_____. 1988. *Rural Poverty in Latin America and the Caribbean.* Rome: FAO.

_____. Various years. *The State of Food and Agriculture*. Rome: FAO.

FEDESARROLLO. 1989. *Coyuntura Social* 1(Decembre), Bogotá.

Fei, John C. H., Gustav Ranis, and Shirley W. Y. Kuo. 1979. *Growth and Equity: The Taiwan Case.* New York: Oxford University Press.

Fields, Gary S. 1989. *A Compendium of Data on Inequality and Poverty for the Developing World.*

Ithaca, NY: Cornell University Department of Economics.

Findlay, Mark, and Ugljesa Zvekic. 1988. *Analysing (In)formal Mechanisms of Crime Control*. United Nations Social Defence Research Institute Publication 31, Rome.

Fordham Institute for Innovation in Social Policy. 1989. *The Index of Social Health 1989, Measuring the Social Well-Being of the Nation*. Fordham University Graduate Center, Tarrytown, NY.

Gertler, Paul, and Jacques van der Gaag. 1988. *Measuring the Willingness to Pay for Social Services in Developing Countries*. World Bank Living Standards Measurement Survey Working Paper 45. Washington, D.C.

Gertler, Paul, and Paul Glewwe. 1989. *The Willingness to Pay for Education in Developing Countries: Evidence from Rural Peru*. World Bank Living Standards Measurement Survey Working Paper 54. Washington, D.C.

Ghai, Dharam. 1989. "Participatory Development: Some Perspectives from Grass-roots Experiences." *Journal of Development Planning* 19:215-46.

Ghai, Dharam, and Cynthia Hewitt de Alcantara. 1989. "The Crisis of the 1980s in Africa, Latin America and the Caribbean: Economic Impact, Social Change and Political Implications." Presented at the Workshop on Economic Crisis and Third World Countries: Impact and Policy Response. University of the West Indies and United Nations Research Institute for Social Development, 3-6 April, Kingston, Jamaica.

Griffin, Keith, and John Knight, eds. 1989. "Human Development in the 1980s and Beyond." Special number of the *Journal of Development Planning* 19.

Gunnerson, Charles G. 1987. "Resource Recovery and Utilization in Shanghai." UNDP/World Bank Global Program of Resource Recovery, Washington, D.C.

Halstead, Scott B., Julia A. Walsh, and Kenneth S. Warren, eds. 1985. *Good Health at Low Cost*. New York: Rockefeller Foundation.

Haq, Mahbub ul. 1976. *The Poverty Curtain: Choices for the Third World*. New York: Columbia University Press.

_____. 1984. "Proposal for an International Debt Refinancing Facility." Presented at the United Nations Economic and Social Council annual session, 1 July, Geneva.

_____. 1988. "People in Development." UNDP Paul G. Hoffman Lecture, New York.

Hardoy, Jorge and David Satterthwaite, eds. 1986. *Small and Intermediate Urban Centers: Their Role in Regional and National Development in the Third World*. Boulder, CO: Westview.

Harpham, Trudy, Patrick Vaughan, and Susan Rifkin. 1985. *Health and the Urban Poor in Developing Countries*. EPC Publication 5. London School of Hygiene and Tropical Medicine, London.

Harrington, Michael. 1968. *The Other America*. New York: Macmillan.

IFAD. 1989. "Poverty Alleviation with Sustainable Growth in the 1990s." Presented at the United Nations Committee on Development Planning Working Group III, International Development Strategies for the 1990s, 22-24 February, New York.

ILO. 1972. *Employment, Incomes and Equality: A Strategy for Increasing Productive Employment in Kenya*. Geneva.

_____. 1987. "Background Document for the High-Level Meeting on Employment and Structural Adjustment." Geneva.

_____. 1988. "Meeting the Social Debt." Programa Regional del Empleo para America Latina y el Caribe, Chile.

_____. Various years. *World Labour Report*. Geneva.

IMF. Various years. *Government Finance Statistics Yearbook*. Washington, D.C.

Jacobson, Jodi L. 1988. *Environmental Refugees: A Yardstick of Habitability*. Worldwatch Paper 86. Washington, D.C.

Jamal, Vali, and J. Weeks. 1988. "The Vanishing Rural-Urban Gap in Sub-Saharan Africa." *International Labour Review* 127(3):271-92.

Jamison, Dean T. 1985. "China's Health Care System: Policies, Organization, Inputs and Finance." In Scott B. Halstead, Julia A. Walsh, and Kenneth S. Warren, eds., *Good Health at Low Cost*. New York: Rockefeller Foundation.

Jamison, Dean T., and L. J. Lau. 1983. *Review of Comparative Agricultural Performance in East and West Punjab*. World Bank South Asia Project Department, Washington, D.C.

Jamison, Dean T., and others. 1984. *China: The Health Sector*. Washington, D.C.: World Bank.

Jolly, Richard. 1989. "A Future for United Nations Aid and Technical Assistance." North-South Roundtable, 6-8 September, Uppsala, Sweden.

Kanbur, Ravi. 1988. *The Principles of Targeting*. University of Warwick Development Economics Research Centre, Warwick.

Kenya, Republic of. 1987. *Economic Survey*. Ministry of Planning and National Development, Central Bureau of Statistics, Nairobi.

Khan, Akhtar Hameed. 1983. "Orangi Project: A Task Bigger Than Colombo." *Pakistan and Gulf Economist* 2(24):12-18.

Kynch, Jocelyn, and Amartya K. Sen. 1983. "Indian Women: Wellbeing and Survival." *Cambridge Journal of Economics* 7(3/4):363-80.

Leonard, Hugh J., and contributors. 1989. *Environment and the Poor*. Overseas Development Counsel, U.S.-Third World Policy Perspectives 11. New Brunswick and Oxford: Transaction Books.

Linn, Johannes F. 1983. *Cities in the Developing World: Policies for their Equitable and Efficient Growth*. New York: Oxford University Press.

McGranahan, Donald V., and Eduardo Pizarro. 1985. *Measurement and Analysis of Socio-Economic Development*. Geneva: UNRISD.

Meerman, Jacob. 1979. *Public Expenditure in Malaysia: Who Benefits and Why.* New York: Oxford University Press.

Moon, Ok Ryun. 1986. "Towards Equity in Health Care." *World Health* (May): 20-21.

Morris, Morris D. 1979. *Measuring the Condition of the World's Poor: The Physical Quality of Life Index.* New York: Pergamon.

Mouly, Jean. 1989. "Reviving the World's Economic Growth: Chances and Risks." Presented at the United Nations Committee for Development Planning Working Group III, International Development Strategies for the 1990s, 22-24 February, New York.

Nadelmann, Ethan A. 1989. "Drug Prohibition in the United States: Costs, Consequences, and Alternatives." *Science* 245(4921):939-46.

Naoroji, Dadabhai. 1901. *Poverty and UnBritish Rule in India.* New Delhi: Government of India Publications Division.

Nelson, Joan M., and contributors. 1989. *Fragile Coalitions: The Politics of Economic Adjustment.* U.S.-Third World Policy Perspectives 12. New Brunswick: Transaction Books.

Newman, Graeme. 1989. "Report on Crime and the Human Condition." United Nations Centre for Social Development and Humanitarian Affairs, Crime Prevention and Criminal Justice Branch, Vienna.

Pantumvanit, Dhira, and Wanai Liengcharernsit. 1989. "Coming to Terms with Bangkok's Environmental Problems." *Environment and Urbanisation* 1(1):31-39.

Patel, Matesh S. 1989. "Eliminating Social Distance Between North and South: Cost-Effective Goals for the 1990s." UNICEF Staff Working Paper 5. New York.

Phua, Kai Hong. 1986. "Singapore's Family Savings Scheme." *World Health* (May):11-12.

Pinstrup-Andersen, Per. 1988. *Food Subsidies in Developing Countries: Costs, Benefits and Policy Options*. Baltimore: Johns Hopkins University Press.

Potter, Joseph E. 1978. "Demographic Factors and Income Distribution in Latin America." Presented at the International Union for the Scientific Study of Population, 28 August-1 September, Helsinki.

Preble, Elizabeth A. 1989. "Projected Impact of HIV/AIDS on Children in Central and East Africa." Presented at the UNICEF Conference on the Implication of AIDS for Mothers and Children, 27-30 November, Paris.

Psacharopoulos, George. 1980. "Returns to Education: An Updated International Comparison." In *Education and Income*, World Bank Staff Working Paper 402. Washington, D.C.

____. 1989. *Recovering Growth with Equity: World Bank Poverty Alleviation Activities in Latin America*. World Bank Internal Discussion Paper IDP-0033. Washington, D.C.

Quinn, Victoria, Mark Cohen, John Mason, and G.N. Kgosidintsi. 1988. "Crisis-proofing the Economy: The Response of Botswana to Economic Recession and Drought." In Giovanni Andrea Cornia, Richard Jolly, and Frances Stewart, eds., *Adjustment with a Human Face,* vol. II. Oxford: Clarendon Press.

Raczynski, Dagmar. 1988. "Social Policy, Poverty, and Vulnerable Groups: Children in Chile." In Giovanni Andrea Cornia, Richard Jolly, and Frances Stewart, eds.,*Adjustment with a Human Face,* vol.II. Oxford: Clarendon Press.

Reid, Elizabeth V. 1988. *AIDS and Development*. Dossier 24, Australian Council for Overseas Aid Development. Canberra.

Rodgers, Gerry, ed. 1989. *Urban Poverty and the Labour Market: Access to Jobs and Incomes in Asian and Latin American Cities*. Geneva: ILO.

Rosero-Bixby, Luis. 1985. "Infant Mortality Decline in Costa Rica." In Scott Halstead, Julia A. Walsh, and Kenneth S. Warren, eds., *Good Health at Low Cost.* New York: Rockefeller Foundation.

Roth, Gabriel. 1987. *The Private Provision of Public Services in Developing Countries*. New York: Oxford University Press.

Roussel, Louis. 1986. "Evolution Récente de la Structure des Ménages dans Quelques Pays Industriels." *Population* 41(6):913-34.

Rowntree, Seebohm. 1901. *Poverty: A Study of Town Life.* London: Longmans.

Sahota, Gian Singh. 1980. *The Distribution of the Benefits of Public Expenditure in Nigeria*. Washington, D.C.: World Bank.

Sen, Amartya K. 1981a. "Public Action and the Quality of Life in Developing Countries." *Oxford Bulletin of Economics and Statistics* 43(4):287-319.

____. 1981b. *Poverty and Famines: An Essay on Entitlement and Deprivation*. Oxford: Clarendon Press.

____. 1985. *Commodities and Capabilities.* Amsterdam: North-Holland.

Sène, Pape Marcel. 1986. "Community Financing in Senegal." *World Health* (May):4-6.

Serageldin, Ismail. 1989. *Poverty, Adjustment, and Growth in Africa.* Washington, D.C.: World Bank.

Sivaramakrishnan, K. C., and Leslie Green. 1986. *Metropolitan Management: The Asian Experience.* New York: Oxford University Press.

Stelcner, Morton, Ana-Maria Arriagada, and Peter Moock. 1987. *Wage Determinants and School Attainment among Men in Peru*. World Bank Living Standards Measurement Survey Working Paper 38. Washington, D.C.

Stewart, Frances. 1985. *Basic Needs in Developing Countries.* Baltimore: Johns Hopkins University Press.

____. 1988. "Monitoring and Statistics for Adjustment with a Human Face." In Giovanni Andrea Cornia, Richard Jolly, and Frances Stewart, eds., *Adjustment with a Human Face,* vol. II. Oxford: Clarendon Press.

Suarez-Berenguela, Rubin M. 1987. *Peru Informal Sector, Labor Markets, and Returns to Education*. World Bank Living Standards Measurement Survey Working Paper 32. Washington, D.C.

Sul, Sang-Mok, and David Williamson. 1988. "The Impact of Adjustment and Stabilization Policies on Social Welfare: The South Korean Experiences during 1978-1985." In Giovanni Andrea Cornia, Richard Jolly, and Frances Stewart, eds., *Adjustment with a Human Face,* vol. II. Oxford: Clarendon Press.

Taylor, John L. 1987. "Evaluation of the Jakarta Kampung Improvement Program." In Reinhard J. Skinner, John L. Taylor, and Emiel A. Wegelin, eds., *Shelter Upgrading for the Urban Poor: Evaluation of Third World Experience*. Manila: Island Publishing House.

Tibaijuka, Anna K. 1988. "The Need to Monitor the Welfare Implications of Structural Adjustment Programmes in Tanzania." Presented at the Fifth Economic Policy Workshop, 23-25 May, Dar-es-Salaam.

Tiecouta, Ngom. 1989. "Appropriate Standards for Infrastructure in Dakar." In Richard E. Stren and Rodney R. White, eds., *African Cities in Crisis*. Boulder, CO: Westview.

Townsend, Peter. 1979. *Poverty in the United Kingdom.* London: Penguin.

United Nations. 1980. *Patterns of Urban and Rural Population Growth.* ST/ESA/SER.A/68 and Corr. 1. UN Publication Sales No. 79.XIII.9. New York.

_____. 1985. *New Dimensions of Criminality and Crime Prevention in the Context of Development: Challenges for the Future*. Report of the Secretary General. A/CONF. 121/18. Vienna.

_____. 1986a. *Compendium of Statistics and Indicators on the Situation of Women*. ST/ESA/STAT/SER.K/5. UN Publication Sales No. 88.XVII.6. New York.

_____. 1986b. *Living Conditions in Developing Countries in the Mid-1980s: Supplement to the 1985 Report on the World Social Situation.* ST/ESA/165/Add.1, UN Publication Sales No. E.85.IV.3. New York.

_____. 1986c. *Situation and Trends in Drug Abuse and the Illicit Traffic: Review of Drug Abuse and Measures to Reduce Illicit Traffic*. E/CN.7/1987/9. Report of the Secretary General. Vienna.

_____. 1986d. *World Comparisons of Purchasing Power and Real Product for 1980. Part I.* ST/ESA/STAT/Ser.F/42, Pt. I. UN Publication Sales No. 86.XVII.9. United Nations Commission of the European Communities and United Nations Statistical Office. New York.

_____. 1987. *Development and International Economic Cooperation: Human Settlements, International Year of Shelter for the Homeless.* Report of the Secretary General. A/42/378. New York.

_____. 1988a. *Overall Socio-economic Perspective of the World Economy to the Year 2000.* Report of the Secretary General. A/43/554. New York.

_____. 1988b. *Population Growth and Policies in Mega-cities: Karachi*. Population Policy Paper 13, ST/ESA/SER.R/77. New York.

_____. 1989a. *External Debt Crisis and Development*. Report of the Secretary General. A/44/628. New York.

_____. 1989b. *National Household Survey Capability Programme, Household Income and Expenditure Surveys: A Technical Study*. DP/UN/INT-88-X01/6E. New York.

_____. 1989c. *Population Growth and Policies in Mega-cities: Jakarta*. Population Policy Paper 18, ST/ESA/SER.R/86. New York.

_____. 1989d. *Report of the United Nations Committee for Development Planning*. E/1989/29. New York.

_____. 1989e. *Report on the World Social Situation.* ST/ESA/213 E/CN.5/1989/2. UN Publication Sales No. 89.IV.1. New York.

_____. 1989f. *World Population Prospects 1988.* Population Studies No. 106, United Nations Department of International, Economic and Social Affairs, ST/ESA/SER. A/106. New York.

_____. 1989g. *World Population Trends and Policies: 1989 Monitoring Report.* ESA/P/WP.107 1. New York.

United Nations Centre for Human Settlements. 1987. *Global Report on Human Settlements*. New York: Oxford University Press.

_____. 1988. *Global Shelter Strategy for the Year 2000*. Report of the Executive Director HC/C/11/3. New Delhi.

_____. 1989. *Municipal Resource Management Subcomponent: Colombian Case Study*. Nairobi.

United Nations Centre on Transnational Corporations. 1989. *International Debt Restructuring: Substantive Issues and Techniques.* ST/CTC/SER.B/4, Publication Sales No. 89.IIA.10. New York.

UNCTAD. 1988a. *The Least Developed Countries 1988 Report.* TD/B/1202. UN Publication Sales No. E.89.II.D.3. New York.

_____. 1988b. *Trade and Development Report 1988.* UN Publications Sales No. E.88.II.D.8. New York.

UNDP. 1988a. *The Amman Statement on Human Development: Goals and Strategies for the Year 2000*. UNDP Study Programme and North-South Roundtable of the Society for International Development. New York.

_____. 1988b. *Conferencia Regional Sobre la Pobreza en America Latina y el Caribe*. PNUD Proyecto RLA/86/004. Bogotá.

_____. 1988c *Regional Conference in Latin America and the Caribbean: Bases for Strategy and Regional Action Programme*. Bogotá.

_____. 1989. *Urban Transition in Developing*

Countries: Policy Issues and Implications for Technical Cooperation in the 1990s. Programme Advisory Note, Technical Advisory Division, Bureau of Programme Policy and Evaluation. New York.

UNDP Development Study Programme and UN Conference on Trade and Development. 1988. *The Role of the Services Sector in the Development Process*. Report of the Schloss Fuschl Roundtable, 1-3 July, Austria. New York.

UNDRO and UNEP. 1986. *Social and Sociological Aspects; Disaster Prevention and Mitigation: A Compendium of Current Knowledge*, vol. 12. New York.

United Nations Economic Commission for Africa. 1989. *African Alternative Framework to Structural Adjustment Programmes for Socio-Economic Recovery and Transformation (AAF-SAP)*. E/ECA/CM.15/6/Rev.3 and A/44/315. Adopted by the Conference of Ministers of the Economic Commission for Africa at its 15th meeting, 10 April, Addis Ababa.

UNICEF. 1988. "Redirecting Adjustment Programmes towards Growth and the Protection of the Poor: The Philippine Case." In Giovanni Andrea Cornia, Richard Jolly, and Frances Stewart, eds., *Adjustment with a Human Face*, vol. II. Oxford: Clarendon Press.

_____. 1989a. *The Social Consequences of Adjustment and Dependency on Primary Commodities in Sub-Saharan Africa*. New York.

_____. 1989b. *The State of the World's Children 1989.* New York: Oxford University Press.

_____. 1989c. *Strategies for Children in the 1990s*. A UNICEF Policy Paper. New York.

UNICEF, Colombo. 1988. "Sri Lanka: The Social Impact of Economic Policies during the Last Decade." In Giovanni Andrea Cornia, Richard Jolly, and Frances Stewart, eds., *Adjustment with a Human Face,* vol. II. Oxford: Clarendon Press.

UNRISD. 1972. *Contents and Measurement of Socio-Economic Development: A Staff Study.* New York: Praeger.

United Nations Social Defence Research Institute. 1984. *Juvenile Social Maladjustment*. Publication 22. Rome.

USAID. 1989. *Development and the National Interest: U.S. Economic Assistance in the 21st Century*. Washington, D.C.

United States Census Bureau and Centre on Budget and Policy Priorities. 1989. New York.

United States House of Representatives. 1989. *U.S. Children and their Families: Current Conditions and Recent Trends*. U.S. House of Representatives Select Committee on Children, Youth, and Families, Washington, D.C.

Uphoff, Norman. 1986. *Local Institutional Development: An Analytical Sourcebook with Cases*. West Hartford, CT: Kumarian.

van Ginneken, Wouter. 1976. *Rural and Urban Income Inequalities in Indonesia, Mexico, Pakistan, Tanzania, and Tunisia.* Geneva: ILO.

_____. 1987. *Trends in Employment and Labour Incomes: Case Studies on Developing Countries.* Geneva: ILO.

Watts, H. 1968. "An Economic Definition of Poverty." In Daniel P. Moynihan, ed., *On Understanding Poverty.* New York: Basic Books.

WHO. 1987a. "Economic Support for National Health for All Strategies." Presented at the 40th World Health Assembly, WHO, 7 May, Geneva.

_____. 1987b. *Health Care: Who Pays?* Geneva.

_____. 1988a. *The Use of Essential Drugs: Third Report of the WHO Expert Committee*. Technical Report Series 770. Geneva.

_____. 1988b. *World Drug Situation.* Geneva.

_____. 1989a. *Contribution to Preparation of the First Report on the State of Human Development*. Rome.

_____. 1989b. *Global Strategy for Health for All by the Year 2000*. Rome.

_____. 1989c. *Global Strategy for the Prevention and Control of AIDS*. Report by the Director General. Rome.

_____. 1989d. *Special Programme for Research and Training in Tropical Diseases (TDR): Progress in Research and Transfer of Technology to National Health Services*. Report by the Director General to the Executive Board's 85th Session, 27 October, Geneva.

_____. 1989e. *Tropical Diseases: Progress in International Research, 1987-88*. Ninth Programme Report of the UNDP/World Bank/WHO Special Programme for Research and Training in Tropical Diseases. Geneva.

Wilson, Gail. 1986. *Money in the Family.* London: Gower.

Woodfield, Anthony. 1989. *Housing and Economic Adjustment*. New York: Taylor and Francis.

World Bank. 1982. *World Development Report 1982.* New York: Oxford University Press.

_____. 1983. *Zimbabwe: Population, Health, and Nutrition Sector Review*. Report 4212-Zim. World Bank Population, Health and Nutrition Department. Washington, D.C.

_____. 1986a. *Financing Education in Developing Countries: An Exploration of Policy Options.* Washington, D.C.

_____. 1986b. *Poverty and Hunger: Issues and Options for Food Security in Developing Countries*. A World Bank Policy Study. Washington, D.C.

_____. 1986c. *Poverty in Latin America: The Impact of Depression*. Washington, D.C.

_____. 1987. *Financing Health Services in Developing Countries: An Agenda for Reform.* A World Bank Policy Study. Washington, D.C.

_____. 1988a. *Brazil: Finance of Public Education*. World Bank Country Study. Washington, D.C.

_____. 1988b. *Brazil: Public Spending on Social Programs, Issues and Options*. vol. I. World Bank Report 7086-BR. Washington, D.C.

_____. 1989. *World Development Report 1989.* New York: Oxford University Press.
World Commission on Environment and Development. 1987. *Our Common Future.* New York: Oxford University Press.
World Food Council. 1987. *The Global State of Hunger and Malnutrition and the Impact of Economic Adjustment on Food and Hunger Problems.* WFC/1987/2. Rome.
Zuckerman, Elaine. 1988. *Poverty and Adjustment Issues and Practices*. World Bank Country Economics Department, Washington, D.C.
Zvekic, Ugljesa, and Aurelio Mattie. 1987. *Research and International Cooperation in Criminal Justice.* United Nations Social Defence Research Institute Publication 29. Rome.

HUMAN
DEVELOPMENT
INDICATORS

undp

Index to countries

9	Afghanistan
84	Albania
57	Algeria
22	Angola
99	Argentina
124	Australia
113	Austria
23	Bangladesh
116	Belgium
10	Benin
12	Bhutan
49	Bolivia
58	Botswana
80	Brazil
104	Bulgaria
3	Burkina Faso
11	Burundi
41	Cameroon
126	Canada
16	Central African Rep.
5	Chad
107	Chile
66	China
86	Colombia
33	Congo
103	Costa Rica
32	Côte d'Ivoire
92	Cuba
106	Czechoslovakia
122	Denmark
63	Dominican Rep.
75	Ecuador
45	Egypt
59	El Salvador
19	Ethiopia
120	Finland
123	France
47	Gabon
110	German Dem. Rep.
119	Germany, Fed. Rep.
30	Ghana
109	Greece
55	Guatemala
6	Guinea
29	Haiti
51	Honduras
108	Hong Kong
101	Hungary
37	India
54	Indonesia
61	Iran, Islamic Rep.
76	Iraq
14	Ireland
111	Israel
117	Italy
87	Jamaica
130	Japan
73	Jordan
40	Kampuchea, Dem.
42	Kenya
82	Korea, Dem. Rep.
97	Korea, Rep.
88	Kuwait
46	Lao PDR
69	Lebanon
53	Lesotho
26	Liberia
67	Libyan Arab Jamahiriya
38	Madagascar
14	Malawi
85	Malaysia
2	Mali
8	Mauritania
81	Mauritius
91	Mexico
70	Mongolia
44	Morocco
13	Mozambique
50	Myanmar
34	Namibia
17	Nepal
127	Netherlands
118	New Zealand
71	Nicaragua
1	Niger
24	Nigeria
125	Norway
48	Oman
36	Pakistan
93	Panama
39	Papua New Guinea
77	Paraguay
74	Peru
65	Philippines
98	Poland
95	Portugal
90	Romania
21	Rwanda
64	Saudi Arabia
18	Senegal
4	Sierra Leone
96	Singapore
7	Somalia
68	South Africa
115	Spain
83	Sri Lanka
15	Sudan
129	Sweden
128	Switzerland
62	Syrian Arab Rep.
35	Tanzania, United Rep.
78	Thailand
27	Togo
94	Trinidad and Tobago
60	Tunisia
72	Turkey
28	Uganda
105	USSR
77	United Arab Emirates
121	United Kingdom
112	USA
102	Uruguay
89	Venezuela
56	Viet Nam
25	Yemen Arab Rep.
31	Yemen, PDR
100	Yugoslavia
20	Zaire
43	Zambia
52	Zimbabwe

Key

In the Human Development Indicators the countries are ranked in ascending order of their human development index (HDI). The reference numbers indicating that rank are in the alphabetical list of countries provided here.

Official government data received by the responsible United Nations or other international agencies have been used whenever possible. In some cases where there are no reliable official figures, estimates by the responsible agency have been used when available. In other cases, the UNDP has made its own estimates based on field information or comparable country data. Generally, only comprehensive or representative national data have been used. If the data refer to only a part of the country, this is indicated. The data in the Human Development Indicators, derived from so many sources, inevitably cover a wide range of data reliability.

Unless otherwise stated, the summary measures for the various human development, income and regional groups of countries are the appropriately weighted values for each group (see the lists following the indicators for the composition of each group). Where the summary measure is a total, the letter T appears after the figure. The following signs have been used:

.. Data not available
T Total
r Rural only
u Urban only
Italicised figures are UNDP estimates.

Contents

1 Human development index

	Life expectancy at birth (years) 1987	Adult literacy rate (%) 1985	Real GDP per capita (PPP $)		Deprivation				Human development index	GNP per capita rank 1987	HDI rank minus GNP rank
			Actual 1987	Log 1987	Life expectancy	Literacy	North minimum purchasing power	Average of the three			
Low human development											
1 Niger	45	14	452	2.66	0.90	0.98	0.77	0.884	0.116	20	-19
2 Mali	45	17	543	2.73	0.92	0.94	0.71	0.857	0.143	15	-13
3 Burkina Faso	48	14	*500*	*2.70*	0.83	0.99	0.73	0.850	0.150	13	-10
4 Sierra Leone	42	30	480	2.68	1.00	0.80	0.75	0.850	0.150	27	-23
5 Chad	46	26	*400*	*2.60*	0.88	0.85	0.81	0.843	0.157	4	1
6 Guinea	43	29	*500*	*2.70*	0.97	0.81	0.73	0.838	0.162	31	-25
7 Somalia	46	12	*1,000*	*3.00*	0.89	1.00	0.51	0.800	0.200	23	-16
8 Mauritania	47	17	840	2.92	0.86	0.95	0.57	0.792	0.208	40	-32
9 Afghanistan	42	24	*1,000*	*3.00*	0.99	0.87	0.51	0.788	0.212	17	-8
10 Benin	47	27	665	2.82	0.85	0.84	0.64	0.776	0.224	28	-18
11 Burundi	50	35	450	2.65	0.78	0.74	0.77	0.765	0.235	18	-7
12 Bhutan	49	*25*	*700*	*2.85*	0.81	0.86	0.63	0.764	0.236	3	9
13 Mozambique	47	39	*500*	*2.70*	0.85	0.70	0.73	0.761	0.239	10	3
14 Malawi	48	42	476	2.68	0.84	0.66	0.75	0.750	0.250	7	7
15 Sudan	51	23	750	2.88	0.76	0.87	0.60	0.745	0.255	32	-17
16 Central African Rep.	46	41	591	2.77	0.88	0.67	0.68	0.742	0.258	29	-13
17 Nepal	52	26	722	2.86	0.72	0.84	0.62	0.727	0.273	8	9
18 Senegal	47	28	1,068	3.03	0.87	0.82	0.49	0.726	0.274	43	-25
19 Ethiopia	42	66	454	2.66	1.00	0.39	0.77	0.718	0.282	1	18
20 Zaire	53	62	220	2.34	0.69	0.43	1.00	0.706	0.294	5	15
21 Rwanda	49	47	571	2.76	0.80	0.60	0.69	0.696	0.304	26	-5
22 Angola	45	41	*1,000*	*3.00*	0.90	0.67	0.51	0.696	0.304	58	-36
23 Bangladesh	52	33	883	2.95	0.73	0.76	0.55	0.682	0.318	6	17
24 Nigeria	51	43	668	2.82	0.74	0.65	0.64	0.678	0.322	36	-12
25 Yemen Arab Rep.	52	25	*1,250*	*3.10*	0.72	0.86	0.44	0.672	0.328	47	-22
26 Liberia	55	35	696	2.84	0.63	0.74	0.63	0.667	0.333	42	-16
27 Togo	54	41	670	2.83	0.67	0.68	0.64	0.663	0.337	24	3
28 Uganda	52	58	511	2.71	0.73	0.48	0.73	0.646	0.354	21	7
29 Haiti	55	38	775	2.89	0.63	0.71	0.59	0.644	0.356	34	-5
30 Ghana	55	54	481	2.68	0.64	0.53	0.75	0.640	0.360	37	-7
31 Yemen,PDR	52	42	*1,000*	*3.00*	0.73	0.66	0.51	0.631	0.369	39	-8
32 Côte d'Ivoire	53	42	1,123	3.05	0.69	0.66	0.47	0.607	0.393	52	-20
33 Congo	49	63	756	2.88	0.80	0.42	0.60	0.605	0.395	59	-26
34 Namibia	56	*30*	*1,500*	*3.18*	0.61	0.80	0.38	0.596	0.404	60	-26
35 Tanzania, United Rep.	54	*75*	405	2.61	0.67	0.29	0.80	0.587	0.413	12	23
36 Pakistan	58	30	1,585	3.20	0.57	0.80	0.36	0.577	0.423	33	3
37 India	59	43	1,053	3.02	0.53	0.66	0.49	0.561	0.439	25	12
38 Madagascar	54	68	634	2.80	0.66	0.36	0.66	0.560	0.440	14	24
39 Papua New Guinea	55	45	1,843	3.27	0.64	0.63	0.31	0.529	0.471	50	-11
40 Kampuchea, Dem.	49	75	*1,000*	*3.00*	0.79	0.28	0.51	0.529	0.471	2	38
41 Cameroon	52	*61*	1,381	3.14	0.73	0.44	0.41	0.526	0.474	64	-23
42 Kenya	59	60	794	2.90	0.52	0.46	0.58	0.519	0.481	30	12
43 Zambia	54	76	717	2.86	0.66	0.28	0.62	0.519	0.481	19	24
44 Morocco	62	34	1,761	3.25	0.46	0.75	0.33	0.511	0.489	48	-4
Medium human development											
45 Egypt	62	45	1,357	3.13	0.46	0.63	0.41	0.499	0.501	49	-4
46 Lao PDR	49	84	*1,000*	*3.00*	0.79	0.18	0.51	0.494	0.506	9	37
47 Gabon	52	62	2,068	3.32	0.71	0.44	0.27	0.475	0.525	93	-46
48 Oman	57	30	*7,750*	*3.89*	0.60	0.80	0.00	0.465	0.535	104	-56
49 Bolivia	54	75	1,380	3.14	0.66	0.29	0.41	0.452	0.548	44	5
50 Myanmar	61	*79*	752	2.88	0.48	0.24	0.60	0.439	0.561	11	39
51 Honduras	65	59	1,119	3.05	0.37	0.46	0.47	0.437	0.563	53	-2
52 Zimbabwe	59	74	1,184	3.07	0.52	0.30	0.45	0.424	0.576	45	7
53 Lesotho	57	73	1,585	3.20	0.59	0.31	0.36	0.420	0.580	35	18
54 Indonesia	57	74	1,660	3.22	0.59	0.30	0.35	0.409	0.591	41	13
55 Guatemala	63	55	1,957	3.29	0.42	0.51	0.29	0.408	0.592	63	-8
56 Viet Nam	62	*80*	*1,000*	*3.00*	0.44	0.23	0.51	0.392	0.608	16	40
57 Algeria	63	50	2,633	3.42	0.41	0.57	0.20	0.391	0.609	91	-34
58 Botswana	59	71	2,496	3.40	0.52	0.33	0.21	0.354	0.646	69	-11
59 El Salvador	64	72	1,733	3.24	0.40	0.32	0.33	0.349	0.651	56	3
60 Tunisia	66	55	2,741	3.44	0.33	0.51	0.18	0.343	0.657	70	-10
61 Iran, Islamic Rep.	66	51	*3,300*	*3.52*	0.34	0.56	0.12	0.340	0.660	97	-36
62 Syrian Arab Rep.	66	60	*3,250*	*3.51*	0.34	0.46	0.13	0.309	0.691	79	-17
63 Dominican Rep.	67	78	*1,750*	*3.24*	0.32	0.25	0.33	0.301	0.699	51	12
64 Saudi Arabia	64	*55*	8,320	3.92	0.38	0.51	0.00	0.298	0.702	107	-43
65 Philippines	64	86	1,878	3.27	0.39	0.16	0.31	0.286	0.714	46	19
66 China	70	69	2,124	3.33	0.23	0.36	0.27	0.284	0.716	22	44

		Life expectancy at birth (years) 1987	Adult literacy rate (%) 1985	Real GDP per capita (PPP $) Actual 1987	Log 1987	Deprivation: Life expectancy	Literacy	North minimum purchasing power	Average of the three	Human development index	GNP per capita rank 1987	HDI rank minus GNP rank
67	Libyan Arab Jamahiriya	62	66	*7,250*	*3.86*	0.46	0.39	0.00	0.281	0.719	103	-36
68	South Africa	61	*70*	4,981	3.70	0.46	0.34	0.00	0.269	0.731	82	-14
69	Lebanon	68	78	*2,250*	*3.35*	0.30	0.25	0.25	0.265	0.735	78	-9
70	Mongolia	64	*90*	*2,000*	*3.30*	0.39	0.11	0.29	0.263	0.737	57	13
71	Nicaragua	64	88	2,209	3.34	0.38	0.14	0.25	0.257	0.743	54	17
72	Turkey	65	74	3,781	3.58	0.37	0.30	0.08	0.249	0.751	71	1
73	Jordan	67	75	3,161	3.50	0.32	0.29	0.14	0.248	0.752	76	-3
74	Peru	63	85	3,129	3.50	0.43	0.17	0.14	0.247	0.753	74	0
75	Ecuador	66	83	2,687	3.43	0.34	0.19	0.19	0.242	0.758	68	7
76	Iraq	65	89	*2,400*	*3.38*	0.37	0.13	0.23	0.241	0.759	96	-20
77	United Arab Emirates	71	*60*	12,191	4.09	0.20	0.46	0.00	0.218	0.782	127	-50
78	Thailand	66	91	2,576	3.41	0.34	0.10	0.20	0.217	0.783	55	23
79	Paraguay	67	88	2,603	3.42	0.31	0.14	0.20	0.216	0.784	65	14
80	Brazil	65	78	4,307	3.63	0.35	0.26	0.04	0.216	0.784	85	-5
81	Mauritius	69	83	2,617	3.42	0.24	0.19	0.20	0.212	0.788	75	6
82	Korea, Dem. Rep.	70	*90*	*2,000*	*3.30*	0.23	0.11	0.29	0.211	0.789	67	15
83	Sri Lanka	71	87	2,053	3.31	0.21	0.15	0.28	0.211	0.789	38	45
84	Albania	72	*85*	*2,000*	*3.30*	0.17	0.17	0.29	0.210	0.790	61	23
High human development												
85	Malaysia	70	74	3,849	3.59	0.23	0.30	0.07	0.200	0.800	80	5
86	Colombia	65	*88*	3,524	3.55	0.36	0.14	0.10	0.199	0.801	72	14
87	Jamaica	74	*82*	2,506	3.40	0.11	0.21	0.21	0.176	0.824	62	25
88	Kuwait	73	70	13,843	4.14	0.14	0.34	0.00	0.161	0.839	122	-34
89	Venezuela	70	87	4,306	3.63	0.23	0.15	0.04	0.139	0.861	95	-6
90	Romania	71	*96*	*3,000*	*3.48*	0.21	0.05	0.15	0.137	0.863	84	6
91	Mexico	69	90	4,624	3.67	0.24	0.11	0.01	0.124	0.876	81	10
92	Cuba	74	96	*2,500*	*3.40*	0.11	0.05	0.21	0.123	0.877	66	26
93	Panama	72	89	4,009	3.60	0.17	0.13	0.06	0.117	0.883	88	5
94	Trinidad and Tobago	71	96	3,664	3.56	0.21	0.05	0.09	0.115	0.885	100	-6
95	Portugal	74	85	5,597	3.75	0.13	0.17	0.00	0.101	0.899	94	1
96	Singapore	73	86	12,790	4.11	0.14	0.16	0.00	0.101	0.899	110	-14
97	Korea, Rep.	70	*95*	4,832	3.68	0.23	0.06	0.00	0.097	0.903	92	5
98	Poland	72	*98*	*4,000*	*3.60*	0.19	0.02	0.06	0.090	0.910	83	15
99	Argentina	71	96	4,647	3.67	0.21	0.05	0.01	0.090	0.910	89	10
100	Yugoslavia	72	92	*5,000*	*3.70*	0.16	0.10	0.00	0.087	0.913	90	10
101	Hungary	71	*98*	*4,500*	*3.65*	0.21	0.02	0.02	0.085	0.915	87	14
102	Uruguay	71	95	5,063	3.70	0.20	0.06	0.00	0.084	0.916	86	16
103	Costa Rica	75	93	3,760	3.58	0.10	0.07	0.08	0.084	0.916	77	26
104	Bulgaria	72	*93*	*4,750*	*3.68*	0.16	0.08	0.01	0.082	0.918	99	5
105	USSR	70	*99*	*6,000*	*3.78*	0.23	0.01	0.00	0.080	0.920	101	4
106	Czechoslovakia	72	*98*	*7,750*	*3.89*	0.19	0.02	0.00	0.069	0.931	102	4
107	Chile	72	98	4,862	3.69	0.18	0.02	0.00	0.069	0.931	73	34
108	Hong Kong	76	88	13,906	4.14	0.05	0.14	0.00	0.064	0.936	111	-3
109	Greece	76	93	*5,500*	*3.74*	0.07	0.08	0.00	0.051	0.949	98	11
110	German Dem. Rep.	74	*99*	*8,000*	*3.90*	0.13	0.01	0.00	0.047	0.953	115	-5
111	Israel	76	95	9,182	3.96	0.07	0.06	0.00	0.043	0.957	108	3
112	USA	76	96	17,615	4.25	0.07	0.05	0.00	0.039	0.961	129	-17
113	Austria	74	*99*	12,386	4.09	0.11	0.01	0.00	0.039	0.961	118	-5
114	Ireland	74	*99*	8,566	3.93	0.11	0.01	0.00	0.039	0.961	106	8
115	Spain	77	95	8,989	3.95	0.04	0.06	0.00	0.035	0.965	105	10
116	Belgium	75	*99*	13,140	4.12	0.09	0.01	0.00	0.034	0.966	116	0
117	Italy	76	97	10,682	4.03	0.07	0.03	0.00	0.034	0.966	112	5
118	New Zealand	75	*99*	10,541	4.02	0.09	0.01	0.00	0.034	0.966	109	9
119	Germany, Fed. Rep.	75	*99*	14,730	4.17	0.09	0.01	0.00	0.033	0.967	120	-1
120	Finland	75	*99*	12,795	4.11	0.09	0.01	0.00	0.033	0.967	121	-1
121	United Kingdom	76	*99*	12,270	4.09	0.08	0.01	0.00	0.030	0.970	113	8
122	Denmark	76	*99*	15,119	4.18	0.07	0.01	0.00	0.029	0.971	123	-1
123	France	76	*99*	13,961	4.14	0.07	0.01	0.00	0.026	0.974	119	4
124	Australia	76	*99*	11,782	4.07	0.06	0.01	0.00	0.022	0.978	114	10
125	Norway	77	*99*	15,940	4.20	0.04	0.01	0.00	0.017	0.983	128	-3
126	Canada	77	*99*	16,375	4.21	0.04	0.01	0.00	0.017	0.983	124	2
127	Netherlands	77	*99*	12,661	4.10	0.04	0.01	0.00	0.016	0.984	117	10
128	Switzerland	77	*99*	15,403	4.19	0.03	0.01	0.00	0.014	0.986	130	-2
129	Sweden	77	*99*	13,780	4.14	0.03	0.01	0.00	0.013	0.987	125	4
130	Japan	78	*99*	13,135	4.12	0.00	0.01	0.00	0.004	0.996	126	4

2 Profile of human development

	Life expectancy at birth (years) 1987	Population with access to health services (%) 1985-87	Population with access to safe water (%) 1985-87	Population with access to sanitation (%) 1985-87	Daily calorie supply (as % of requirements) 1984-86	Adult literacy rate (%) 1985	GNP per capita (US $) 1987	Real GDP per capita (PPP$) 1987
Low human development	55	47	48	14	95	41	300	970
Excluding India	52	..	39	22	91	40	300	880
1 Niger	45	41	47	..	100	14	260	450
2 Mali	45	15	17	19	86	17	210	540
3 Burkina Faso	48	49	67	9	86	14	190	..
4 Sierra Leone	42	..	25	25	81	30	300	480
5 Chad	46	30	..	..	69	26	150	..
6 Guinea	43	32	19	..	77	29	..	..
7 Somalia	46	27	34	18	90	12	290	..
8 Mauritania	47	30	..	..	92	17	440	840
9 Afghanistan	42	29	21	..	94	24	..	..
10 Benin	47	18	52	35	95	27	310	670
11 Burundi	50	61	26	58	97	35	250	450
12 Bhutan	49	65	..	..	..	..	150	..
13 Mozambique	47	39	16	21	69	39	170	..
14 Malawi	48	80	56	..	102	42	160	480
15 Sudan	51	51	21	..	88	23	330	750
16 Central African Rep.	46	45	..	..	86	41	330	590
17 Nepal	52	..	29	2	93	26	160	720
18 Senegal	47	40	53	..	99	28	520	1,070
19 Ethiopia	42	46	16	..	71	66	130	450
20 Zaire	53	26	33	..	98	62	150	220
21 Rwanda	49	27	50	57	81	47	300	570
22 Angola	45	30	30	19	82	41	470	..
23 Bangladesh	52	45	46	6	83	33	160	880
24 Nigeria	51	40	46	..	90	43	370	670
25 Yemen Arab Rep.	52	35	42	..	94	25	590	..
26 Liberia	55	39	55	..	102	35	450	700
27 Togo	54	61	55	14	97	41	290	670
28 Uganda	52	61	20	30	95	58	260	510
29 Haiti	55	70	38	21	84	38	360	780
30 Ghana	55	60	56	30	76	54	390	480
31 Yemen, PDR	52	30	54	..	96	42	420	..
32 Côte d'Ivoire	53	30	19	..	110	42	740	1,120
33 Congo	49	83	21	..	117	63	870	760
34 Namibia	56	..	..	..	..	..	..	..
35 Tanzania, United Rep.	54	76	56	68	96	..	180	410
36 Pakistan	58	55	44	20	97	30	350	1,590
37 India	59	..	57	10	100	43	300	1,050
38 Madagascar	54	56	32	..	106	68	210	630
39 Papua New Guinea	55	..	27	45	96	45	700	1,840
40 Kampuchea, Dem.	49	53	3	..	98	75	..	..
41 Cameroon	52	41	33	46	88	..	970	1,380
42 Kenya	59	..	30	..	92	60	330	790
43 Zambia	54	75	59	56	92	76	250	720
44 Morocco	62	70	60	..	118	34	610	1,760
Medium human development	67	75	59	50	113	71	690	2,370
Excluding China	63	..	..	..	115	73	1,250	2,730
45 Egypt	62	..	73	..	132	45	680	1,360
46 Lao PDR	49	67	21	..	104	84	170	..
47 Gabon	52	90	92	..	107	62	2,700	2,070
48 Oman	57	91	53	31	..	30	5,810	..
49 Bolivia	54	63	44	21	89	75	580	1,380
50 Myanmar	61	33	27	24	119	..	200	750
51 Honduras	65	73	50	30	92	59	810	1,120
52 Zimbabwe	59	71	..	..	89	74	580	1,180
53 Lesotho	57	80	36	15	101	73	370	1,590
54 Indonesia	57	80	38	37	116	74	450	1,660
55 Guatemala	63	34	38	24	105	55	950	1,960
56 Viet Nam	62	80	46	..	105	..	..	..
57 Algeria	63	88	68	57	112	50	2,680	2,630
58 Botswana	59	89	54	42	96	71	1,050	2,500
59 El Salvador	64	56	52	60	94	72	860	1,730
60 Tunisia	66	90	68	52	123	55	1,180	2,740
61 Iran, Islamic Rep.	66	78	76	..	138	51	..	..
62 Syrian Arab Rep.	66	76	76	..	131	60	1,640	..
63 Dominican Rep.	67	80	63	28	109	78	730	..
64 Saudi Arabia	64	97	97	..	125	..	6,200	8,320
65 Philippines	64	..	52	67	104	86	590	1,880
66 China	70	..	..	..	111	69	290	2,120

	Life expectancy at birth (years) 1987	Population with access to health services (%) 1985-87	Population with access to safe water (%) 1985-87	Population with access to sanitation (%) 1985-87	Daily calorie supply (as % of requirements) 1984-86	Adult literacy rate (%) 1985	GNP per capita (US $) 1987	Real GDP per capita (PPP$) 1987
67 Libyan Arab Jamahiriya	62	..	97	..	153	66	5,460	..
68 South Africa	61	..	..	..	120	..	1,890	4,980
69 Lebanon	68	..	93	..	125	78	..	..
70 Mongolia	64	..	..	..	116	..	..	..
71 Nicaragua	64	83	49	27	110	88	830	2,210
72 Turkey	65	..	78	..	125	74	1,210	3,780
73 Jordan	67	97	96	61	121	75	1,560	3,160
74 Peru	63	75	55	50	93	85	1,470	3,130
75 Ecuador	66	62	58	67	89	83	1,040	2,690
76 Iraq	65	93	87	75	124	89	3,020	..
77 United Arab Emirates	71	90	..	..	..	..	15,830	12,190
78 Thailand	66	70	64	53	105	91	850	2,580
79 Paraguay	67	61	29	86	123	88	990	2,600
80 Brazil	65	..	78	64	111	78	2,020	4,310
81 Mauritius	69	100	100	92	121	83	1,490	2,620
82 Korea, Dem. Rep.	70	..	..	..	135	..	..	..
83 Sri Lanka	71	93	40	45	110	87	400	2,050
84 Albania	72	..	..	..	114	..	..	..
High human development	73	..	..	..	131	..	9,250	11,860
85 Malaysia	70	..	84	76	121	74	1,810	3,850
86 Colombia	65	60	92	70	110	..	1,240	3,520
87 Jamaica	74	90	96	91	116	..	940	2,510
88 Kuwait	73	100	..	..	..	70	14,610	13,840
89 Venezuela	70	..	90	51	102	87	3,230	4,310
90 Romania	71	..	..	..	127	..	2,560	..
91 Mexico	69	..	77	58	135	90	1,830	4,620
92 Cuba	74	..	..	..	135	96	..	..
93 Panama	72	80	83	81	107	89	2,240	4,010
94 Trinidad and Tobago	71	99	98	98	126	96	4,210	3,660
95 Portugal	74	..	..	..	128	85	2,830	5,600
96 Singapore	73	100	100	99	124	86	7,940	12,790
97 Korea, Rep.	70	93	77	100	122	..	2,690	4,830
98 Poland	72	..	..	..	126	..	2,070	..
99 Argentina	71	71	56	69	136	96	2,390	4,650
100 Yugoslavia	72	..	..	..	139	92	2,480	..
101 Hungary	71	..	..	..	135	..	2,240	..
102 Uruguay	71	82	85	59	100	95	2,190	5,060
103 Costa Rica	75	80	91	94	124	93	1,610	3,760
104 Bulgaria	72	..	..	..	145	..	4,150	..
105 USSR	70	..	..	..	133	..	4,550	..
106 Czechoslovakia	72	..	..	..	141	..	5,820	..
107 Chile	72	97	94	85	106	98	1,310	4,860
108 Hong Kong	76	..	..	..	121	88	8,070	13,910
109 Greece	76	..	..	..	147	93	4,020	..
110 German Dem. Rep.	74	..	..	..	145	..	7,180	..
111 Israel	76	..	..	..	118	95	6,800	9,180
112 USA	76	..	..	..	138	96	18,530	17,620
113 Austria	74	..	..	..	130	..	11,980	12,390
114 Ireland	74	..	..	..	146	..	6,120	8,570
115 Spain	77	..	..	..	137	95	6,010	8,990
116 Belgium	75	..	..	..	146	..	11,480	13,140
117 Italy	76	..	..	..	139	97	10,350	10,680
118 New Zealand	75	..	..	..	129	..	7,750	10,540
119 Germany, Fed. Rep.	75	..	..	..	130	..	14,400	14,730
120 Finland	75	..	..	..	113	..	14,470	12,800
121 United Kingdom	76	..	..	..	128	..	10,420	12,270
122 Denmark	76	..	..	..	131	..	14,930	15,120
123 France	76	..	..	..	130	..	12,790	13,960
124 Australia	76	..	..	..	125	..	11,100	11,780
125 Norway	77	..	..	..	120	..	17,190	15,940
126 Canada	77	..	..	..	129	..	15,160	16,380
127 Netherlands	77	..	..	..	121	..	11,860	12,660
128 Switzerland	77	..	..	..	128	..	21,330	15,400
129 Sweden	77	..	..	..	113	..	15,550	13,780
130 Japan	78	..	..	..	122	..	15,760	13,140
All developing countries	62	61	55	32	107	60	650	1,970
Least developed countries	50	46	35	20	89	37	210	690
Sub-Saharan Africa	51	45	37	..	91	48	440	990
Industrial countries	74	..	..	..	132	..	10,760	14,260
World	65	..	..	..	113	..	3,100	4,110

Note: Summary data for regional and income groups are given in tables 23 and 24.

3 Profile of human deprivation

	Without access to health services 1985-87 (millions)	Without access to safe water 1985-87 (millions)	Without access to adequate sanitation 1985-87 (millions)	Malnourished (underweight) children 1980-88 (millions)	Illiterate adults 1985 (millions)	Out-of-school children 1986-88 (millions)	Population below poverty line 1977-86 (millions)
Low human development	775 T	750 T	1,300 T	90 T	500 T	140 T	700 T
Excluding India	..	..	..	..	..	..	..
1 Niger	4.0	3.6	..	0.6	3.1	1.7	1.9 r
2 Mali	7.5	7.3	7.1	0.5	3.9	2.2	3.9
3 Burkina Faso	4.3	2.8	7.7	..	4.2	1.9	..
4 Sierra Leone	..	2.9	2.9	0.2	1.6	0.7	1.8 r
5 Chad	3.8	..	..	..	2.3	1.1	2.6
6 Guinea	4.4	5.3	..	..	2.6	1.5	..
7 Somalia	5.2	4.7	5.8	..	3.3	1.8	4.2
8 Mauritania	1.3	..	..	0.1	0.9	0.4	..
9 Afghanistan	11.0	12.0	..	..	7.0	3.4	1.2 u
10 Benin	3.6	2.1	2.9	..	1.7	0.7	1.8 r
11 Burundi	2.0	3.8	2.1	0.3	1.8	1.1	4.3
12 Bhutan	0.5	..	..	..	..	0.3	..
13 Mozambique	9.0	12.0	12.0	..	5.1	2.8	..
14 Malawi	1.6	3.5	..	0.4	2.5	1.2	6.1
15 Sudan	12.0	19.0	..	..	10.0	4.5	16.0 r
16 Central African Rep.	1.5	..	..	..	0.9	0.5	1.4 r
17 Nepal	..	13.0	18.0	2.1	7.8	2.2	11.0
18 Senegal	4.2	3.3	..	0.3	2.8	1.2	..
19 Ethiopia	24.0	38.0	..	2.9	8.3	10.0	29.0
20 Zaire	25.0	23.0	..	1.8	6.9	4.8	17.0 r
21 Rwanda	5.0	3.4	2.9	0.4	1.8	1.2	5.8
22 Angola	6.7	6.7	7.7	..	3.1	1.4	..
23 Bangladesh	60.0	59.0	103.0	11.0	41.0	19.0	94.0
24 Nigeria	63.0	57.0	..	..	31.0	11.0	..
25 Yemen Arab Rep.	4.9	4.4	..	0.9	2.9	0.9	..
26 Liberia	1.5	1.1	..	0.1	0.9	0.5	0.3 r
27 Togo	1.2	1.4	2.8	0.2	1.1	0.3	0.3 u
28 Uganda	6.7	14.0	12.0	0.3	3.8	2.8	..
29 Haiti	1.9	3.9	5.0	0.3	2.4	0.7	4.7
30 Ghana	5.6	6.2	9.9	0.7	3.6	1.8	6.2
31 Yemen, PDR	1.6	1.1	..	0.1	0.7	0.3	0.3 r
32 Côte d'Ivoire	8.1	9.4	..	0.3	3.4	1.9	3.2
33 Congo	0.3	1.5	..	0.1	0.4	..	..
34 Namibia	..	..	..	..	..	..	..
35 Tanzania, United Rep.	6.1	11.0	8.1	2.5	..	4.2	..
36 Pakistan	52.0	64.0	92.0	11.0	44.0	23.0	34.0
37 India	..	352.0	737.0	46.0	297.0	10.0	394.0
38 Madagascar	4.9	7.6	..	0.7	2.0	0.9	5.6
39 Papua New Guinea	..	2.8	2.1	0.2	1.2	0.6	2.5
40 Kampuchea, Dem.	3.7	7.7	..	..	1.3	..	..
41 Cameroon	6.3	7.2	5.8	0.3	..	0.9	3.0
42 Kenya	..	16.0	..	..	4.5	..	10.0
43 Zambia	2.0	3.2	3.5	..	1.0	0.7	1.1 u
44 Morocco	7.2	9.6	..	0.6	9.3	3.2	8.9
Medium human development	475 T	750 T	1,100 T	55 T	370 T	100 T	450 T
Excluding China	..	..	..	..	..	..	..
45 Egypt	..	14.0	..	1.3	17.0	2.7	12.0
46 Lao PDR	1.3	3.0	..	0.3	0.4	0.3	..
47 Gabon	0.1	0.1	..	..	0.3	..	..
48 Oman	0.1	0.7	1.0	..	0.5	0.1	..
49 Bolivia	2.6	3.9	5.5	0.2	1.0	0.5	2.9 r
50 Myanmar	27.0	29.0	30.0	2.1	..	4.2	16.0
51 Honduras	1.3	2.4	3.4	0.2	1.1	0.4	1.8
52 Zimbabwe	2.6	..	..	0.2	1.3	0.2	..
53 Lesotho	0.3	1.1	1.4	..	0.3	0.1	0.9
54 Indonesia	35.0	109.0	110.0	11.0	29.0	17.0	68.0
55 Guatemala	5.7	5.4	6.6	0.5	2.1	1.3	6.2
56 Viet Nam	13.0	35.0	..	4.7	..	5.5	3.0 r
57 Algeria	2.9	7.6	10.0	..	6.6	1.7	2.1 u
58 Botswana	0.1	0.6	0.7	0.0	0.2	0.1	0.6
59 El Salvador	2.2	2.4	2.0	..	0.8	0.5	1.4
60 Tunisia	0.8	2.5	3.7	0.2	2.2	0.5	1.4
61 Iran, Islamic Rep.	12.0	13.0	..	4.1	15.0	2.5	..
62 Syrian Arab Rep.	2.8	2.8	..	..	2.4	0.5	..
63 Dominican Rep.	1.4	2.6	5.0	0.1	0.9	0.1	3.0
64 Saudi Arabia	0.4	0.4	..	..	..	1.5	..
65 Philippines	..	29.0	20.0	2.9	5.0	1.5	35.0
66 China	..	..	..	..	251.0	40.0	87.0 r

	Without access to health services 1985-87	Without access to safe water 1985-87	Without access to adequate sanitation 1985-87	Malnourished (underweight) children 1980-88	Illiterate adults 1985	Out-of-school children 1986-88	Population below poverty line 1977-86
	Millions						
67 Libyan Arab Jamahiriya	..	0.1	..	..	0.8	..	..
68 South Africa	..	..	..	..	..	..	..
69 Lebanon	..	0.2	..	0.1	0.4	0.1	..
70 Mongolia	..	..	..	..	..	0.0	..
71 Nicaragua	0.6	1.8	2.6	0.1	0.2	0.3	0.7
72 Turkey	..	12.0	..	..	9.1	2.9	..
73 Jordan	0.1	0.2	1.5	..	0.5	0.0	0.6
74 Peru	5.3	9.6	11.0	0.4	1.9	0.1	7.2 u
75 Ecuador	3.9	4.3	3.4	0.3	1.0	0.3	5.2
76 Iraq	1.2	2.3	4.4	..	1.0	1.2	..
77 United Arab Emirates	0.2	..	..	..	..	0.1	0.0 r
78 Thailand	16.0	19.0	25.0	1.5	3.2	5.2	16.0
79 Paraguay	1.6	2.8	0.6	..	0.3	0.4	1.4
80 Brazil	..	32.0	52.0	2.4	21.0	5.0	..
81 Mauritius	0.0	0.0	0.1	0.0	0.1	0.0	0.1
82 Korea, Dem. Rep.	..	..	..	..	..	1.5	..
83 Sri Lanka	1.2	10.0	9.2	0.7	1.5	0.6	..
84 Albania	..	..	..	..	..	0.1	..
High human development	..	..	..	..	..	..	..
85 Malaysia	..	2.7	4.0	..	2.7	0.9	4.6
86 Colombia	12.0	2.4	9.2	0.5	..	1.3	13.0
87 Jamaica	0.2	0.1	0.2	0.0	..	0.1	1.0 r
88 Kuwait	0.0	..	..	0.0	0.4	0.1	..
89 Venezuela	..	1.9	9.2	0.3	1.5	0.8	..
90 Romania	..	..	..	..	..	0.3	..
91 Mexico	..	20.0	36.0	..	5.2	3.5	..
92 Cuba	..	..	..	..	0.3	0.1	..
93 Panama	0.5	0.4	0.4	0.0	0.2	0.1	0.6
94 Trinidad and Tobago	0.0	0.0	0.0	0.0	0.0	0.0	0.2 r
95 Portugal	..	..	..	..	1.2	0.2	..
96 Singapore	0.0	0.0	0.0	0.0	0.3	0.0	..
97 Korea, Rep.	3.0	9.8	0.0	..	..	0.2	6.8
98 Poland	..	..	..	..	..	0.4	..
99 Argentina	9.1	14.0	9.8	..	1.0	0.3	..
100 Yugoslavia	..	..	..	..	1.5	0.6	..
101 Hungary	..	..	..	..	..	0.2	..
102 Uruguay	0.6	0.5	1.3	..	0.1	0.0	0.6 u
103 Costa Rica	0.6	0.3	0.2	0.0	0.1	0.2	..
104 Bulgaria	..	..	..	..	..	0.1	..
105 USSR	..	..	..	..	..	0.0	..
106 Czechoslovakia	..	..	..	..	..	0.6	..
107 Chile	0.4	0.8	1.9	0.0	0.2	0.2	..
108 Hong Kong	..	..	..	..	0.5	0.1	..
100 Greece	..	..	..	..	0.6	0.0	..
110 German Dem. Rep.	..	..	..	..	..	0.3	..
111 Israel	..	..	..	..	0.2	0.0	..
112 USA	..	..	..	..	..	0.0	..
113 Austria	..	..	..	..	..	0.1	..
114 Ireland	..	..	..	..	..	0.2	..
115 Spain	..	..	..	..	1.7	0.0	..
116 Belgium	..	..	..	..	..	0.0	..
117 Italy	..	..	..	..	1.4	0.7	..
118 New Zealand	..	..	..	..	..	0.0	..
119 Germany, Fed. Rep.	..	..	..	..	..	0.3	..
120 Finland	..	..	..	..	..	0.0	..
121 United Kingdom	..	..	..	..	..	0.5	..
122 Denmark	..	..	..	..	..	0.0	..
123 France	..	..	..	..	..	0.0	..
124 Australia	..	..	..	..	..	0.0	..
125 Norway	..	..	..	..	..	0.0	..
126 Canada	..	..	..	..	..	0.0	..
127 Netherlands	..	..	..	..	..	0.0	..
128 Switzerland	..	..	..	..	..	0.0	..
129 Sweden	..	..	..	..	..	0.1	..
130 Japan	..	..	..	..	..	0.0	..
All developing countries	1,500 T	1,750 T	2,800 T	150 T	870 T	240 T	1,150 T
Least developed countries	225 T	250 T	350 T	35 T	150 T	70 T	250 T
Sub-Saharan Africa	..	..	..	..	..	..	..
Industrial countries	..	..	..	..	..	..	..
World	..	..	..	..	..	..	..

Note: Summary data for regional and income groups are given in tables 23 and 24.

4 Trends in human development

	Life expectancy (years)			Under-five mortality rate (per 1,000)		Population with access to safe water (%)		Daily calorie supply (as % of requirements)		Adult literacy rate (%)		GNP per capita (US$)	
	1960	1975	1987	1960	1988	1975	1985-87	1964-66	1984-86	1970	1985	1976	1987
Low human development	42	49	55	285	170	31	48	89	95	29	41	180	300
Excluding India	40	46	52	287	186	30	39	88	91	23	40	220	300
1 Niger	35	40	45	320	228	27	47	85	100	4	14	160	260
2 Mali	35	39	45	370	292	..	17	83	86	8	17	100	210
3 Burkina Faso	36	42	48	362	233	25	67	91	86	8	14	110	190
4 Sierra Leone	32	36	42	386	266	..	25	79	81	13	30	200	300
5 Chad	35	40	46	326	223	26	..	99	69	11	26	120	150
6 Guinea	33	37	43	346	248	14	19	81	77	14	29	150	..
7 Somalia	36	42	46	294	221	38	34	92	90	3	12	110	290
8 Mauritania	35	41	47	320	220	..	..	88	92	..	17	340	440
9 Afghanistan	33	39	42	380	300	9	21	90	94	8	24	160	..
10 Benin	35	41	47	310	185	34	52	88	95	16	27	130	310
11 Burundi	42	45	50	258	188	..	26	103	97	20	35	120	250
12 Bhutan	38	43	49	297	197	..	..	..	..	..	..	70	150
13 Mozambique	37	43	47	330	298	..	16	86	69	22	39	170	170
14 Malawi	38	42	48	364	262	..	56	91	102	30	42	140	160
15 Sudan	39	44	51	293	181	..	21	79	88	17	23	290	330
16 Central African Rep.	37	42	46	308	223	..	..	91	86	16	41	230	330
17 Nepal	38	45	52	297	197	8	29	87	93	13	26	120	160
18 Senegal	37	41	47	313	136	..	53	104	99	12	28	390	520
19 Ethiopia	36	42	42	294	259	8	16	77	71	..	66	100	130
20 Zaire	42	47	53	251	138	19	33	98	98	42	62	140	150
21 Rwanda	42	45	49	248	206	68	50	73	81	32	47	110	300
22 Angola	33	39	45	346	292	..	30	81	82	12	41	330	470
23 Bangladesh	40	46	52	262	188	56	46	91	83	24	33	110	160
24 Nigeria	40	46	51	318	174	..	46	95	90	25	43	380	370
25 Yemen Arab Rep.	37	45	52	378	190	..	42	80	94	5	25	250	590
26 Liberia	41	49	55	258	147	..	55	94	102	18	35	450	450
27 Togo	39	47	54	305	153	16	55	101	97	17	41	260	290
28 Uganda	43	48	52	224	169	35	20	96	95	41	58	240	260
29 Haiti	42	50	55	294	171	12	38	88	84	22	38	200	360
30 Ghana	45	51	55	224	146	35	56	87	76	31	54	580	390
31 Yemen, PDR	37	45	52	378	197	..	54	84	96	20	42	280	420
32 Côte d'Ivoire	39	47	53	264	142	..	19	102	110	18	42	610	740
33 Congo	38	44	49	241	114	38	21	101	117	35	63	520	870
34 Namibia	42	50	56	262	176	..	..	..	..	..	..	..	..
35 Tanzania, United Rep.	41	48	54	248	176	39	56	85	96	33	..	180	180
36 Pakistan	43	50	58	277	166	25	44	76	97	21	30	170	350
37 India	44	52	59	282	149	31	57	89	100	34	43	150	300
38 Madagascar	41	48	54	364	184	25	32	108	106	50	68	200	210
39 Papua New Guinea	41	49	55	247	81	20	27	72	96	32	45	490	700
40 Kampuchea, Dem.	42	36	49	218	199	..	3	98	98	..	75	..	..
41 Cameroon	40	46	52	275	153	..	33	89	88	33	..	290	970
42 Kenya	45	52	59	208	113	17	30	98	92	32	60	240	330
43 Zambia	42	48	54	228	127	42	59	91	92	52	76	440	250
44 Morocco	47	54	62	265	119	..	60	92	118	22	34	540	610
Medium human development	48	61	67	209	72	33	59	88	113	57	71	540	690
Excluding China	48	56	63	214	94	..	..	92	115	..	73	740	1,250
45 Egypt	46	54	62	300	125	..	73	97	132	35	45	280	680
46 Lao PDR	40	42	49	232	159	..	21	86	104	33	84	90	170
47 Gabon	41	46	52	288	169	..	92	81	107	33	62	..	2,700
48 Oman	40	48	57	378	64	..	53	..	..	..	30	..	5,810
49 Bolivia	43	48	54	282	172	34	44	77	89	57	75	390	580
50 Myanmar	44	54	61	229	95	17	27	89	119	71	..	120	200
51 Honduras	46	56	65	232	107	41	50	87	92	53	59	390	810
52 Zimbabwe	45	53	59	182	113	..	..	87	89	55	74	550	580
53 Lesotho	42	50	57	208	136	17	36	89	101	62	73	170	370
54 Indonesia	41	49	57	235	119	11	38	81	116	54	74	240	450
55 Guatemala	46	55	63	230	99	39	38	93	105	44	55	630	950
56 Viet Nam	44	53	62	233	88	..	46	97	105	..	..	..	..
57 Algeria	47	55	63	270	107	77	68	72	112	25	50	990	2,680
58 Botswana	46	52	59	174	92	..	54	88	96	41	71	..	1,050
59 El Salvador	50	58	64	206	84	53	52	80	94	57	72	490	860
60 Tunisia	48	58	66	255	83	..	68	94	123	31	55	840	1,180
61 Iran, Islamic Rep.	50	57	66	254	90	51	76	87	138	29	51	1,930	..
62 Syrian Arab Rep.	50	59	66	218	64	..	76	89	131	40	60	780	1,640
63 Dominican Rep.	52	61	67	200	81	55	63	85	109	67	78	780	730
64 Saudi Arabia	44	56	64	292	98	64	97	79	125	9	..	4,480	6,200
65 Philippines	53	59	64	135	73	40	52	82	104	83	86	410	590
66 China	47	65	70	202	43	..	..	86	111	..	69	410	290

	Life expectancy (years)			Under-five mortality rate (per 1,000)		Population with access to safe water (%)		Daily calorie supply (as % of requirements)		Adult literacy rate (%)		GNP per capita (US$)	
	1960	1975	1987	1960	1988	1975	1985-87	1964-66	1984-86	1970	1985	1976	1987
67 Libyan Arab Jamahiriya	47	54	62	268	119	87	97	83	153	37	66	6,310	5,460
68 South Africa	49	55	61	192	95	..	..	107	120	..	..	1,340	1,890
69 Lebanon	60	65	68	92	51	..	93	99	125	69	78	..	..
70 Mongolia	52	61	64	158	59	..	..	106	116	81	..	860	..
71 Nicaragua	47	55	64	210	95	46	49	107	110	58	88	750	830
72 Turkey	50	59	65	258	93	68	78	105	125	52	74	990	1,210
73 Jordan	47	59	67	218	57	..	96	93	121	47	75	610	1,560
74 Peru	48	56	63	233	123	47	55	98	93	71	85	800	1,470
75 Ecuador	53	60	66	183	87	36	58	83	89	72	83	640	1,040
76 Iraq	48	59	65	222	94	66	87	89	124	34	89	1,390	3,020
77 United Arab Emirates	53	65	71	239	32	..	..	..	..	16	..	..	15,830
78 Thailand	52	60	66	149	49	25	64	95	105	79	91	380	850
79 Paraguay	64	66	67	134	62	13	29	112	123	80	88	640	990
80 Brazil	55	61	65	160	85	..	78	100	111	66	78	1,140	2,020
81 Mauritius	59	64	69	104	29	..	100	103	121	68	83	..	1,490
82 Korea, Dem. Rep.	54	64	70	120	33	..	..	99	135	..	..	470	..
83 Sri Lanka	62	66	71	113	43	19	40	100	110	77	87	200	400
84 Albania	62	68	72	151	34	..	..	102	114	..	..	540	..
High human development	68	71	73	67	27	..	..	121	131	..	..	4,350	9,250
85 Malaysia	54	64	70	106	32	34	84	101	121	60	74	860	1,810
86 Colombia	55	61	65	148	68	64	92	94	110	78	..	630	1,240
87 Jamaica	63	68	74	88	22	86	96	100	116	97	..	1,070	940
88 Kuwait	60	68	73	128	22	89	..	..	..	54	70	15,480	14,610
89 Venezuela	60	67	70	114	44	..	90	94	102	75	87	2,570	3,230
90 Romania	65	69	71	82	28	..	..	114	127	94	..	1,450	2,560
91 Mexico	57	64	69	140	68	62	77	111	135	74	90	1,090	1,830
92 Cuba	63	72	74	87	18	..	..	102	135	87	96	860	..
93 Panama	61	68	72	105	34	77	83	98	107	81	89	1,310	2,240
94 Trinidad and Tobago	64	67	71	67	23	93	98	103	126	92	96	2,240	4,210
95 Portugal	63	69	74	112	17	..	..	107	128	72	85	1,690	2,830
96 Singapore	64	70	73	50	12	..	100	87	124	74	86	2,700	7,940
97 Korea, Rep.	54	64	70	120	33	66	77	96	122	88	..	670	2,690
98 Poland	67	71	72	70	18	..	..	123	126	98	..	2,860	2,070
99 Argentina	65	68	71	75	37	66	56	119	136	93	96	1,550	2,390
100 Yugoslavia	63	69	72	113	28	..	..	128	139	84	92	1,680	2,480
101 Hungary	68	70	71	57	19	..	..	122	135	98	..	2,280	2,240
102 Uruguay	68	69	71	56	31	..	85	106	100	93	95	1,390	2,190
103 Costa Rica	62	69	75	121	22	72	91	104	124	88	93	1,040	1,610
104 Bulgaria	68	71	72	69	20	..	..	137	145	92	..	2,310	4,150
105 USSR	68	68	70	53	32	..	..	126	133	98	..	2,760	4,550
106 Czechoslovakia	70	70	72	32	15	..	..	139	141	..	..	3,840	5,820
107 Chile	57	65	72	142	26	70	94	108	106	89	98	1,050	1,310
108 Hong Kong	66	73	76	65	10	..	..	..	121	77	88	2,110	8,070
109 Greece	69	73	76	64	18	..	..	124	147	85	93	2,590	4,020
110 German Dem. Rep.	70	71	74	44	12	..	..	122	145	..	..	4,220	7,180
111 Israel	69	72	76	40	14	..	..	109	118	88	95	3,920	6,800
112 USA	70	72	76	30	13	..	..	126	138	..	96	7,890	18,530
113 Austria	69	71	74	43	10	..	..	126	130	..	..	5,330	11,980
114 Ireland	70	72	74	36	9	..	..	138	146	..	..	2,560	6,120
115 Spain	69	74	77	56	12	100	..	117	137	90	95	2,920	6,010
116 Belgium	70	72	75	35	13	..	..	128	146	99	..	6,780	11,480
117 Italy	69	73	76	50	11	..	..	124	139	94	97	3,050	10,350
118 New Zealand	71	72	75	27	12	..	..	127	129	..	..	4,250	7,750
119 Germany, Fed. Rep.	69	71	75	40	10	..	..	118	130	..	..	7,380	14,400
120 Finland	68	71	75	28	7	..	..	116	113	..	..	5,620	14,470
121 United Kingdom	71	72	76	27	11	..	..	133	128	..	..	4,020	10,420
122 Denmark	72	74	76	25	11	..	..	127	131	..	..	7,450	14,930
123 France	70	73	76	34	10	..	..	135	130	99	..	6,550	12,790
124 Australia	71	73	76	25	10	..	..	120	125	..	..	6,100	11,100
125 Norway	73	75	77	23	10	..	..	115	120	..	..	7,420	17,190
126 Canada	71	74	77	33	8	..	..	122	129	..	..	7,510	15,160
127 Netherlands	73	75	77	22	8	..	..	123	121	..	..	6,200	11,860
128 Switzerland	71	75	77	27	8	..	..	127	128	..	..	8,880	21,330
129 Sweden	73	75	77	20	7	..	..	117	113	..	..	8,670	15,550
130 Japan	68	74	78	40	8	..	..	112	122	99	..	4,910	15,760
All developing countries	46	57	62	243	121	35	55	90	107	43	60	450	650
Least developed countries	39	45	50	288	205	31	35	87	89	25	37	140	210
Sub-Saharan Africa	40	46	51	284	183	24	37	92	91	26	48	350	440
Industrial countries	69	71	74	46	18	..	..	124	132	..	..	4,850	10,760
World	53	61	65	218	108	..	..	100	113	..	..	1,800	3,100

Note: Summary data for regional and income groups are given in tables 23 and 24.

5 Human capital formation

		Adult literacy rate (percentage of those 15 years and older)						Scientists and technicians (per 1,000 people) 1970-87
		1970			1985			
		Male	Female	Total	Male	Female	Total	
Low human development		40	17	29	54	29	41	2.9
Excluding India		32	13	23	51	29	40	1.9
1	Niger	6	2	4	19	9	14	..
2	Mali	11	4	8	23	11	17	..
3	Burkina Faso	13	3	8	21	6	14	..
4	Sierra Leone	18	8	13	38	21	30	..
5	Chad	20	2	11	40	11	26	..
6	Guinea	21	7	14	40	17	29	..
7	Somalia	5	1	3	18	6	12	..
8	Mauritania	..	..	..	..	..	17	..
9	Afghanistan	13	2	8	39	8	24	..
10	Benin	23	8	16	37	16	27	..
11	Burundi	29	10	20	43	27	35	..
12	Bhutan	..	..	..	..	..	..	..
13	Mozambique	29	14	22	55	22	39	..
14	Malawi	42	18	30	52	31	42	2.8
15	Sudan	28	6	17	33	15	23	0.8
16	Central African Rep.	26	6	16	53	29	41	..
17	Nepal	23	3	13	39	12	26	0.9
18	Senegal	18	5	12	37	19	28	..
19	Ethiopia	8	..	..	..	..	66	..
20	Zaire	61	22	42	79	45	62	..
21	Rwanda	43	21	32	61	33	47	0.4
22	Angola	16	7	12	49	33	41	..
23	Bangladesh	36	12	24	43	22	33	1.1
24	Nigeria	35	14	25	54	31	43	1.7
25	Yemen Arab Rep.	9	1	5	42	7	25	0.4
26	Liberia	27	8	18	47	23	35	..
27	Togo	27	7	17	53	28	41	0.3
28	Uganda	52	30	41	70	45	58	..
29	Haiti	26	17	22	40	35	38	6.0
30	Ghana	43	18	31	64	43	54	2.6
31	Yemen, PDR	31	9	20	59	25	42	..
32	Côte d'Ivoire	26	10	18	53	31	42	..
33	Congo	50	19	35	71	55	63	3.9
34	Namibia	..	..	..	..	..	..	..
35	Tanzania, United Rep.	48	18	33	93	88	..	..
36	Pakistan	30	11	21	40	19	30	..
37	India	47	20	34	57	29	43	3.3
38	Madagascar	56	43	50	74	62	68	..
39	Papua New Guinea	39	24	32	55	35	45	4.7
40	Kampuchea, Dem.	..	23	..	85	65	75	..
41	Cameroon	47	19	33	68	*49*	..	5.8
42	Kenya	44	19	32	70	49	60	3.7
43	Zambia	66	37	52	84	67	76	7.6
44	Morocco	34	10	22	45	22	34	..
Medium human development		67	48	57	81	61	71	11.8
Excluding China		..	..	..	80	66	73	19.6
45	Egypt	50	20	35	59	30	45	47.0
46	Lao PDR	37	28	33	92	76	84	..
47	Gabon	43	22	33	70	53	62	..
48	Oman	..	..	..	47	12	30	..
49	Bolivia	68	46	57	84	65	75	..
50	Myanmar	85	57	71	..	..	..	2.1
51	Honduras	55	50	53	61	58	59	3.2
52	Zimbabwe	63	47	55	81	67	74	..
53	Lesotho	49	74	62	62	84	73	..
54	Indonesia	66	42	54	83	65	74	13.0
55	Guatemala	51	37	44	63	47	55	2.1
56	Viet Nam	..	..	..	88	80	..	..
57	Algeria	39	11	25	63	37	50	..
58	Botswana	37	44	41	73	69	71	2.5
59	El Salvador	61	53	57	75	69	72	1.8
60	Tunisia	44	17	31	68	41	55	2.0
61	Iran, Islamic Rep.	40	17	29	62	39	51	12.0
62	Syrian Arab Rep.	60	20	40	76	43	60	7.2
63	Dominican Rep.	69	65	67	78	77	78	6.2
64	Saudi Arabia	15	2	9	71	31	..	..
65	Philippines	84	81	83	86	85	86	..
66	China	..	..	..	82	56	69	7.0

		Adult literacy rate (percentage of those 15 years and older)						Scientists and technicians (per 1,000 people)
		1970			1985			
		Male	Female	Total	Male	Female	Total	1970-87
67	Libyan Arab Jamahiriya	60	13	37	81	50	66	17.0
68	South Africa	..	..	..	..	..	..	..
69	Lebanon	79	58	69	86	69	78	..
70	Mongolia	87	74	81	95	*87*	..	1.6
71	Nicaragua	58	57	58	..	..	88	..
72	Turkey	69	34	52	86	62	74	34.0
73	Jordan	64	29	47	87	63	75	6.6
74	Peru	81	60	71	91	78	85	..
75	Ecuador	75	68	72	85	80	83	13.0
76	Iraq	50	18	34	90	87	89	7.3
77	United Arab Emirates	24	7	16	..	..	..	..
78	Thailand	86	72	79	94	88	91	1.6
79	Paraguay	85	75	80	91	85	88	..
80	Brazil	69	63	66	79	76	78	36.0
81	Mauritius	77	59	68	89	77	83	25.0
82	Korea, Dem. Rep.	..	..	..	..	..	..	..
83	Sri Lanka	85	69	77	91	83	87	4.7
84	Albania	..	..	..	..	..	..	..
High human development		..	..	..	..	..	..	135.5
85	Malaysia	71	48	60	81	66	74	6.9
86	Colombia	79	76	78	82	*88*	..	..
87	Jamaica	96	97	97	..	..	..	8.0
88	Kuwait	65	42	54	76	63	70	76.0
89	Venezuela	79	71	75	88	85	87	125.0
90	Romania	96	91	94	..	..	..	..
91	Mexico	78	69	74	92	88	90	..
92	Cuba	86	87	87	96	96	96	52.0
93	Panama	81	81	81	89	88	89	11.0
94	Trinidad and Tobago	95	89	92	97	95	96	11.0
95	Portugal	78	65	72	89	81	85	..
96	Singapore	92	55	74	93	79	86	26.0
97	Korea, Rep.	94	81	88	96	*91*	..	..
98	Poland	98	97	98	..	..	..	168.0
99	Argentina	94	92	93	96	96	96	..
100	Yugoslavia	92	76	84	97	87	92	192.0
101	Hungary	98	98	98	..	..	..	251.0
102	Uruguay	93	93	93	..	..	95	..
103	Costa Rica	88	87	88	94	93	93	77.0
104	Bulgaria	94	89	92	..	..	..	113.0
105	USSR	98	97	98	..	..	..	128.0
106	Czechoslovakia	..	..	..	..	..	..	130.0
107	Chile	90	88	89	97	97	98	27.0
108	Hong Kong	90	64	77	95	81	88	214.0
109	Greece	93	76	85	97	89	93	166.0
110	German Dem. Rep.	..	..	..	..	..	..	101.0
111	Israel	93	83	88	97	93	95	82.0
112	USA	..	..	..	..	..	96	55.0
113	Austria	..	..	..	..	..	..	268.0
114	Ireland	..	..	..	..	..	..	244.0
115	Spain	93	87	90	97	92	95	130.0
116	Belgium	99	99	99	..	..	..	..
117	Italy	95	93	94	98	96	97	83.0
118	New Zealand	..	..	..	..	..	..	49.0
119	Germany, Fed. Rep.	..	..	..	..	..	..	139.0
120	Finland	..	..	..	..	..	..	376.0
121	United Kingdom	..	..	..	..	..	..	..
122	Denmark	..	..	..	..	..	..	63.0
123	France	99	98	99	..	..	..	83.0
124	Australia	..	..	..	..	..	..	157.0
125	Norway	..	..	..	..	..	..	103.0
126	Canada	..	..	..	..	..	..	257.0
127	Netherlands	..	..	..	..	..	..	219.0
128	Switzerland	..	..	..	..	..	..	202.0
129	Sweden	..	..	..	..	..	..	262.0
130	Japan	99	99	99	..	..	..	317.0
All developing countries		53	33	43	71	50	60	9.7
Least developed countries		33	16	25	47	27	37	1.4
Sub-Saharan Africa		34	17	26	59	38	48	2.6
Industrial countries		..	..	..	..	..	..	140.5
World		..	..	..	..	..	..	44.4

Note: Summary data for regional and income groups are given in tables 23 and 24.

6 South-North human gaps

Index: North = 100 (see footnote)

		Life expectancy 1987	Adult literacy rate 1985	Female literacy rate 1985	Under-five mortality rate 1988	Maternal mortality rate 1980-87	Combined primary and secondary enrolment ratio 1986-88	Doctors (per 1,000 people) 1984	Scientists and technicians (per 1,000 people) 1970-85	GNP per capita 1987	Real GDP per capita 1985
Low human development		69	41	30	10	4	43	3	2	3	5
Excluding India		..	..	..	..	..	..	..	..	..	..
1	Niger	61	15	10	8	6	17	1	..	2	3
2	Mali	60	18	12	6	..	15	2	..	2	4
3	Burkina Faso	65	14	7	8	3	20	1	..	2	..
4	Sierra Leone	56	32	23	7	5	39	3	..	3	3
5	Chad	62	27	12	8	3	30	1	..	1	..
6	Guinea	58	30	18	7	..	20	1	..	..	..
7	Somalia	62	13	7	8	2	14	3	..	3	..
8	Mauritania	63	18	..	8	..	37	4	..	4	6
9	Afghanistan	57	26	9	6	4	17	..	..	..	..
10	Benin	64	28	17	10	..	43	3	..	3	5
11	Burundi	67	37	29	10	..	34	2	..	2	3
12	Bhutan	66	..	..	9	1	18	2	..	1	..
13	Mozambique	64	41	24	6	..	35	1	..	2	..
14	Malawi	64	44	34	7	..	49	4	2	1	3
15	Sudan	68	25	16	10	4	37	5	1	3	5
16	Central African Rep.	62	44	32	8	4	41	2	..	3	4
17	Nepal	70	28	13	9	3	58	1	1	1	5
18	Senegal	63	30	21	14	4	39	3	..	5	7
19	Ethiopia	56	70	..	7	..	29	1	..	1	3
20	Zaire	72	66	49	13	..	53	..	..	1	2
21	Rwanda	66	50	36	9	12	46	1	0	3	4
22	Angola	61	44	36	6	..	50	..	..	4	..
23	Bangladesh	69	35	24	10	4	43	7	1	1	6
24	Nigeria	69	45	34	11	3	66	6	1	3	5
25	Yemen Arab Rep.	70	27	8	10	..	65	7	0	5	..
26	Liberia	74	37	25	13	..	31	5	..	4	5
27	Togo	72	43	30	12	..	66	5	0	3	5
28	Uganda	70	61	49	11	8	48	2	..	2	4
29	Haiti	75	40	38	11	11	60	6	4	3	5
30	Ghana	74	57	47	13	2	58	3	2	4	3
31	Yemen, PDR	70	45	27	9	..	53	11	..	4	..
32	Côte d'Ivoire	72	45	34	13	..	48	..	..	7	8
33	Congo	66	67	60	16	2	..	6	3	8	5
34	Namibia	75	..	..	10	..	..	..	..	..	..
35	Tanzania, United Rep.	72	..	96	10	7	43	..	..	2	3
36	Pakistan	77	32	21	11	5	30	16	..	3	11
37	India	79	45	32	12	7	98	18	2	3	7
38	Madagascar	73	72	67	10	10	75	5	..	2	4
39	Papua New Guinea	74	47	38	23	3	43	7	3	7	13
40	Kampuchea, Dem.	67	80	71	9	..	..	..	..	..	..
41	Cameroon	70	..	53	12	8	72	..	4	9	10
42	Kenya	80	64	53	16	14	..	5	3	3	6
43	Zambia	73	80	73	15	16	71	6	5	2	5
44	Morocco	83	36	24	16	8	53	3	..	6	12
Medium human development		87	79	74	21	22	80	27	5	9	16
Excluding China		..	..	..	..	..	..	..	..	..	..
45	Egypt	83	48	33	15	8	83	58	33	6	10
46	Lao PDR	67	89	83	12	..	69	34	..	2	..
47	Gabon	70	66	58	11	..	..	16	..	25	14
48	Oman	76	32	13	29	..	74	27	..	54	..
49	Bolivia	73	79	71	11	5	78	30	..	5	10
50	Myanmar	82	..	..	19	17	62	12	2	2	5
51	Honduras	87	63	63	17	49	77	30	2	8	8
52	Zimbabwe	80	79	73	16	5	96	7	..	5	8
53	Lesotho	77	77	91	14	..	80	2	..	3	11
54	Indonesia	77	79	71	16	5	66	5	9	4	12
55	Guatemala	85	58	51	19	22	53	21	1	9	14
56	Viet Nam	84	..	87	21	17	71	46	..	..	..
57	Algeria	85	53	40	17	17	78	20	..	25	18
58	Botswana	80	76	75	20	10	84	7	2	10	17
59	El Salvador	86	77	75	22	35	70	16	1	8	12
60	Tunisia	89	59	45	22	8	80	21	1	11	19
61	Iran, Islamic Rep.	89	54	42	21	..	85	17	9	..	..
62	Syrian Arab Rep.	89	64	47	29	9	89	36	5	15	..
63	Dominican Rep.	90	83	84	23	33	95	26	4	7	..
64	Saudi Arabia	87	..	34	19	..	62	66	..	58	58
65	Philippines	86	91	92	25	26	94	7	..	5	13
66	China	94	73	61	43	56	84	46	5	3	15

	Index: North = 100 (see footnote)									
	Life expectancy 1987	Adult literacy rate 1985	Female literacy rate 1985	Under-five mortality rate 1988	Maternal mortality rate 1980-87	Combined primary and secondary enrolment ratio 1986-88	Doctors (per 1,000 people) 1984	Scientists and technicians (per 1,000 people) 1970-85	GNP per capita 1987	Real GDP per capita 1985
67 Libyan Arab Jamahiriya	83	70	54	16	31	..	66	12	51	..
68 South Africa	83	..	..	19	30	..	..	..	18	35
69 Lebanon	91	83	75	36	..	92	..	..	..	..
70 Mongolia	86	..	95	31	24	98	..	1	..	..
71 Nicaragua	87	94	..	19	52	78	31	..	8	15
72 Turkey	87	78	67	20	12	80	33	24	11	27
73 Jordan	90	79	68	32	..	102	40	5	14	22
74 Peru	84	90	85	15	28	101	44	..	14	22
75 Ecuador	89	88	87	21	13	91	55	9	10	19
76 Iraq	87	95	95	20	49	81	26	5	28	..
77 United Arab Emirates	96	..	..	58	..	85	45	..	147	85
78 Thailand	89	97	96	38	..	64	7	1	8	18
79 Paraguay	90	94	92	30	6	70	31	..	9	18
80 Brazil	88	82	83	22	20	88	42	26	19	30
81 Mauritius	94	88	84	64	24	80	24	18	14	18
82 Korea, Dem. Rep.	94	..	..	56	60	75	..	..	..	..
83 Sri Lanka	95	93	90	43	41	87	8	3	4	14
84 Albania	97	..	..	54	..	95	..	..	..	..
High human development	100	99	96	100 +	100 +	96	90	89	57	79
85 Malaysia	94	79	72	58	42	81	24	5	17	27
86 Colombia	88	..	96	27	22	85	38	..	12	25
87 Jamaica	100	..	..	84	22	84	22	6	9	18
88 Kuwait	99	74	68	84	100 +	89	72	54	136	97
89 Venezuela	94	93	92	42	42	86	65	89	30	30
90 Romania	95	..	..	66	16	95	80	..	24	..
91 Mexico	94	96	96	27	30	87	37	..	17	32
92 Cuba	100	102	104	100 +	72	98	..	37	..	..
93 Panama	97	95	96	54	43	85	47	8	21	28
94 Trinidad and Tobago	95	102	103	80	45	96	48	8	39	26
95 Portugal	99	90	88	100 +	100 +	91	112	..	26	39
96 Singapore	99	91	86	100 +	100 +	95	35	19	74	90
97 Korea, Rep.	94	..	99	56	94	99	39	..	25	34
98 Poland	96	..	..	100 +	100 +	98	93	120	19	..
99 Argentina	95	102	104	50	36	99	124	..	22	33
100 Yugoslavia	97	97	95	66	100 +	87	83	137	23	..
101 Hungary	95	..	..	97	94	90	148	179	21	..
102 Uruguay	96	101	..	60	64	95	90	..	20	35
103 Costa Rica	101	99	101	84	68	77	48	55	15	26
104 Bulgaria	98	..	..	92	100 +	98	..	80	39	..
105 USSR	94	..	..	58	51	103	..	91	42	..
106 Czechoslovakia	96	..	..	100 +	100 +	80	..	93	54	..
107 Chile	97	104	105	71	52	96	37	19	12	34
108 Hong Kong	103	93	88	100 +	100 +	90	43	152	75	97
109 Greece	102	98	97	100 +	100 +	100	131	118	37	..
110 German Dem. Rep.	99	..	..	100 +	100 +	89	..	72	67	..
111 Israel	102	101	101	100 +	100 +	102	131	58	63	64
112 USA	102	102	..	100 +	100 +	103	97	39	172	123
113 Austria	100	..	..	100 +	100 +	88	117	191	111	87
114 Ireland	100	..	..	100 +	100 +	79	67	174	57	60
115 Spain	103	101	100	100 +	100 +	103	143	93	56	63
116 Belgium	101	..	..	100 +	100 +	102	139	..	107	92
117 Italy	102	103	104	100 +	100 +	95	199	59	96	75
118 New Zealand	101	..	..	100 +	100 +	97	79	35	72	74
119 Germany, Fed. Rep.	101	..	..	100 +	100 +	99	121	99	134	103
120 Finland	101	..	..	100 +	100 +	103	104	268	134	90
121 United Kingdom	102	..	..	100 +	100 +	96	..	..	97	86
122 Denmark	102	..	..	100 +	100 +	103	115	45	139	106
123 France	102	..	..	100 +	100 +	103	143	59	119	98
124 Australia	103	..	..	100 +	100 +	103	104	112	103	83
125 Norway	104	..	..	100 +	100 +	98	102	73	160	112
126 Canada	104	..	..	100 +	100 +	103	90	183	141	115
127 Netherlands	104	..	..	100 +	100 +	103	102	156	110	89
128 Switzerland	104	..	..	100 +	100 +	103	65	144	198	108
129 Sweden	104	..	..	100 +	100 +	98	117	187	145	97
130 Japan	106	..	..	100 +	100 +	..	69	226	146	92
All developing countries	80	66	56	16	10	71	16	4	6	12
Least developed countries	66	35	24	9	4	42	3	1	2	4
Sub-Saharan Africa	69	44	34	10	6	45	3	2	3	5
Industrial countries	100	100	100	100	100	100	100	100	100	100
World	..	..	..	..	..	..	..	..	..	..

Note: All figures are expressed in relation to the North average, which is indexed to equal 100. The smaller the figure the bigger the gap, the closer the figure is to 100 the smaller the gap, and a figure above 100 indicates that the country has overtaken the North average. All aggregates in this table are medians. Summary data for regional and income groups are given in tables 23 and 24.

7 South-North real expenditure

		Index of real per capita expenditure: North = 100 (see footnote)							
		All food 1985	Meat 1985	Dairy, oils, fats 1985	Cereals, bread 1985	Health services 1985	Pharma-ceuticals 1985	Education services 1985	Books 1985
Low human development		19	16	10	63	3	3	21	3
Excluding India		..	..	..	..	..	..	..	..
1	Niger	..	..	..	..	..	..	..	..
2	Mali	15	23	2	52	0	2	9	1
3	Burkina Faso	..	..	..	..	..	..	..	..
4	Sierra Leone	..	..	..	..	..	..	..	..
5	Chad	..	..	..	..	..	..	..	..
6	Guinea	..	..	..	..	..	..	..	..
7	Somalia	..	..	..	..	..	..	..	..
8	Mauritania	..	..	..	..	..	..	..	..
9	Afghanistan	..	..	..	..	..	..	..	..
10	Benin	..	..	..	..	..	..	..	..
11	Burundi	..	..	..	..	..	..	..	..
12	Bhutan	..	..	..	..	..	..	..	..
13	Mozambique	..	..	..	..	..	..	..	..
14	Malawi	19	10	3	100	1	3	12	0
15	Sudan	..	..	..	..	..	..	..	..
16	Central African Rep.	..	..	..	..	..	..	..	..
17	Nepal	..	..	..	..	..	..	..	..
18	Senegal	30	16	22	80	1	3	21	3
19	Ethiopia	7	5	4	26	3	1	10	7
20	Zaire	..	..	..	..	..	..	..	..
21	Rwanda	..	..	..	..	..	..	..	..
22	Angola	..	..	..	..	..	..	..	..
23	Bangladesh	..	..	..	..	..	..	..	..
24	Nigeria	25	16	15	41	4	3	17	6
25	Yemen Arab Rep.	..	..	..	..	..	..	..	..
26	Liberia	..	..	..	..	..	..	..	..
27	Togo	..	..	..	..	..	..	..	..
28	Uganda	..	..	..	..	..	..	..	..
29	Haiti	..	..	..	..	..	..	..	..
30	Ghana	..	..	..	..	..	..	..	..
31	Yemen, PDR	..	..	..	..	..	..	..	..
32	Côte d'Ivoire	27	21	16	62	1	2	21	6
33	Congo	..	..	..	..	..	..	..	..
34	Namibia	..	..	..	..	..	..	..	..
35	Tanzania, United Rep.	13	6	3	47	2	2	21	0
36	Pakistan	42	21	49	128	23	2	26	3
37	India	19	2	16	63	3	7	21	2
38	Madagascar	25	25	5	93	0	1	22	2
39	Papua New Guinea	..	..	..	..	..	..	..	..
40	Kampuchea, Dem.	..	..	..	..	..	..	..	..
41	Cameroon	13	11	9	39	6	14	23	4
42	Kenya	17	5	11	69	4	2	39	3
43	Zambia	17	15	8	41	3	7	15	0
44	Morocco	41	27	27	119	3	8	36	3
Medium human development		44	31	25	103	16	26	48	7
Excluding China		..	..	..	..	..	..	..	..
45	Egypt	..	..	..	..	..	..	..	..
46	Lao PDR	..	..	..	..	..	..	..	..
47	Gabon	..	..	..	..	..	..	..	..
48	Oman	..	..	..	..	..	..	..	..
49	Bolivia	47	70	25	41	18	21	19	1
50	Myanmar	..	..	..	..	..	..	..	..
51	Honduras	27	24	28	32	19	22	17	3
52	Zimbabwe	14	12	10	42	4	4	20	10
53	Lesotho	..	..	..	..	..	..	..	..
54	Indonesia	31	6	5	114	3	1	14	2
55	Guatemala	47	49	39	103	61	39	67	29
56	Viet Nam	..	..	..	..	..	..	..	..
57	Algeria	..	..	..	..	..	..	..	..
58	Botswana	29	27	13	117	10	8	48	1
59	El Salvador	26	13	23	62	16	39	32	5
60	Tunisia	59	25	57	170	8	26	58	10
61	Iran, Islamic Rep.	..	..	..	..	..	..	..	..
62	Syrian Arab Rep.	..	..	..	..	..	..	..	..
63	Dominican Rep.	59	49	44	110	29	43	57	16
64	Saudi Arabia	..	..	..	..	..	..	..	..
65	Philippines	44	31	10	175	32	7	48	2
66	China	..	..	..	..	..	..	..	..

		Index of real per capita expenditure: North = 100 (see footnote)							
		All food 1985	Meat 1985	Dairy, oils, fats 1985	Cereals, bread 1985	Health services 1985	Pharma-ceuticals 1985	Education services 1985	Books 1985
67	Libyan Arab Jamahiriya	..	..	..	..	..	..	..	..
68	South Africa	..	..	..	..	..	..	..	..
69	Lebanon	..	..	..	..	..	..	..	..
70	Mongolia	..	..	..	..	..	..	..	..
71	Nicaragua	..	..	..	..	..	..	..	..
72	Turkey	..	..	..	..	..	..	..	..
73	Jordan	..	..	..	..	..	..	..	..
74	Peru	53	48	28	96	12	30	56	7
75	Ecuador	73	31	18	62	17	41	61	9
76	Iraq	..	..	..	..	..	..	..	..
77	United Arab Emirates	..	..	..	..	..	..	..	..
78	Thailand	..	..	..	..	..	..	..	..
79	Paraguay	43	48	40	49	4	24	50	18
80	Brazil	96	102	62	231	16	97	36	13
81	Mauritius	..	..	..	..	..	..	..	..
82	Korea, Dem. Rep.	..	..	..	..	..	..	..	..
83	Sri Lanka	33	3	7	144	8	35	27	7
84	Albania	..	..	..	..	..	..	..	..
High human development		95	94	72	104	66	88	94	62
85	Malaysia	..	..	..	..	..	..	..	..
86	Colombia	52	82	22	57	26	50	53	11
87	Jamaica	..	..	..	..	..	..	..	..
88	Kuwait	..	..	..	..	..	..	..	..
89	Venezuela	69	18	64	135	28	81	107	1
90	Romania	..	..	..	..	..	..	..	..
91	Mexico	..	..	..	..	..	..	..	..
92	Cuba	..	..	..	..	..	..	..	..
93	Panama	68	41	27	61	26	26	93	12
94	Trinidad and Tobago	..	..	..	..	..	..	..	..
95	Portugal	109	102	166	106	49	69	51	39
96	Singapore	..	..	..	..	..	..	..	..
97	Korea, Rep.	51	17	10	193	1	48	56	6
98	Poland	67	81	73	84	52	79	70	35
99	Argentina	95	129	34	64	8	49	49	7
100	Yugoslavia	54	49	57	93	63	85	48	65
101	Hungary	71	75	68	108	91	114	68	32
102	Uruguay	102	124	60	164	25	92	100	14
103	Costa Rica	58	25	57	116	24	104	95	5
104	Bulgaria	..	..	..	..	..	..	..	..
105	USSR	..	..	..	..	..	..	..	..
106	Czechoslovakia	..	..	..	..	..	..	..	..
107	Chile	63	47	38	114	22	38	102	2
108	Hong Kong	92	113	42	126	55	85	54	53
109	Greece	114	123	104	80	39	89	54	36
110	German Dem. Rep.	..	..	..	..	..	..	..	..
111	Israel	96	66	71	151	95	53	148	43
112	USA	111	115	86	89	114	119	133	120
113	Austria	97	91	103	90	135	117	91	58
114	Ireland	78	106	65	83	69	100	105	122
115	Spain	142	163	155	124	59	174	47	85
116	Belgium	123	163	104	120	103	185	148	100
117	Italy	135	154	115	145	59	146	106	95
118	New Zealand	..	..	..	..	..	..	..	..
119	Germany, Fed. Rep.	111	119	96	160	100	246	70	125
120	Finland	91	0	0	0	137	88	85	80
121	United Kingdom	91	98	77	101	96	92	126	147
122	Denmark	100	93	106	99	102	208	169	163
123	France	125	152	119	121	132	268	105	132
124	Australia	..	..	..	..	..	..	..	..
125	Norway	97	80	111	75	183	62	87	100
126	Canada	126	161	132	87	113	129	100	107
127	Netherlands	111	94	97	114	139	83	114	114
128	Switzerland	..	..	..	..	..	..	..	..
129	Sweden	..	..	..	..	..	..	..	..
130	Japan	72	24	23	98	136	74	144	144
All developing countries		41	24	22	87	8	21	36	5
Least developed countries		..	..	..	..	..	..	..	..
Sub-Saharan Africa		17	15	9	52	3	3	21	3
Industrial countries		100	100	100	100	100	100	100	100
World		..	..	..	..	..	..	..	..

Note: All figures are expressed in relation to the North average, which is indexed to equal 100. The smaller the figure the bigger the gap, the closer the figure is to 100 the smaller the gap, and a figure above 100 indicates that the country has overtaken the North average. All aggregates in this table are medians. Summary data for regional and income groups are given in tables 23 and 24.

8 Rural-urban gaps

	Rural population (%) 1988	Rural population with access to services (%) 1985-87			Urban population with access to services (%) 1985-87			Rural-urban disparity (100 = rural-urban parity)		
		Health	Water	Sanitation	Health	Water	Sanitation	Health	Water	Sanitation
Low human development	73	37	38	6	81	74	39	46	53	15
Excluding India	74	..	23	11	..	73	56	..	38	20
1 Niger	82	30	49	..	99	35	..	30	140	..
2 Mali	81	..	10	3	..	46	90	..	22	3
3 Burkina Faso	91	48	69	6	51	43	44	94	160	14
4 Sierra Leone	69	..	7	10	..	68	60	..	10	17
5 Chad	69	..	..	..	..	..	..	..	..	..
6 Guinea	76	..	12	..	..	41	..	..	29	..
7 Somalia	65	15	22	5	50	58	44	30	38	11
8 Mauritania	61	..	..	..	..	73	8	..	..	..
9 Afghanistan	79	17	17	..	80	38	5	21	45	..
10 Benin	60	..	34	20	..	80	58	..	43	34
11 Burundi	93	..	21	56	..	98	84	..	21	67
12 Bhutan	95	..	19	..	..	..	..	..	..	..
13 Mozambique	76	30	9	12	100	38	53	30	24	23
14 Malawi	86	..	50	..	..	97	..	..	52	..
15 Sudan	78	40	10	..	90	60	..	44	17	..
16 Central African Rep.	55	..	..	..	..	13	..	..	..	..
17 Nepal	91	..	25	1	..	70	17	..	36	6
18 Senegal	62	..	38	..	..	79	87	..	48	..
19 Ethiopia	88	..	9	..	..	69	96	..	13	..
20 Zaire	62	17	21	9	40	52	..	43	40	..
21 Rwanda	93	25	48	55	60	79	77	42	61	71
22 Angola	73	..	15	16	..	87	29	..	17	55
23 Bangladesh	87	..	..	3	..	24	24	..	..	13
24 Nigeria	66	30	20	5	75	100	..	40	20	..
25 Yemen Arab Rep.	77	24	25	..	75	100	83	32	25	..
26 Liberia	58	30	23	..	50	100	..	60	23	..
27 Togo	76	..	41	9	..	99	31	..	41	29
28 Uganda	90	57	18	30	90	37	32	63	49	94
29 Haiti	71	70	30	13	80	59	42	88	51	31
30 Ghana	68	45	39	16	92	93	61	49	42	26
31 Yemen, PDR	58	..	32	..	..	85	..	..	38	..
32 Côte d'Ivoire	55	11	10	..	61	30	..	18	33	..
33 Congo	59	70	7	..	97	42	..	72	17	..
34 Namibia	45	..	..	..	..	..	..	..	..	..
35 Tanzania, United Rep.	70	72	42	58	99	90	93	73	47	62
36 Pakistan	69	35	27	6	99	83	51	35	33	12
37 India	73	..	50	2	..	76	31	..	66	6
38 Madagascar	76	..	17	..	..	81	..	..	21	..
39 Papua New Guinea	85	..	15	35	..	95	99	..	16	35
40 Kampuchea, Dem.	89	50	2	..	80	10	..	63	20	..
41 Cameroon	53	39	24	1	44	43	100	89	56	1
42 Kenya	78	..	21	..	..	61	..	..	34	..
43 Zambia	47	100	41	34	..	76	76	..	54	45
44 Morocco	53	50	25	16	100	100	..	50	25	..
Medium human development	68	57	45	33	95	80	72	59	56	46
Excluding China	55	..	..	..	..	77	..	..	..	..
45 Egypt	52	..	56	..	..	92	..	..	61	..
46 Lao PDR	82	..	20	..	..	28	..	..	71	..
47 Gabon	56	..	..	..	..	..	..	..	..	..
48 Oman	90	90	49	25	100	90	88	90	54	28
49 Bolivia	50	36	13	10	90	75	33	40	17	30
50 Myanmar	76	11	24	21	100	36	33	11	67	64
51 Honduras	58	65	45	34	85	56	24	76	80	142
52 Zimbabwe	74	62	32	15	100	..	..	62	..	..
53 Lesotho	81	..	30	14	..	65	22	..	46	64
54 Indonesia	73	..	36	38	..	43	33	..	84	115
55 Guatemala	59	25	14	12	47	72	41	53	19	29
56 Viet Nam	79	75	39	..	100	70	..	75	56	..
57 Algeria	56	80	55	40	100	85	80	80	65	50
58 Botswana	78	85	46	28	100	84	93	85	55	30
59 El Salvador	56	40	40	43	80	68	82	50	59	52
60 Tunisia	46	80	31	16	100	100	84	80	31	19
61 Iran, Islamic Rep.	46	60	55	..	95	95	..	63	58	..
62 Syrian Arab Rep.	49	60	54	..	92	98	..	65	55	..
63 Dominican Rep.	41	..	33	10	..	85	41	..	39	24
64 Saudi Arabia	24	88	88	..	100	100	100	88	88	..
65 Philippines	59	..	54	56	..	49	83	..	110	67
66 China	79	..	..	..	..	85	..	..	..	..

	Rural population (%) 1988	Rural population with access to services (%) 1985-87			Urban population with access to services (%) 1985-87			Rural-urban disparity (100 = rural-urban parity)		
		Health	Water	Sanitation	Health	Water	Sanitation	Health	Water	Sanitation
67 Libyan Arab Jamahiriya	32	..	90	..	..	100	..	..	90	..
68 South Africa	42	..	..	..	..	..	..	..	..	..
69 Lebanon	18	..	85	..	..	95	..	..	89	..
70 Mongolia	49	..	..	..	..	..	..	..	..	..
71 Nicaragua	41	60	11	16	100	76	35	60	14	46
72 Turkey	53	..	63	..	..	95	..	..	66	..
73 Jordan	33	95	88	..	98	100	92	97	88	..
74 Peru	31	..	17	12	..	73	67	..	23	18
75 Ecuador	45	30	31	29	90	81	98	33	38	30
76 Iraq	27	70	54	11	97	100	100	72	54	11
77 United Arab Emirates	22	..	..	..	..	..	..	..	..	..
78 Thailand	78	..	66	46	..	56	78	..	118	59
79 Paraguay	54	38	8	83	90	53	89	42	15	93
80 Brazil	25	..	56	1	..	85	86	..	66	1
81 Mauritius	58	100	100	86	100	100	100	100	100	86
82 Korea, Dem. Rep.	34	..	..	..	..	..	..	..	..	..
83 Sri Lanka	79	..	29	39	..	82	65	..	35	60
84 Albania	65	..	..	..	..	..	..	..	..	..
High human development	28	..	..	..	..	..	..	..	..	..
85 Malaysia	59	..	76	60	..	96	100	..	79	60
86 Colombia	31	..	76	13	..	100	96	..	76	14
87 Jamaica	49	..	93	90	..	99	92	..	94	98
88 Kuwait	5	..	..	..	..	97	100	..	..	..
89 Venezuela	11	..	65	5	..	93	57	..	70	9
90 Romania	50	..	..	..	..	..	..	..	..	..
91 Mexico	29	..	47	13	..	89	77	..	53	17
92 Cuba	26	..	..	..	..	..	..	..	..	..
93 Panama	46	64	64	61	95	100	99	67	64	62
94 Trinidad and Tobago	33	..	95	95	..	100	100	..	95	95
95 Portugal	68	..	..	..	..	..	..	..	..	..
96 Singapore	..	..	..	..	100	100	99	..	..	..
97 Korea, Rep.	31	86	48	100	97	90	100	89	53	100
98 Poland	38	..	..	..	..	..	..	..	..	..
99 Argentina	14	21	17	35	80	63	75	26	27	47
100 Yugoslavia	51	..	..	..	..	..	..	..	..	..
101 Hungary	41	..	..	..	..	..	..	..	..	..
102 Uruguay	15	..	27	59	..	95	59	..	28	100
103 Costa Rica	48	63	83	89	100	100	99	63	83	90
104 Bulgaria	31	..	..	..	..	..	..	..	..	..
105 USSR	33	..	..	..	..	..	..	..	..	..
106 Czechoslovakia	33	..	..	..	..	..	..	..	..	..
107 Chile	15	..	71	4	..	98	100	..	72	4
108 Hong Kong	7	..	..	..	..	..	..	..	..	..
109 Greece	38	..	..	..	..	..	..	..	..	..
110 German Dem. Rep.	22	..	..	..	..	..	..	..	..	..
111 Israel	9	..	..	..	..	..	..	..	..	..
112 USA	26	..	..	..	..	..	..	..	..	..
113 Austria	43	..	..	..	..	..	..	..	..	..
114 Ireland	42	..	..	..	..	..	..	..	..	..
115 Spain	23	..	..	..	..	..	..	..	..	..
116 Belgium	3	..	..	..	..	..	..	..	..	..
117 Italy	32	..	..	..	..	..	..	..	..	..
118 New Zealand	16	..	..	..	..	..	..	..	..	..
119 Germany, Fed. Rep.	14	..	..	..	..	..	..	..	..	..
120 Finland	34	..	..	..	..	..	..	..	..	..
121 United Kingdom	8	..	..	..	..	..	..	..	..	..
122 Denmark	14	..	..	..	..	..	..	..	..	..
123 France	26	..	..	..	..	..	..	..	..	..
124 Australia	15	..	..	..	..	..	..	..	..	..
125 Norway	26	..	..	..	..	..	..	..	..	..
126 Canada	24	..	..	..	..	..	..	..	..	..
127 Netherlands	12	..	..	..	..	..	..	..	..	..
128 Switzerland	41	..	..	..	..	..	..	..	..	..
129 Sweden	16	..	..	..	..	..	..	..	..	..
130 Japan	23	..	..	..	..	..	..	..	..	..
All developing countries	67	45	41	14	88	79	61	51	52	23
Least developed countries	82	36	23	15	..	52	49	..	59	31
Sub-Saharan Africa	70	38	24	17	72	74	..	53	32	..
Industrial countries	28	..	..	..	..	..	..	..	..	..
World	58	..	..	..	..	..	..	..	..	..

Note: Summary data for regional and income groups are given in tables 23 and 24.

9 Female-male gaps

	Females as a percentage of males (see footnote)							
	Life expectancy	Literacy rate		Primary enrolment		Secondary enrolment	Labour force	Parliament
	1988	1970	1985	1960	1986-88	1986-88	1988	1988
Low human development	102.1	40	52	46	71	53	39	7.6
Excluding India	104.2	36	54	40	70	52	43	7.0
1 Niger	107.5	33	47	43	54	..	89	..
2 Mali	107.5	36	48	43	59	44	20	3.7
3 Burkina Faso	107.1	23	29	42	59	50	87	..
4 Sierra Leone	108.0	44	55	..	71	48	49	..
5 Chad	107.2	10	28	14	40	20	27	..
6 Guinea	107.8	33	43	36	44	31	67	..
7 Somalia	107.3	20	33	100	..	..	64	3.9
8 Mauritania	107.3	..	..	23	69	39	28	..
9 Afghanistan	102.4	15	21	13	52	50	9	..
10 Benin	107.2	35	43	39	51	39	91	4.0
11 Burundi	107.0	34	60	33	74	50	91	..
12 Bhutan	97.1	..	..	..	65	29	48	..
13 Mozambique	107.2	48	40	60	78	57	92	16.0
14 Malawi	102.9	43	60	..	81	60	72	8.9
15 Sudan	104.9	21	42	40	69	74	27	0.6
16 Central African Rep.	107.2	23	55	23	62	35	86	..
17 Nepal	97.6	13	31	5	45	31	51	5.7
18 Senegal	107.3	28	51	..	69	53	66	11.7
19 Ethiopia	107.9	..	..	27	61	67	61	..
20 Zaire	106.7	36	57	36	81	44	56	3.5
21 Rwanda	107.0	49	54	..	96	71	92	12.9
22 Angola	107.4	44	67	..	..	..	64	15.4
23 Bangladesh	98.6	33	51	39	84	46	8	9.1
24 Nigeria	107.0	40	57	59	..	..	54	..
25 Yemen Arab Rep.	105.9	11	17	..	28	13	15	0.0
26 Liberia	105.6	30	49	40	61	..	44	6.2
27 Togo	106.9	26	53	38	63	33	58	5.2
28 Uganda	106.7	58	64	..	83	56	71	..
29 Haiti	106.2	65	88	84	87	89	73	..
30 Ghana	106.8	42	67	48	81	65	67	..
31 Yemen, PDR	106.0	29	42	25	36	42	13	9.0
32 Côte d'Ivoire	106.7	38	58	35	..	46	52	5.7
33 Congo	107.1	38	77	51	..	..	64	9.8
34 Namibia	104.5	..	..	..	..	..	31	..
35 Tanzania, United Rep.	106.6	38	95	55	99	60	93	..
36 Pakistan	100.0	37	48	28	55	42	14	8.8
37 India	100.3	43	51	50	72	54	34	7.9
38 Madagascar	105.7	77	84	78	95	83	66	1.5
39 Papua New Guinea	103.0	62	64	12	85	56	64	0.0
40 Kampuchea, Dem.	106.2	..	76	..	..	44	64	..
41 Cameroon	108.1	40	66	49	84	63	51	14.1
42 Kenya	107.0	43	70	47	95	70	67	1.7
43 Zambia	104.0	56	80	67	90	..	40	2.9
44 Morocco	105.7	29	49	40	66	70	26	0.0
Medium human development	105.2	69	74	77	90	81	62	15.1
Excluding China	106.3	..	82	..	92	94	45	7.4
45 Egypt	104.5	40	51	65	79	73	11	3.9
46 Lao PDR	106.3	76	83	47	83	70	81	..
47 Gabon	106.6	51	76	..	..	..	61	13.3
48 Oman	104.9	..	26	..	89	63	9	..
49 Bolivia	109.0	68	77	64	88	88	34	3.1
50 Myanmar	106.0	67	..	85	..	..	60	..
51 Honduras	106.7	91	95	99	104	..	22	5.2
52 Zimbabwe	106.4	75	83	..	97	86	54	9.0
53 Lesotho	117.4	151	135	162	125	144	78	..
54 Indonesia	105.1	64	78	67	96	..	45	11.4
55 Guatemala	107.9	73	75	78	85	..	19	7.0
56 Viet Nam	107.4	..	91	..	88	93	88	17.7
57 Algeria	105.2	28	59	67	83	75	10	2.3
58 Botswana	110.8	119	95	137	105	106	55	5.1
59 El Salvador	114.3	87	92	..	105	111	33	3.3
60 Tunisia	102.4	39	60	49	85	74	31	5.6
61 Iran, Islamic Rep.	102.0	43	63	48	86	68	21	1.5
62 Syrian Arab Rep.	106.0	33	57	44	90	70	20	9.2
63 Dominican Rep.	106.6	94	99	99	104	..	17	5.0
64 Saudi Arabia	105.6	13	44	..	83	67	8	..
65 Philippines	106.0	96	99	95	102	100	46	7.8
66 China	104.3	..	68	..	89	74	76	21.2

	Females as a percentage of males (see footnote)							
	Life expectancy	Literacy rate		Primary enrolment		Secondary enrolment	Labour force	Parliament
	1988	1970	1985	1960	1986-88	1986-88	1988	1988
67 Libyan Arab Jamahiriya	105.7	22	62	26	..	..	10	..
68 South Africa	110.3	..	..	90	..	..	55	2.8
69 Lebanon	106.0	73	80	94	90	98	37	0.0
70 Mongolia	106.7	85	95	99	103	109	83	24.8
71 Nicaragua	104.3	98	..	102	111	200	33	13.5
72 Turkey	105.4	49	72	64	93	60	51	1.3
73 Jordan	105.7	45	72	63	101	98	11	0.0
74 Peru	106.4	74	86	75	96	90	32	5.5
75 Ecuador	106.6	91	94	91	98	104	24	1.4
76 Iraq	103.2	36	97	38	87	63	27	13.2
77 United Arab Emirates	106.3	29	..	..	102	120	7	0.0
78 Thailand	106.5	84	94	90	..	..	82	3.4
79 Paraguay	106.6	88	93	86	95	100	26	1.6
80 Brazil	108.5	91	96	96	..	128	38	5.3
81 Mauritius	107.7	77	87	90	102	94	35	5.7
82 Korea, Dem. Rep.	109.7	..	..	..	..	..	85	21.1
83 Sri Lanka	106.1	81	91	90	97	110	37	4.7
84 Albania	107.3	..	..	84	99	89	70	28.8
High human development	110.1	97	97	99	99	103	66	14.8
85 Malaysia	106.1	68	81	77	100	100	54	5.1
86 Colombia	107.3	96	100	100	103	102	28	..
87 Jamaica	107.5	101	..	101	102	108	84	11.6
88 Kuwait	106.0	65	83	78	97	92	17	..
89 Venezuela	109.2	90	97	100	100	123	38	3.9
90 Romania	108.1	95	..	94	..	101	86	34.4
91 Mexico	110.0	88	96	94	97	98	37	10.7
92 Cuba	105.0	101	100	100	93	108	46	33.9
93 Panama	105.6	100	99	96	95	113	37	5.9
94 Trinidad and Tobago	107.5	94	98	98	101	106	43	16.6
95 Portugal	109.7	83	90	98	94	119	58	7.6
96 Singapore	107.9	60	85	93	96	104	48	3.8
97 Korea, Rep.	109.4	86	92	90	100	95	51	2.9
98 Poland	111.8	99	..	97	100	105	84	20.2
99 Argentina	110.0	98	99	101	100	113	39	4.7
100 Yugoslavia	108.4	83	89	96	99	96	63	17.7
101 Hungary	111.2	100	..	97	100	101	81	20.9
102 Uruguay	109.9	100	..	100	98	..	45	0.0
103 Costa Rica	106.4	99	99	98	97	108	28	12.2
104 Bulgaria	108.3	95	..	98	98	101	86	21.0
105 USSR	114.1	99	..	100	..	..	93	34.5
106 Czechoslovakia	110.9	..	..	100	101	181	87	29.5
107 Chile	110.3	98	99	96	98	106	39	..
108 Hong Kong	107.8	71	85	85	99	107	51	..
109 Greece	105.9	82	91	97	100	90	36	4.3
110 German Dem. Rep.	108.2	..	..	102	98	96	84	32.2
111 Israel	104.9	89	96	98	103	110	51	8.3
112 USA	109.9	100	..	..	99	101	71	5.3
113 Austria	110.1	..	..	98	99	104	67	10.9
114 Ireland	107.6	..	..	105	100	111	41	8.4
115 Spain	108.2	94	95	109	100	110	32	6.4
116 Belgium	109.2	100	..	97	101	101	51	8.5
117 Italy	109.2	98	98	97	..	..	47	12.8
118 New Zealand	108.4	..	..	96	99	102	53	14.4
119 Germany, Fed. Rep.	109.0	..	..	..	100	96	60	15.4
120 Finland	110.8	..	..	95	99	116	88	31.5
121 United Kingdom	108.0	..	..	100	101	104	63	6.3
122 Denmark	108.0	..	..	100	101	101	80	29.0
123 France	111.2	99	..	99	99	108	66	6.4
124 Australia	109.1	..	..	100	99	103	61	6.1
125 Norway	109.1	..	..	100	100	105	69	34.4
126 Canada	109.4	..	..	97	98	100	66	9.9
127 Netherlands	109.0	..	..	99	102	98	45	..
128 Switzerland	108.9	..	..	100	..	..	58	14.0
129 Sweden	107.9	..	..	101	..	102	80	30.9
130 Japan	107.6	100	..	99	100	102	61	1.4
All developing countries	104.2	54	67	61	83	71	51	11.8
Least developed countries	104.0	37	51	44	72	53	46	7.6
Sub-Saharan Africa	107.2	42	61	52	77	58	62	6.3
Industrial countries	110.4	98	..	99	100	104	71	16.0
World	105.7	69	..	74	86	77	56	12.9

Note: All figures are expressed in relation to the male average, which is indexed to equal 100. The smaller the figure, the bigger the gap; the closer the figure is to 100, the smaller the gap; and a figure above 100 indicates that the female average is higher than the male. Summary data for regional and income groups are given in tables 23 and 24.

10 Child survival and development

	Births attended by health personnel (%) 1983-88	Low birth-weight babies (%) 1982-88	Infant mortality rate (per 1,000 live births) 1988	Mothers breast-feeding at one year (%) 1980-87	One-year olds immunised (%) 1987	Child malnutrition 1980-88: % of under-fives underweight	Child malnutrition 1980-88: % of 12-23 months wasted	Child malnutrition 1980-88: % of 24-59 months stunted	Under-five mortality rate (per 1,000 live births) 1988
Low human development	30	25	107	72	55	42	16	44	170
Excluding India	27	20	114	..	49	43	..	..	186
1 Niger	47	15	134	15	24	49	23	38	228
2 Mali	27	17	168	82	31	31	16	34	292
3 Burkina Faso	30	..	137	97	46	..	..	..	233
4 Sierra Leone	25	17	153	83	40	23	26	46	266
5 Chad	24	11	131	..	21	..	..	..	223
6 Guinea	25	..	146	40	23	..	..	..	248
7 Somalia	2	..	131	..	28	..	..	..	221
8 Mauritania	20	11	126	67	45	31	..	..	220
9 Afghanistan	8	20	171	..	27	..	..	..	300
10 Benin	45	8	109	76	35	..	14	..	185
11 Burundi	21	9	111	90	54	38	10	60	188
12 Bhutan	7	..	127	..	67	..	..	..	197
13 Mozambique	28	20	172	..	42	..	..	..	298
14 Malawi	45	20	149	96	83	24	8	61	262
15 Sudan	20	..	107	72	58	..	13	..	181
16 Central African Rep.	66	15	131	..	33	..	..	..	223
17 Nepal	6	..	127	82	71	70	..	..	197
18 Senegal	50	11	80	82	57	22	8	28	136
19 Ethiopia	14	..	153	95	18	38	19	43	259
20 Zaire	..	13	83	86	46	28	11	40	138
21 Rwanda	22	17	121	74	82	28	23	45	206
22 Angola	15	17	172	..	28	..	..	..	292
23 Bangladesh	5	28	118	82	18	60	17	59	188
24 Nigeria	40	20	104	60	62	..	21	..	174
25 Yemen Arab Rep.	12	..	115	29	32	61	17	69	190
26 Liberia	87	..	86	70	43	20	7	38	147
27 Togo	15	20	93	90	73	25	..	..	153
28 Uganda	45	..	102	20	52	7	3	32	169
29 Haiti	40	17	116	88	50	37	17	51	171
30 Ghana	40	17	89	72	42	27	28	31	146
31 Yemen, PDR	10	13	118	55	37	26	8	36	197
32 Côte d'Ivoire	20	14	95	78	37	12	4	10	142
33 Congo	..	12	72	95	76	24	13	33	114
34 Namibia	..	..	105	82	..	..	..	..	176
35 Tanzania, United Rep.	60	14	105	70	86	48	17	..	176
36 Pakistan	24	25	108	70	65	52	17	42	166
37 India	33	30	98	..	63	41	..	..	149
38 Madagascar	62	10	119	85	44	33	18	41	184
39 Papua New Guinea	34	25	57	..	55	35	..	58	81
40 Kampuchea, Dem.	47	..	127	..	47	..	..	..	199
41 Cameroon	..	13	93	77	52	17	2	43	153
42 Kenya	28	15	71	67	74	..	10	42	113
43 Zambia	..	14	79	93	84	..	12	41	127
44 Morocco	29	..	80	76	65	16	6	34	119
Medium human development	61	9	51	60	81	32	10	43	72
Excluding China	..	12	66	..	70	..	..	..	94
45 Egypt	47	5	83	81	85	17	3	34	125
46 Lao PDR	..	39	109	93	20	37	20	44	159
47 Gabon	92	..	102	..	76	..	..	..	169
48 Oman	60	6	40	20	90	..	..	..	64
49 Bolivia	36	12	109	..	38	15	1	46	172
50 Myanmar	57	16	69	..	24	38	17	75	95
51 Honduras	50	20	68	24	76	21	2	34	107
52 Zimbabwe	69	15	71	84	81	12	1	29	113
53 Lesotho	40	11	99	..	81	..	7	23	136
54 Indonesia	31	14	84	83	71	51	17	..	119
55 Guatemala	34	10	58	74	49	34	3	68	99
56 Viet Nam	99	18	63	..	58	52	12	60	88
57 Algeria	15	9	73	..	71	..	..	..	107
58 Botswana	77	8	66	73	90	15	19	51	92
59 El Salvador	35	15	58	55	63	..	..	54	84
60 Tunisia	68	7	58	71	88	21	3	45	83
61 Iran, Islamic Rep.	82	5	61	..	81	43	23	55	90
62 Syrian Arab Rep.	37	..	47	41	63	..	..	..	64
63 Dominican Rep.	57	16	64	..	70	13	3	26	81
64 Saudi Arabia	74	6	70	52	88	..	..	..	98
65 Philippines	57	18	44	53	82	33	7	42	73
66 China	..	5	31	..	96	..	..	..	43

	Births attended by health personnel (%) 1983-88	Low birth-weight babies (%) 1982-88	Infant mortality rate (per 1,000 live births) 1988	Mothers breast-feeding at one year (%) 1980-87	One-year olds immunised (%) 1987	Child malnutrition 1980-88			Under-five mortality rate (per 1,000 live births) 1988
						% of under-fives underweight	% of 12-23 months wasted	% of 24-59 months stunted	
67 Libyan Arab Jamahiriya	76	..	80	..	62	..	..	..	119
68 South Africa	..	12	71	..	..	..	..	..	95
69 Lebanon	..	..	39	15	88	16	..	..	51
70 Mongolia	99	10	44	..	67	..	..	..	59
71 Nicaragua	41	15	61	71	70	11	..	22	95
72 Turkey	78	8	74	51	71	..	..	..	93
73 Jordan	83	5	43	50	71	..	..	..	57
74 Peru	44	9	87	37	66	13	3	43	123
75 Ecuador	27	10	62	48	62	17	4	39	87
76 Iraq	56	9	68	..	84	..	..	..	94
77 United Arab Emirates	96	7	25	..	73	..	..	..	32
78 Thailand	40	12	38	68	79	26	10	28	49
79 Paraguay	22	7	42	49	65	..	..	..	62
80 Brazil	95	8	62	34	68	13	2	31	85
81 Mauritius	85	9	22	40	84	24	16	22	29
82 Korea, Dem. Rep.	65	..	24	..	59	..	..	..	33
83 Sri Lanka	87	28	32	81	79	38	19	34	43
84 Albania	..	7	28	..	95	..	..	..	34
High human development	94	8	20	..	76	..	..	..	27
85 Malaysia	82	10	24	..	74	..	12	33	32
86 Colombia	51	15	46	36	85	12	1	27	68
87 Jamaica	89	8	18	43	82	9	5	9	22
88 Kuwait	99	7	19	12	51	6	2	14	22
89 Venezuela	82	9	36	30	62	10	3	7	44
90 Romania	100	6	22	..	93	..	..	..	28
91 Mexico	94	15	46	36	74	..	..	..	68
92 Cuba	..	8	15	..	93	..	1	..	18
93 Panama	89	8	23	53	79	16	7	24	34
94 Trinidad and Tobago	98	..	20	14	78	7	5	4	23
95 Portugal	87	5	14	7	78	..	..	..	17
96 Singapore	100	6	9	..	95	14	..	..	12
97 Korea, Rep.	70	6	24	27	89	..	..	..	33
98 Poland	100	8	16	..	97	..	..	..	18
99 Argentina	..	..	32	14	68	..	..	..	37
100 Yugoslavia	86	7	25	..	90	..	..	..	28
101 Hungary	99	10	17	..	99	..	..	..	19
102 Uruguay	97	8	27	..	84	..	..	16	31
103 Costa Rica	93	10	18	22	89	6	3	8	22
104 Bulgaria	100	6	15	..	99	..	..	..	20
105 USSR	98	6	25	..	83	..	..	..	32
106 Czechoslovakia	100	6	12	..	99	..	..	..	15
107 Chile	98	7	19	17	96	3	1	10	26
108 Hong Kong	92	5	8	..	91	..	..	..	10
109 Greece	97	6	13	..	86	..	..	..	18
110 German Dem. Rep.	99	6	8	..	97	..	..	..	12
111 Israel	100	7	11	..	90	..	..	..	14
112 USA	99	7	10	..	48	..	..	..	13
113 Austria	..	6	8	..	83	..	..	..	10
114 Ireland	..	4	7	..	70	..	..	..	9
115 Spain	96	1	9	..	78	..	..	..	12
116 Belgium	100	5	10	..	83	..	..	..	13
117 Italy	100	7	10	..	59	..	..	..	11
118 New Zealand	99	5	10	..	59	..	..	..	12
119 Germany, Fed. Rep.	100	6	8	..	68	..	..	..	10
120 Finland	100	4	6	..	89	..	..	..	7
121 United Kingdom	100	7	9	..	82	..	..	..	11
122 Denmark	100	6	8	..	90	..	..	..	11
123 France	99	5	8	..	83	..	..	..	10
124 Australia	99	6	9	10	68	..	..	..	10
125 Norway	100	4	8	..	84	..	..	..	10
126 Canada	99	6	7	..	85	..	..	..	8
127 Netherlands	100	..	8	..	96	..	..	..	8
128 Switzerland	99	5	7	..	88	..	..	..	8
129 Sweden	100	4	6	..	76	..	..	..	7
130 Japan	100	5	5	..	84	..	..	..	8
All developing countries	42	17	79	64	68	38	13	42	121
Least developed countries	23	21	124	75	37	46	16	54	205
Sub-Saharan Africa	36	16	110	71	52	29	16	38	183
Industrial countries	99	6	15	..	75	..	..	..	18
World	51	15	71	..	69	..	..	..	108

Note: Summary data for regional and income groups are given in tables 23 and 24.

11 Health profile

	Population with access to health services (%) 1985-87	Maternal mortality rate (per 100,000 live births) 1980-87	Thousands of people: Per doctor 1984	Thousands of people: Per nurse 1984	Nurses per doctor 1984	Health expenditure (% of GNP) 1960	Health expenditure (% of GNP) 1986	Private expenditure on health per capita (PPP$) 1985
Low human development	47	460	8.8	2.6	4.3	0.6	0.8	4.2
Excluding India	..	570	16.8	3.9	7.9	0.7	0.7	7.5
1 Niger	41	420	38.8	0.5	86.2	0.2	0.8	..
2 Mali	15	..	25.4	1.4	18.8	1.0	0.7	0.1
3 Burkina Faso	49	810	57.2	1.7	34.0	0.6	0.9	..
4 Sierra Leone	..	450	13.6	1.1	12.5	..	0.7	..
5 Chad	30	860	38.4	3.4	11.3	0.5	0.6	..
6 Guinea	32	..	57.4	6.4	9.0	1.0	1.0	..
7 Somalia	27	1,100	16.1	1.5	10.5	0.6	0.2	..
8 Mauritania	30	..	12.1	1.2	10.1	0.5	1.9	..
9 Afghanistan	29	690	..	..	..	..	..	..
10 Benin	18	..	15.9	1.8	9.1	1.5	0.8	..
11 Burundi	61	..	21.1	3.0	6.9	0.8	0.7	..
12 Bhutan	65	1,710	23.3	3.0	7.8	..	..	..
13 Mozambique	39	..	38.0	5.8	6.6	..	1.8	..
14 Malawi	80	..	11.6	3.1	3.7	0.2	2.4	0.7
15 Sudan	51	660	10.1	1.3	8.1	1.0	0.2	..
16 Central African Rep.	45	600	23.1	2.2	10.6	1.3	1.2	..
17 Nepal	..	830	32.7	4.7	7.0	0.2	0.9	..
18 Senegal	40	600	13.5	2.1	6.4	1.5	1.1	0.6
19 Ethiopia	46	..	77.4	5.3	14.6	0.7	1.0	2.2
20 Zaire	26	..	..	..	..	..	0.8	..
21 Rwanda	27	210	34.7	3.7	9.5	0.5	0.6	..
22 Angola	30	..	..	..	..	..	1.0	..
23 Bangladesh	45	600	6.7	9.0	0.7	..	0.6	..
24 Nigeria	40	800	8.0	1.0	7.8	0.3	0.4	3.2
25 Yemen Arab Rep.	35	..	6.3	2.7	2.3	..	1.1	..
26 Liberia	39	..	9.2	1.4	6.8	0.8	1.5	..
27 Togo	61	..	8.7	1.2	7.0	1.3	1.6	..
28 Uganda	61	300	21.9	2.1	10.6	0.7	0.2	..
29 Haiti	70	230	7.2	2.3	3.1	1.0	0.9	..
30 Ghana	60	1,000	14.9	0.6	23.3	1.1	0.3	..
31 Yemen, PDR	30	..	4.3	1.1	4.1	..	2.0	..
32 Côte d'Ivoire	30	..	..	..	..	1.5	1.1	1.2
33 Congo	83	1,000	8.1	0.6	14.3	1.6	2.0	..
34 Namibia	..	..	..	..	..	..	..	..
35 Tanzania, United Rep.	76	340	..	..	..	0.5	1.2	1.4
36 Pakistan	55	500	2.9	4.9	0.6	0.3	0.2	20.0
37 India	..	340	2.5	1.7	1.5	0.5	0.9	2.6
38 Madagascar	56	240	10.0	..	..	1.4	1.8	0.4
39 Papua New Guinea	..	900	6.2	0.9	6.9	..	3.3	..
40 Kampuchea, Dem.	53	..	..	..	..	..	..	..
41 Cameroon	41	300	..	..	..	1.0	0.7	5.3
42 Kenya	..	170	10.1	1.0	10.6	1.5	1.1	3.3
43 Zambia	75	150	7.1	0.7	9.6	1.0	2.1	2.7
44 Morocco	70	300	15.6	0.9	17.0	1.0	1.0	2.6
Medium human development	75	130	2.3	1.5	1.9	0.8	1.5	11.8
Excluding China	..	210	3.9	1.1	3.6	0.7	1.5	..
45 Egypt	..	320	0.8	0.8	1.0	0.6	1.0	..
46 Lao PDR	67	..	1.4	0.5	2.6	0.5	..	..
47 Gabon	90	..	2.8	0.3	10.3	0.5	2.0	..
48 Oman	91	..	1.7	0.8	2.2	..	3.3	..
49 Bolivia	63	480	1.5	2.5	0.6	0.4	0.4	16.0
50 Myanmar	33	140	3.7	0.9	4.2	0.7	1.0	..
51 Honduras	73	50	1.5	0.7	2.3	1.0	2.6	17.0
52 Zimbabwe	71	480	6.7	1.0	6.7	1.2	2.3	3.9
53 Lesotho	80	..	18.6	..	..	1.0	1.6	..
54 Indonesia	80	450	9.5	1.3	7.5	0.3	0.7	2.9
55 Guatemala	34	110	2.2	0.9	2.6	0.6	0.7	53.0
56 Viet Nam	80	140	1.0	0.6	1.6	..	..	..
57 Algeria	88	140	2.3	0.3	7.1	1.2	2.2	..
58 Botswana	89	250	6.9	0.7	9.9	1.5	2.9	9.1
59 El Salvador	56	70	2.8	0.9	3.0	0.9	1.0	14.0
60 Tunisia	90	310	2.2	0.4	5.8	1.6	2.7	6.7
61 Iran, Islamic Rep.	78	..	2.7	1.1	2.6	0.8	1.8	..
62 Syrian Arab Rep.	76	280	1.3	1.4	0.9	0.4	0.8	..
63 Dominican Rep.	80	74	1.8	1.2	1.5	1.3	1.4	25.0
64 Saudi Arabia	97	..	0.7	0.3	2.2	0.6	4.0	..
65 Philippines	..	93	6.7	2.7	2.4	0.4	0.7	28.0
66 China	..	44	1.0	1.7	0.6	1.3	1.4	..

		Population with access to health services (%) 1985-87	Maternal mortality rate (per 100,000 live births) 1980-87	Thousands of people		Nurses per doctor 1984	Health expenditure (% of GNP)		Private expenditure on health per capita (PPP$) 1985
				Per doctor 1984	Per nurse 1984		1960	1986	
67	Libyan Arab Jamahiriya	..	80	0.7	0.4	2.0	1.3	3.0	..
68	South Africa	..	83	..	..	..	0.5	0.6	..
69	Lebanon	..	..	..	..	..	..	..	..
70	Mongolia	..	100	..	..	..	..	1.3	..
71	Nicaragua	83	47	1.5	0.5	2.8	0.4	6.6	..
72	Turkey	..	210	1.4	1.0	1.3	0.8	0.5	..
73	Jordan	97	..	1.1	1.3	0.9	0.6	1.9	..
74	Peru	75	88	1.0	..	..	1.1	1.0	11.0
75	Ecuador	62	190	0.8	0.6	1.3	0.4	1.1	15.0
76	Iraq	93	50	1.7	1.7	1.0	1.0	0.8	..
77	United Arab Emirates	90	..	1.0	0.4	2.6	..	1.0	..
78	Thailand	70	..	6.3	0.7	8.9	0.4	1.3	..
79	Paraguay	61	380	1.5	1.0	1.5	0.5	0.4	3.9
80	Brazil	..	120	1.1	1.2	0.9	0.6	1.3	14.0
81	Mauritius	100	100	1.9	0.6	3.3	1.5	1.8	..
82	Korea, Dem. Rep.	..	41	..	..	..	0.5	1.0	..
83	Sri Lanka	93	60	5.5	1.3	4.3	2.0	1.3	6.6
84	Albania	..	..	..	..	..	..	2.3	..
High human development		..	37	0.6	0.3	3.6	2.2	4.6	82.5
85	Malaysia	..	59	1.9	1.0	1.9	1.1	1.8	..
86	Colombia	60	110	1.2	0.6	1.9	0.4	0.8	23.0
87	Jamaica	90	110	2.1	0.5	4.2	2.0	2.8	..
88	Kuwait	100	6	0.6	0.2	3.2	..	2.9	..
89	Venezuela	..	59	0.7	..	..	2.6	2.7	24.0
90	Romania	..	150	0.6	0.3	2.0	2.0	1.9	..
91	Mexico	..	82	1.2	0.9	1.4	1.9	1.7	..
92	Cuba	..	34	..	..	..	3.0	3.2	..
93	Panama	80	57	1.0	0.4	2.5	3.0	5.4	22.0
94	Trinidad and Tobago	99	54	1.0	0.3	3.7	1.7	3.0	..
95	Portugal	..	12	0.4	..	..	0.8	5.7	43.0
96	Singapore	100	5	1.3	..	..	1.0	1.2	..
97	Korea, Rep.	93	26	1.2	0.6	2.0	0.2	0.3	..
98	Poland	..	11	0.5	0.2	2.6	3.5	4.0	45.0
99	Argentina	71	69	0.4	1.0	0.4	1.3	1.6	6.9
100	Yugoslavia	..	22	0.6	0.3	2.1	3.2	4.3	55.0
101	Hungary	..	26	0.3	0.2	1.8	2.6	3.2	79.0
102	Uruguay	82	38	0.5	..	..	2.6	2.7	22.0
103	Costa Rica	80	36	1.0	0.5	2.1	3.0	5.4	21.0
104	Bulgaria	..	13	..	..	..	2.0	3.2	..
105	USSR	..	48	..	..	..	3.5	3.2	..
106	Czechoslovakia	..	10	..	..	..	2.9	4.2	..
107	Chile	97	47	1.2	0.4	3.3	2.0	2.1	20.0
108	Hong Kong	..	5	1.1	0.2	4.5	..	..	48.0
109	Greece	..	9	0.4	0.5	0.8	1.7	3.5	34.0
110	German Dem. Rep.	..	16	..	..	..	2.2	2.6	..
111	Israel	..	5	0.4	0.1	3.2	1.0	2.1	83.0
112	USA	..	8	0.5	0.1	6.7	1.3	4.5	100.0
113	Austria	..	7	0.4	0.2	2.2	2.9	5.3	118.0
114	Ireland	..	12	0.7	0.1	4.9	3.0	7.8	61.0
115	Spain	..	11	0.3	0.3	1.2	2.0	4.3	52.0
116	Belgium	..	9	0.3	0.1	3.0	2.7	5.6	90.0
117	Italy	..	10	0.2	..	..	3.2	4.5	51.0
118	New Zealand	..	6	0.6	0.1	7.3	2.7	5.6	..
119	Germany, Fed. Rep.	..	11	0.4	0.2	1.7	3.1	6.3	88.0
120	Finland	..	6	0.4	0.1	7.3	2.3	6.0	120.0
121	United Kingdom	..	9	..	0.1	..	3.4	5.3	84.0
122	Denmark	..	4	0.4	0.1	6.7	3.2	5.3	89.0
123	France	..	14	0.3	0.1	2.9	2.5	6.6	115.0
124	Australia	..	8	0.4	0.1	4.0	2.5	5.1	..
125	Norway	..	2	0.5	0.1	7.5	2.8	5.5	160.0
126	Canada	..	3	0.5	0.1	4.3	2.4	6.6	99.0
127	Netherlands	..	5	0.5	0.2	2.6	1.3	7.5	122.0
128	Switzerland	..	5	0.7	0.1	5.4	2.5	6.8	..
129	Sweden	..	5	0.4	0.1	3.9	3.4	8.0	..
130	Japan	..	16	0.7	0.2	3.7	1.3	4.9	119.0
All developing countries		61	290	4.8	1.9	2.8	0.9	1.4	7.1
Least developed countries		46	520	23.3	4.7	8.4	0.7	0.9	..
Sub-Saharan Africa		45	530	24.6	2.2	12.4	0.7	0.8	2.6
Industrial countries		..	24	0.5	0.1	4.2	2.2	4.7	90.9
World		..	250	4.0	1.6	3.1	2.0	4.2	33.0

Note: Summary data for regional and income groups are given in tables 23 and 24.

		Food production per capita index (1979-81=100) 1985-87	Daily calorie supply per capita 1986	Food import dependency ratio 1979-81	Food import dependency ratio 1984-86	Food aid in cereals (1,000 metric tons) 1981-82	Food aid in cereals (1,000 metric tons) 1987-88
Low human development		105	2,190	6.9	7.7	4,630 T	6,390 T
	Excluding India	97	2,130	12.2	12.9	..	..
1	Niger	98	2,430	7.3	12.6	71.4	18.9
2	Mali	108	2,070	7.1	16.2	66.4	25.9
3	Burkina Faso	121	2,140	6.6	9.7	80.9	38.2
4	Sierra Leone	89	1,860	18.6	15.3	28.9	57.9
5	Chad	105	1,720	2.3	7.5	28.6	14.5
6	Guinea	90	1,780	11.9	14.4	38.6	26.2
7	Somalia	98	2,140	33.6	26.4	185.9	152.4
8	Mauritania	91	2,320	53.0	63.3	86.4	51.4
9	Afghanistan	..	..	..	..	81.8	104.0
10	Benin	112	2,180	6.7	6.8	8.3	11.3
11	Burundi	102	2,340	2.1	2.4	9.0	4.3
12	Bhutan	118	..	..	..	1.1	2.1
13	Mozambique	84	1,600	..	..	148.5	339.6
14	Malawi	83	2,310	3.3	1.8	2.0	108.8
15	Sudan	100	2,210	9.6	12.1	194.1	604.2
16	Central African Rep.	83	1,950	2.8	4.9	2.0	6.4
17	Nepal	102	2,050	1.7	3.1	23.2	20.5
18	Senegal	103	2,350	34.8	31.8	82.7	109.0
19	Ethiopia	94	1,750	..	..	189.7	825.3
20	Zaire	94	2,160	5.6	4.5	97.5	176.5
21	Rwanda	75	1,830	1.6	4.0	12.6	7.8
22	Angola	85	..	..	..	74.5	102.9
23	Bangladesh	88	1,930	9.5	16.9	1,005.5	1,394.6
24	Nigeria	96	2,150	15.0	9.1	1.4	..
25	Yemen Arab Rep.	105	2,320	39.7	60.2	12.9	159.7
26	Liberia	93	2,380	25.2	23.4	42.4	55.7
27	Togo	87	2,210	10.9	14.5	4.6	16.3
28	Uganda	82	2,340	2.2	1.0	48.5	29.3
29	Haiti	89	1,900	19.7	20.9	89.9	153.5
30	Ghana	108	1,760	8.8	5.3	43.1	109.6
31	Yemen, PDR	86	2,300	62.4	71.4	25.3	31.2
32	Côte d'Ivoire	94	2,560	18.3	18.1	0.9	0.9
33	Congo	94	2,620	19.6	29.1	0.4	0.7
34	Namibia	..	..	..	..	..	..
35	Tanzania, United Rep.	86	2,190	4.7	4.9	307.5	71.5
36	Pakistan	104	2,320	10.2	11.9	347.4	657.4
37	India	112	2,240	2.7	3.4	337.6	223.0
38	Madagascar	91	2,440	8.6	7.0	87.1	75.8
39	Papua New Guinea	97	2,210	25.3	19.7	..	..
40	Kampuchea, Dem.	141	..	..	..	49.9	6.2
41	Cameroon	95	2,030	7.6	10.0	10.5	2.3
42	Kenya	90	2,060	10.2	12.8	127.2	118.8
43	Zambia	92	..	22.1	14.1	100.0	140.4
44	Morocco	122	2,920	38.2	30.8	477.5	339.6
Medium human development		118	2,650	8.8	7.7	3,190 T	5,790 T
	Excluding China	108	2,680	15.0	14.4	..	..
45	Egypt	120	3,340	40.6	46.9	1,956.6	1,737.8
46	Lao PDR	110	2,390	..	..	1.3	21.4
47	Gabon	82	2,520	21.9	31.9	..	..
48	Oman	..	..	..	..	..	..
49	Bolivia	95	2,140	20.9	14.9	44.2	290.4
50	Myanmar	124	2,610	0.9	0.6	5.0	..
51	Honduras	87	2,070	14.8	12.0	33.8	145.6
52	Zimbabwe	97	2,130	4.2	7.6	..	13.9
53	Lesotho	86	2,300	46.9	59.2	34.2	55.3
54	Indonesia	119	2,580	8.4	3.9	106.6	319.1
55	Guatemala	96	2,310	11.2	10.3	10.6	320.1
56	Viet Nam	114	2,300	9.6	3.6	43.8	65.0
57	Algeria	95	2,720	64.5	69.1	5.4	3.8
58	Botswana	75	2,200	65.8	79.5	6.5	52.8
59	El Salvador	88	2,160	..	..	129.1	177.3
60	Tunisia	90	2,990	42.8	41.6	96.0	393.1
61	Iran, Islamic Rep.	89	3,310	..	..	..	..
62	Syrian Arab Rep.	95	3,260	22.6	35.2	9.3	26.3
63	Dominican Rep.	91	2,480	31.2	30.0	57.1	278.1
64	Saudi Arabia	236	3,000	89.7	78.4	..	..
65	Philippines	88	2,370	7.6	8.5	54.5	470.9
66	China	127	2,630	4.0	2.4	78.4	347.1

	Food production per capita index (1979-81=100) 1985-87	Daily calorie supply per capita 1986	Food import dependency ratio		Food aid in cereals (1,000 metric tons)	
			1979-81	1984-86	1981-82	1987-88
67 Libyan Arab Jamahiriya	90	3,600	69.7	78.2	..	..
68 South Africa	88	2,920	4.7	15.8	..	..
69 Lebanon	..	..	..	..	9.0	58.3
70 Mongolia	95	..	27.4	2.6	..	..
71 Nicaragua	65	2,500	..	..	103.6	86.6
72 Turkey	98	3,230	1.9	5.2	0.3	0.6
73 Jordan	117	2,990	91.0	97.3	72.5	28.8
74 Peru	103	2,250	30.1	27.6	76.2	355.4
75 Ecuador	92	2,060	15.1	16.1	8.3	32.6
76 Iraq	102	2,930	..	..	..	..
77 United Arab Emirates	..	3,730	114.7	98.0	..	..
78 Thailand	104	2,330	3.5	2.5	2.3	97.2
79 Paraguay	117	2,850	3.5	2.3	1.1	1.9
80 Brazil	111	2,660	9.5	8.2	3.0	20.7
81 Mauritius	99	2,750	76.0	63.6	42.5	31.5
82 Korea, Dem. Rep.	109	..	7.1	3.5	..	..
83 Sri Lanka	88	2,400	26.3	27.5	202.5	360.5
84 Albania	95	..	4.5	4.4	..	..
High human development	100	3,280	24.4	22.9	..	..
85 Malaysia	138	2,730	42.0	48.5	..	..
86 Colombia	101	2,540	12.2	14.0	2.6	89.9
87 Jamaica	98	2,590	57.3	63.5	82.8	208.2
88 Kuwait	..	3,020	110.9	103.9	..	..
89 Venezuela	92	2,490	49.3	44.0	..	..
90 Romania	110	3,370	10.6	4.3	..	..
91 Mexico	93	3,130	18.0	14.8	..	32.1
92 Cuba	105	..	49.5	47.2	..	..
93 Panama	93	2,450	25.6	25.7	3.1	0.1
94 Trinidad and Tobago	71	3,080	82.6	92.0	..	..
95 Portugal	102	3,150	58.9	54.9	..	..
96 Singapore	83	2,840	157.4	179.5	..	..
97 Korea, Rep.	100	2,910	44.2	46.9	429.2	..
98 Poland	102	3,340	14.7	6.4	417.4	0.7
99 Argentina	96	3,210	1.1	0.5	..	..
100 Yugoslavia	98	3,560	8.5	4.9	..	..
101 Hungary	106	3,570	6.5	2.2	..	..
102 Uruguay	102	2,650	14.8	11.3	..	..
103 Costa Rica	86	2,800	25.4	18.0	45.2	235.1
104 Bulgaria	102	..	11.9	15.4	..	..
105 USSR	109	..	15.8	16.5	..	..
106 Czechoslovakia	122	..	19.6	9.8	..	..
107 Chile	108	2,580	35.2	18.0	18.3	13.9
108 Hong Kong	54	2,860	105.7	120.3	..	..
109 Greece	100	3,690	17.4	15.9	..	..
110 German Dem. Rep.	117	..	12.5	12.4	..	..
111 Israel	100	3,060	71.1	71.9	0.2	2.1
112 USA	85	3,650	5.1	4.3	..	..
113 Austria	104	3,430	12.9	12.1	..	..
114 Ireland	98	3,630	21.8	21.1	..	..
115 Spain	110	3,360	24.2	16.5	..	..
116 Belgium	116	..	128.9	119.8	..	..
117 Italy	98	3,520	29.5	27.8	..	..
118 New Zealand	111	3,460	25.4	24.1	..	..
119 Germany, Fed. Rep.	112	3,530	35.2	34.3	..	..
120 Finland	99	3,120	21.9	13.5	..	..
121 United Kingdom	105	3,260	37.3	31.7	..	..
122 Denmark	119	3,630	15.7	15.3	..	..
123 France	100	3,340	15.8	14.3	..	..
124 Australia	97	3,330	4.0	5.5	..	..
125 Norway	109	3,220	49.4	39.4	..	..
126 Canada	93	3,460	15.5	12.6	..	..
127 Netherlands	111	3,330	114.5	112.0	..	..
128 Switzerland	108	3,440	50.1	47.1	..	..
129 Sweden	92	3,060	19.4	19.8	..	..
130 Japan	98	2,860	55.5	54.5	..	..
All developing countries	111	2,480	9.6	9.2	8,410 T	12,770 T
Least developed countries	96	2,070	9.1	12.8	2,800 T	4,410 T
Sub-Saharan Africa	93	2,160	11.1	10.8	2,070 T	2,850 T
Industrial countries	101	3,390	23.1	21.4	..	..
World	109	2,650	12.9	12.2	..	..

Note: Summary data for regional and income groups are given in tables 23 and 24.

13 Education profile

	Gross primary enrolment ratio		Net primary enrolment ratio		Gross secondary enrolment ratio		Gross tertiary enrolment ratio		Radios (per 1,000 people)	Tele-visions (per 1,000 people)
	Male 1986-88	Female 1986-88	Male 1986-88	Female 1986-88	Male 1986-88	Female 1986-88	Male 1986-88	Female 1986-88	1986-87	1986-87
Low human development	92	67	55	38	37	20	9.3	3.8	92	8
Excluding India	70	51	..	..	23	12	4.8	1.7	108	10
1 Niger	37	20	..	..	..	..	1.0	0.2	62	3
2 Mali	29	17	23	14	9	4	1.4	0.2	37	0
3 Burkina Faso	41	24	34	20	8	4	1.0	0.3	24	5
4 Sierra Leone	68	48	..	..	23	11	..	..	216	8
5 Chad	73	29	52	23	10	2	0.8	0.1	237	1
6 Guinea	41	18	31	15	13	4	1.9	0.2	33	2
7 Somalia	..	..	19	10	..	..	5.0	1.1	38	0
8 Mauritania	61	42	..	..	23	9	6.0	0.9	139	1
9 Afghanistan	27	14	..	..	10	5	..	..	102	8
10 Benin	84	43	66	34	23	9	4.4	0.8	75	4
11 Burundi	68	50	46	37	6	3	1.2	0.4	56	0
12 Bhutan	31	20	..	..	7	2	0.3	0.1	15	..
13 Mozambique	76	59	49	41	7	4	0.3	0.1	38	1
14 Malawi	73	59	50	47	5	3	0.9	0.3	197	..
15 Sudan	59	41	..	..	23	17	2.5	1.5	229	52
16 Central African Rep.	82	51	59	39	17	6	..	..	60	2
17 Nepal	104	47	76	35	35	11	7.6	1.8	31	1
18 Senegal	71	49	59	41	19	10	4.6	1.2	103	32
19 Ethiopia	46	28	32	22	18	12	1.6	0.3	193	2
20 Zaire	84	68	..	..	32	14	..	..	98	1
21 Rwanda	69	66	65	63	7	5	0.6	0.1	54	..
22 Angola	..	..	..	..	..	..	..	..	49	5
23 Bangladesh	76	64	67	44	24	11	7.8	1.9	40	3
24 Nigeria	..	..	..	..	..	..	..	..	163	6
25 Yemen Arab Rep.	141	40	..	..	46	6	..	..	34	8
26 Liberia	82	50	..	..	..	..	4.0	1.2	224	18
27 Togo	124	78	87	59	36	12	..	..	178	5
28 Uganda	76	63	43	38	16	9	1.2	0.4	96	6
29 Haiti	83	72	45	42	19	17	1.6	0.8	41	4
30 Ghana	78	63	..	..	49	32	2.4	0.6	292	13
31 Yemen, PDR	96	35	..	..	26	11	..	..	154	21
32 Côte d'Ivoire	..	..	..	..	26	12	3.9	1.0	131	54
33 Congo	..	..	..	..	..	..	12.3	2.2	120	3
34 Namibia	..	..	..	..	..	..	..	..	123	11
35 Tanzania, United Rep.	67	66	50	51	5	3	0.4	0.1	16	1
36 Pakistan	51	28	..	..	26	11	6.8	3.1	86	14
37 India	113	81	..	..	50	27	12.3	5.2	77	7
38 Madagascar	97	92	89	..	23	19	4.5	3.0	193	6
39 Papua New Guinea	75	64	..	..	16	9	2.8	1.1	64	2
40 Kampuchea, Dem.	..	..	..	..	45	20	..	..	106	8
41 Cameroon	119	100	..	..	32	20	..	..	125	12
42 Kenya	98	93	..	..	27	19	1.9	0.7	90	6
43 Zambia	102	92	..	..	..	..	2.3	0.6	73	15
44 Morocco	85	56	68	46	43	30	13.2	6.5	206	56
Medium human development	127	114	98	91	51	41	6.9	4.4	197	48
Excluding China	111	103	95	90	53	46	16.4	11.3	213	85
45 Egypt	100	79	..	..	79	58	25.5	14.0	310	83
46 Lao PDR	102	85	..	..	23	16	2.1	1.2	123	2
47 Gabon	..	..	..	..	..	..	7.6	2.8	119	23
48 Oman	103	92	83	77	46	29	3.0	1.2	649	739
49 Bolivia	97	85	88	78	40	35	..	..	527	77
50 Myanmar	..	..	..	..	..	..	..	..	79	1
51 Honduras	104	108	..	..	..	..	11.7	7.2	376	67
52 Zimbabwe	130	126	100	100	49	42	4.7	2.8	85	22
53 Lesotho	102	127	..	..	18	26	1.4	2.2	68	1
54 Indonesia	120	115	99	97	..	..	8.8	4.2	145	40
55 Guatemala	82	70	..	..	..	..	..	..	65	37
56 Viet Nam	107	94	..	..	44	41	..	..	99	34
57 Algeria	105	87	97	81	61	46	..	..	227	70
58 Botswana	111	117	85	93	31	33	..	..	130	7
59 El Salvador	77	81	61	62	27	30	20.8	14.8	401	82
60 Tunisia	126	107	100	89	46	34	7.0	4.4	171	68
61 Iran, Islamic Rep.	122	105	98	89	57	39	7.1	2.7	236	53
62 Syrian Arab Rep.	115	104	100	94	69	48	22.2	13.1	231	58
63 Dominican Rep.	99	103	80	78	..	..	..	..	164	79
64 Saudi Arabia	78	65	64	48	52	35	15.0	11.4	272	268
65 Philippines	105	107	94	94	66	66	35.5	40.3	135	36
66 China	140	124	99	91	50	37	2.3	1.1	184	17

	Gross primary enrolment ratio		Net primary enrolment ratio		Gross secondary enrolment ratio		Gross tertiary enrolment ratio		Radios (per 1,000 people)	Tele-visions (per 1,000 people)
	Male 1986-88	Female 1986-88	Male 1986-88	Female 1986-88	Male 1986-88	Female 1986-88	Male 1986-88	Female 1986-88	1986-87	1986-87
67 Libyan Arab Jamahiriya	..	..	..	..	..	..	15.2	5.5	221	63
68 South Africa	..	..	..	..	..	..	..	..	319	97
69 Lebanon	105	95	..	..	57	56	35.5	20.2	772	302
70 Mongolia	100	103	..	..	88	96	17.4	26.0	128	31
71 Nicaragua	94	104	74	79	29	58	7.6	9.2	237	60
72 Turkey	121	113	..	..	57	34	13.4	7.1	160	172
73 Jordan	98	99	88	88	80	78	..	..	237	69
74 Peru	125	120	..	..	68	61	..	..	241	84
75 Ecuador	118	116	..	..	55	57	34.8	23.7	292	81
76 Iraq	105	91	91	82	60	38	14.8	10.1	199	64
77 United Arab Emirates	98	100	88	89	55	66	4.9	16.3	319	106
78 Thailand	..	..	..	..	..	..	..	..	174	103
79 Paraguay	104	99	86	84	30	30	..	..	165	24
80 Brazil	..	..	..	..	32	41	..	..	368	191
81 Mauritius	105	107	93	95	53	50	1.6	0.9	263	188
82 Korea, Dem. Rep.	..	..	..	..	..	..	49.0	22.9	110	12
83 Sri Lanka	105	102	100	100	63	69	4.7	3.2	187	31
84 Albania	100	99	..	..	80	71	7.5	8.1	167	83
High human development	106	106	96	96	81	83	30.2	30.9	910	413
85 Malaysia	102	102	..	..	59	59	7.6	6.1	436	140
86 Colombia	112	115	72	74	55	56	14.3	13.4	167	108
87 Jamaica	104	106	..	..	62	67	4.4	3.5	400	108
88 Kuwait	95	92	81	77	86	79	13.2	21.1	327	261
89 Venezuela	107	107	..	..	48	59	27.5	25.5	395	142
90 Romania	..	..	..	..	79	80	12.2	10.4	288	166
91 Mexico	119	116	..	..	54	53	19.3	12.1	241	120
92 Cuba	107	100	95	94	85	92	19.4	26.0	334	193
93 Panama	109	104	90	89	56	63	21.7	30.2	220	163
94 Trinidad and Tobago	99	100	87	88	80	85	4.7	3.6	457	290
95 Portugal	131	123	..	..	47	56	12.1	14.1	212	159
96 Singapore	118	113	100	100	70	73	13.3	10.3	306	..
97 Korea, Rep. of	104	104	100	99	91	86	..	..	986	194
98 Poland	101	101	99	99	78	82	15.3	20.4	289	263
99 Argentina	110	110	..	..	69	78	35.9	41.5	659	217
100 Yugoslavia	95	94	..	..	82	79	19.1	18.0	344	175
101 Hungary	97	97	94	96	69	70	13.8	16.7	586	402
102 Uruguay	111	109	..	..	..	..	20.9	26.6	594	173
103 Costa Rica	100	97	85	85	40	43	..	..	258	79
104 Bulgaria	105	103	..	..	75	76	19.6	25.8	357	189
105 USSR	..	..	..	..	..	..	19.9	25.4	685	314
106 Czechoslovakia	95	96	..	..	27	49	18.2	14.0	577	281
107 Chile	103	101	..	..	72	76	19.6	16.0	335	163
108 Hong Kong	106	105	95	95	71	76	16.8	9.3	633	241
109 Greece	106	106	91	92	89	80	25.9	26.5	411	175
110 German Dem. Rep.	107	105	92	91	79	76	29.9	34.2	663	754
111 Israel	94	97	..	..	79	87	35.7	32.9	470	264
112 USA	101	100	97	97	98	99	55.5	63.7	2,119	811
113 Austria	102	101	..	..	78	81	31.3	27.4	561	480
114 Ireland	100	100	..	..	91	101	27.0	21.4	580	..
115 Spain	113	113	98	98	97	107	29.6	30.3	295	368
116 Belgium	99	100	82	83	99	100	33.8	31.6	465	320
117 Italy	..	..	97	98	..	..	25.4	23.2	786	..
118 New Zealand	107	106	100	100	84	86	37.1	35.6	923	369
119 Germany, Fed. Rep.	101	101	..	..	96	92	34.2	25.6	954	385
120 Finland	102	101	..	..	98	114	36.7	38.5	991	..
121 United Kingdom	105	106	97	97	82	85	23.3	20.1	1,145	434
122 Denmark	98	99	..	..	106	107	29.1	30.2	956	386
123 France	114	113	100	100	89	96	29.7	32.0	893	..
124 Australia	106	105	97	98	96	99	28.8	28.8	1,270	483
125 Norway	95	95	97	97	92	97	34.1	36.7	790	348
126 Canada	106	104	97	97	104	104	53.1	63.5	953	577
127 Netherlands	114	116	85	88	105	103	35.7	26.8	908	469
128 Switzerland	..	..	..	..	..	..	31.7	15.4	834	405
129 Sweden	..	..	..	..	90	92	28.8	33.7	875	39
130 Japan	102	102	100	100	95	97	35.0	21.4	863	587
All developing countries	110	92	87	79	45	33	8.9	5.1	172	40
Least developed countries	69	50	53	38	19	9	3.9	1.0	85	6
Sub-Saharan Africa	75	61	48	37	21	13	2.1	0.7	139	14
Industrial countries	103	103	97	97	90	93	32.0	33.3	1,013	477
World	109	93	89	82	51	40	15.1	12.7	369	134

Note: Summary data for regional and income groups are given in tables 23 and 24.

Education imbalances

	Dropout rate (% of grade 1 enrolment not completing primary school) 1985-87	Primary pupil to teacher ratio 1985-88	Combined primary and secondary enrolment ratio 1986-88	Secondary technical enrolment (as % of total secondary enrolment) 1986-88	Education expenditure (as % of GNP) 1960	Education expenditure (as % of GNP) 1986	Primary education expenditure (as % of total education expenditure) 1985-88	Educated unemployed (as % of total unemployed) 1982-87
Low human development	50	45	68	3.0	2.1	3.2	43.2	..
Excluding India	..	44	43	4.7	1.8	3.1	43.0	..
1 Niger	25	38	17	1.3	0.5	4.0	36.8	..
2 Mali	61	38	15	9.1	2.0	3.2	48.4	..
3 Burkina Faso	26	68	19	6.6	1.5	2.5	38.1	..
4 Sierra Leone	..	34	38	1.6	..	3.0	..	..
5 Chad	83	71	29	7.0	0.9	2.0	82.9	..
6 Guinea	30	40	19	6.7	1.5	3.0	30.8	..
7 Somalia	67	20	14	16.1	0.9	6.0	..	..
8 Mauritania	8	50	36	5.3	2.1	6.0	30.0	..
9 Afghanistan	37	37	17	9.1	..	..	43.2	..
10 Benin	64	33	42	6.2	2.5	3.5	46.2	..
11 Burundi	13	62	33	19.1	2.4	2.8	45.0	..
12 Bhutan	..	37	18	37.0	..	..	..	..
13 Mozambique	61	63	34	6.7	..	..	..	..
14 Malawi	67	63	48	2.2	2.1	3.7	47.0	..
15 Sudan	39	35	36	4.6	1.9	4.0	55.5	..
16 Central African Rep.	83	63	40	4.1	2.0	5.3	51.2	0.9
17 Nepal	73	35	56	6.4	0.4	2.8	35.7	..
18 Senegal	17	54	38	67.3	2.4	4.6	56.8	..
19 Ethiopia	59	49	28	..	0.8	4.2	52.8	1.5
20 Zaire	40	37	52	10.0	2.4	0.4	47.1	..
21 Rwanda	51	57	45	26.4	0.3	3.2	68.0	..
22 Angola	..	46	49	2.3	0.3	3.4	94.6	..
23 Bangladesh	80	59	42	0.6	0.6	2.2	39.1	..
24 Nigeria	37	44	64	2.5	1.5	1.4	17.2	..
25 Yemen Arab Rep.	85	54	63	1.6	..	5.1	58.8	..
26 Liberia	..	36	30	4.3	0.7	5.0	17.6	..
27 Togo	41	52	64	5.8	1.9	5.5	34.0	..
28 Uganda	24	30	47	2.9	3.2	1.1	20.1	..
29 Haiti	85	38	58	1.7	1.4	1.2	56.8	..
30 Ghana	..	24	56	2.2	3.8	3.5	29.3	..
31 Yemen, PDR	60	26	52	12.5	..	6.0	63.2	14.6
32 Côte d'Ivoire	32	36	47	7.1	4.6	5.0	46.8	..
33 Congo	25	64	..	10.5	2.5	5.0	30.0	..
34 Namibia	..	..	..	..	..	..	..	..
35 Tanzania, United Rep.	24	33	42	1.5	2.1	4.2	59.1	..
36 Pakistan	51	41	29	1.7	1.1	2.2	40.2	..
37 India	..	46	95	1.2	2.3	3.4	43.3	4.6
38 Madagascar	70	40	73	1.8	2.3	3.5	42.3	..
39 Papua New Guinea	33	31	42	14.3	2.5	5.6	..	..
40 Kampuchea, Dem.	50	..	..	..	..	..	..	..
41 Cameroon	33	50	70	26.3	1.7	2.8	77.7	..
42 Kenya	38	34	..	1.7	4.6	5.0	61.9	..
43 Zambia	9	47	69	2.1	1.6	4.4	44.2	..
44 Morocco	31	26	52	1.3	3.1	5.9	36.5	..
Medium human development	35	27	79	9.4	2.5	4.0	43.8	..
Excluding China	37	29	76	11.7	2.7	4.3	49.7	..
45 Egypt	36	30	81	22.9	4.1	4.8	67.3	..
46 Lao PDR	86	25	67	6.0	..	1.2	..	..
47 Gabon	41	47	..	20.0	2.1	4.8	..	..
48 Oman	11	26	72	5.5	..	6.6	52.2	..
49 Bolivia	..	27	76	..	1.5	2.4	71.9	..
50 Myanmar	73	45	60	1.4	2.2	2.1	..	..
51 Honduras	57	39	75	25.2	2.2	5.0	46.6	..
52 Zimbabwe	26	39	93	0.1	0.5	7.9	66.0	..
53 Lesotho	48	53	78	3.4	3.2	3.5	39.1	..
54 Indonesia	20	28	64	11.0	2.5	3.5	..	14.4
55 Guatemala	64	35	52	17.3	1.4	1.8	38.2	12.3
56 Viet Nam	50	34	69	1.6	..	..	..	..
57 Algeria	10	28	76	4.9	5.6	6.1	28.5	..
58 Botswana	11	32	82	6.1	2.7	9.1	36.8	1.9
59 El Salvador	69	45	68	67.5	2.3	2.3	60.3	..
60 Tunisia	23	31	78	17.3	3.3	5.0	45.0	..
61 Iran, Islamic Rep.	17	26	83	5.6	2.4	3.5	41.5	..
62 Syrian Arab Rep.	33	26	87	6.2	2.0	5.7	44.8	27.8
63 Dominican Rep.	65	41	92	4.6	2.1	1.6	44.4	..
64 Saudi Arabia	10	16	60	1.9	3.2	10.6	..	..
65 Philippines	25	32	91	..	2.3	1.7	61.8	5.9
66 China	32	24	82	7.0	1.8	2.7	28.5	..

	Dropout rate (% of grade 1 enrolment not completing primary school) 1985-87	Primary pupil to teacher ratio 1985-88	Combined primary and secondary enrolment ratio 1986-88	Secondary technical enrolment (as % of total secondary enrolment) 1986-88	Education expenditure (as % of GNP) 1960	1986	Primary education expenditure (as % of total education expenditure) 1985-88	Educated unemployed (as % of total unemployed) 1982-87
67 Libyan Arab Jamahiriya	18	17	..	8.3	2.8	10.1	..	..
68 South Africa	..	..	..	..	3.0	4.6	..	..
69 Lebanon	..	17	90	10.9	..	..	..	..
70 Mongolia	..	32	95	7.2	..	..	..	..
71 Nicaragua	80	32	76	22.3	1.5	6.1	36.7	..
72 Turkey	15	31	78	21.5	2.6	2.1	55.8	2.5
73 Jordan	4	30	99	9.3	3.0	5.1	89.9	..
74 Peru	49	35	98	4.3	2.3	1.6	31.1	12.5
75 Ecuador	50	31	89	14.8	1.9	3.6	37.1	..
76 Iraq	29	25	79	11.8	5.8	3.7	46.5	..
77 United Arab Emirates	18	18	83	0.9	..	2.2	..	..
78 Thailand	36	20	62	16.3	2.3	4.1	59.0	1.1
79 Paraguay	50	25	68	6.4	1.3	1.4	36.6	..
80 Brazil	78	24	86	15.9	1.9	3.4	52.3	..
81 Mauritius	4	22	78	1.4	3.0	3.3	44.2	1.2
82 Korea, Dem. Rep.	1	36	73	..	..	..	43.6	..
83 Sri Lanka	12	32	85	0.4	3.8	3.6	93.5	..
84 Albania	..	20	92	72.5	..	..	..	..
High human development	15	21	95	21.4	3.7	5.1	34.1	12.0
85 Malaysia	3	22	79	1.3	2.9	7.9	37.9	4.5
86 Colombia	43	29	83	20.8	1.7	2.8	39.9	..
87 Jamaica	..	34	82	3.6	2.3	5.6	34.5	3.8
88 Kuwait	9	18	87	0.3	..	4.6	48.7	..
89 Venezuela	27	26	84	5.1	3.7	6.6	24.5	6.7
90 Romania	..	21	92	..	2.9	1.8	..	..
91 Mexico	29	32	85	12.7	1.2	2.8	26.5	..
92 Cuba	8	13	95	27.5	5.0	6.2	20.8	..
93 Panama	18	22	83	26.8	3.6	5.5	36.3	7.9
94 Trinidad and Tobago	16	24	93	11.3	2.8	5.8	47.5	3.8
95 Portugal	..	17	89	1.4	1.8	4.3	48.5	..
96 Singapore	5	27	92	4.8	2.8	5.2	28.7	8.7
97 Korea, Rep.	1	..	96	16.8	2.0	4.9	..	8.1
98 Poland	6	16	95	77.0	3.8	4.5	32.8	..
99 Argentina	..	19	96	60.5	2.1	3.3	37.7	..
100 Yugoslavia	2	23	85	24.6	2.3	3.8	74.5	21.2
101 Hungary	8	14	88	73.9	3.2	3.8	36.6	..
102 Uruguay	14	22	92	12.0	3.7	6.6	36.1	7.8
103 Costa Rica	19	31	75	22.7	4.1	4.7	34.4	5.2
104 Bulgaria	10	18	95	56.8	2.6	4.4	47.2	..
105 USSR	20	17	100	14.7	4.7	5.2	39.1	..
106 Czechoslovakia	7	21	78	58.9	3.4	3.6	43.3	..
107 Chile	67	29	93	18.7	2.7	5.2	51.9	4.1
108 Hong Kong	2	27	88	7.1	..	..	31.4	6.0
109 Greece	1	23	97	12.6	1.5	2.5	31.6	12.0
110 German Dem. Rep.	..	17	87	25.3	4.0	3.8	55.7	..
111 Israel	..	16	99	38.5	8.0	7.3	33.3	12.4
112 USA	..	21	100	..	3.8	5.3	36.7	11.6
113 Austria	5	11	86	25.6	2.9	6.0	17.5	8.5
114 Ireland	..	27	77	6.1	3.1	6.9	28.6	4.0
115 Spain	4	26	100	25.7	1.1	3.2	58.9	4.2
116 Belgium	23	15	99	45.6	4.6	5.6	25.0	10.6
117 Italy	1	14	92	33.9	3.2	4.0	21.3	..
118 New Zealand	..	21	94	0.7	2.2	4.8	34.5	8.5
119 Germany, Fed. Rep.	5	17	96	12.2	2.8	4.5	13.8	26.3
120 Finland	2	14	100	27.8	4.7	5.9	31.3	11.9
121 United Kingdom	..	..	93	8.3	3.5	5.3	24.4	..
122 Denmark	1	11	100	30.9	3.1	7.5	43.8	14.4
123 France	5	19	100	26.0	2.4	5.9	19.2	..
124 Australia	..	17	100	..	2.8	5.1	60.3	6.3
125 Norway	1	16	95	27.6	4.1	6.8	44.3	11.3
126 Canada	..	17	100	..	4.2	7.4	61.7	14.8
127 Netherlands	6	17	100	44.0	4.6	6.6	22.6	..
128 Switzerland	..	..	100	6.7	3.0	4.8	73.6	..
129 Sweden	..	16	95	32.8	4.3	7.6	48.1	16.4
130 Japan	1	23	..	12.6	4.0	5.0	27.1	..
All developing countries	39	35	75	6.9	2.4	3.9	41.1	6.1
Least developed countries	60	48	40	4.2	1.5	3.3	46.2	..
Sub-Saharan Africa	41	43	49	6.3	2.3	3.6	44.3	..
Industrial countries	11	19	97	23.0	3.7	5.2	34.1	13.3
World	35	33	78	9.0	3.5	5.0	35.0	8.0

Note: Summary data for regional and income groups are given in tables 23 and 24.

15 Employment

	Labour force (as % of total population) 1988	Women in labour force (as % of total labour force) 1988	Percentage of labour force in: Agriculture 1965	Agriculture 1985-87	Industry 1965	Industry 1985-87	Services 1965	Services 1985-87	Earnings per employee (annual growth rate) 1970-80	Earnings per employee (annual growth rate) 1980-86
Low human development	37.3	26.5	75.3	64.7	10.3	10.3	14.4	25.0	..	..
Excluding India	36.5	27.6	78.2	67.0	8.2	9.8	13.6	23.3	..	..
1 Niger	51.4	47.0	95.0	85.0	1.0	2.7	4.0	12.3	..	..
2 Mali	31.8	16.4	90.0	85.5	1.0	2.0	8.0	12.5	-8.4	..
3 Burkina Faso	46.8	46.5	89.0	86.6	3.0	4.3	7.0	9.1	..	2.6
4 Sierra Leone	35.5	33.1	78.0	69.6	11.0	14.1	11.0	16.4	..	..
5 Chad	35.1	21.3	92.0	83.2	3.0	4.6	5.0	12.0	..	..
6 Guinea	45.7	40.2	87.0	80.7	6.0	9.0	7.0	10.3	..	..
7 Somalia	29.4	39.0	81.0	75.6	6.0	8.4	13.0	16.0	-6.4	-8.6
8 Mauritania	33.4	21.7	89.0	69.4	3.0	8.9	8.0	21.7	..	..
9 Afghanistan	36.4	8.4	69.0	61.0	11.0	14.0	20.0	25.0	..	..
10 Benin	47.1	47.8	83.0	70.2	5.0	6.6	12.0	23.1	..	..
11 Burundi	52.3	47.7	94.0	92.9	2.0	1.6	4.0	5.5	-7.8	..
12 Bhutan	46.1	32.4	95.0	92.5	2.0	2.8	4.0	4.7	..	..
13 Mozambique	54.6	47.9	87.0	84.5	6.0	7.4	7.0	8.1	..	..
14 Malawi	42.0	41.8	92.0	83.4	3.0	7.4	5.0	9.3	..	..
15 Sudan	32.1	21.5	82.0	64.9	5.0	3.9	14.0	31.2	..	..
16 Central African Rep.	48.4	46.2	88.0	83.7	3.0	2.8	9.0	13.5	..	..
17 Nepal	40.5	33.8	94.0	93.0	2.0	0.6	4.0	6.5	..	..
18 Senegal	43.9	39.7	83.0	80.6	6.0	6.2	11.0	13.1	-4.8	-0.2
19 Ethiopia	45.4	37.8	86.0	79.8	5.0	7.9	9.0	12.3	-4.6	-3.1
20 Zaire	36.9	35.9	82.0	71.5	9.0	12.9	9.0	15.6	..	..
21 Rwanda	49.2	48.0	94.0	92.8	2.0	3.0	3.0	4.3	..	..
22 Angola	41.4	39.0	79.0	73.8	8.0	9.5	13.0	16.7	..	..
23 Bangladesh	28.8	7.1	84.0	56.6	5.0	10.4	11.0	33.0	-2.9	-3.7
24 Nigeria	37.5	35.1	72.0	68.1	10.0	11.7	18.0	20.2	0.0	..
25 Yemen Arab Rep.	24.4	13.4	79.0	68.8	7.0	9.2	14.0	22.1	..	..
26 Liberia	36.2	30.6	79.0	74.2	10.0	9.4	11.0	16.4	..	1.6
27 Togo	41.0	36.8	78.0	64.3	9.0	6.3	13.0	29.4	..	..
28 Uganda	44.6	41.4	91.0	85.9	3.0	4.4	6.0	9.7	..	..
29 Haiti	48.0	42.1	77.0	50.4	7.0	5.7	16.0	43.9	-3.3	-0.5
30 Ghana	38.1	40.0	61.0	59.3	15.0	11.1	24.0	29.6	..	..
31 Yemen, PDR	26.0	11.8	54.0	41.0	12.0	17.5	33.0	41.2	..	..
32 Côte d'Ivoire	37.6	34.4	81.0	65.2	5.0	8.3	15.0	26.5	-0.9	..
33 Congo	39.8	39.0	66.0	62.4	11.0	11.9	23.0	25.6	..	..
34 Namibia	29.1	23.8	..	43.5	..	21.9	..	34.8	..	..
35 Tanzania, United Rep.	46.8	48.3	92.0	85.6	3.0	4.5	6.0	9.9	..	-11.4
36 Pakistan	28.0	12.1	60.0	48.7	18.0	13.3	22.0	38.0	3.4	8.8
37 India	37.9	25.6	73.0	62.6	12.0	10.8	15.0	26.6	-0.2	5.6
38 Madagascar	42.6	39.7	85.0	80.9	4.0	6.0	11.0	13.2	-0.9	-12.9
39 Papua New Guinea	47.0	39.1	87.0	76.3	6.0	10.2	7.0	13.5	2.9	0.1
40 Kampuchea, Dem.	47.0	39.2	80.0	74.4	4.0	6.7	16.0	18.9	..	..
41 Cameroon	39.3	33.7	86.0	74.0	4.0	4.5	9.0	21.5	..	..
42 Kenya	40.3	40.3	86.0	81.0	5.0	6.8	9.0	12.1	-3.4	-3.7
43 Zambia	31.5	28.7	79.0	37.9	8.0	7.8	13.0	54.9	-3.3	0.2
44 Morocco	30.8	20.3	61.0	45.6	15.0	25.0	24.0	29.4	..	..
Medium human development	49.1	36.8	72.8	59.8	10.8	14.1	16.5	26.1	2.8	2.5
Excluding China	37.3	29.2	61.5	43.7	14.7	14.7	24.1	41.6	..	..
45 Egypt	27.0	9.8	55.0	38.2	15.0	13.3	30.0	48.5	4.0	1.6
46 Lao PDR	55.4	44.7	81.0	75.7	5.0	7.1	15.0	17.2	..	..
47 Gabon	48.2	37.7	..	75.5	..	10.8	..	13.7	..	..
48 Oman	28.1	8.1	62.0	50.0	15.0	21.8	23.0	28.6	..	..
49 Bolivia	31.2	25.2	54.0	46.5	20.0	19.7	26.0	33.9	2.5	4.4
50 Myanmar	44.2	37.4	64.0	63.9	14.0	9.1	23.0	27.0	..	..
51 Honduras	30.3	18.3	68.0	60.4	12.0	16.1	20.0	23.4	-0.4	..
52 Zimbabwe	40.6	35.0	79.0	72.8	8.0	10.5	13.0	16.7	1.6	6.1
53 Lesotho	46.3	43.8	92.0	23.3	3.0	33.1	6.0	43.6	..	..
54 Indonesia	39.0	31.2	71.0	53.5	9.0	9.7	21.0	36.8	4.7	9.2
55 Guatemala	28.5	15.9	64.0	49.8	15.0	12.3	21.0	37.9	-3.2	-0.1
56 Viet Nam	48.6	46.9	79.0	67.5	6.0	11.8	15.0	20.7	..	..
57 Algeria	22.7	9.3	57.0	25.7	17.0	15.3	26.0	59.0	0.1	-3.9
58 Botswana	34.9	35.5	89.0	43.2	4.0	4.9	8.0	51.9	10.4	-4.2
59 El Salvador	40.1	25.1	59.0	43.2	16.0	19.4	26.0	37.5	2.4	..
60 Tunisia	31.3	23.9	49.0	23.4	21.0	17.5	29.0	59.1	4.2	-4.9
61 Iran, Islamic Rep.	27.1	17.5	49.0	36.4	26.0	32.8	25.0	30.8	..	..
62 Syrian Arab Rep.	24.8	16.8	52.0	24.9	20.0	16.0	28.0	59.1	2.2	-1.8
63 Dominican Rep.	30.0	14.5	59.0	45.7	14.0	15.5	27.0	38.8	-1.0	-4.8
64 Saudi Arabia	29.1	7.1	68.0	48.5	11.0	14.4	21.0	37.2	..	..
65 Philippines	36.0	31.6	58.0	43.4	16.0	9.7	26.0	46.9	-3.0	..
66 China	59.3	43.2	81.0	73.7	8.0	13.6	11.0	12.7	..	..

	Labour force (as % of total population) 1988	Women in labour force (as % of total labour force) 1988	Percentage of labour force in Agriculture		Industry		Services		Earnings per employee (annual growth rate)	
			1965	1985-87	1965	1985-87	1965	1985-87	1970-80	1980-86
67 Libyan Arab Jamahiriya	23.7	8.7	41.0	18.1	21.0	28.9	38.0	53.0	..	..
68 South Africa	34.9	35.4	32.0	13.6	30.0	24.4	39.0	62.0	2.7	0.4
69 Lebanon	30.1	27.2	29.0	14.3	24.0	27.4	47.0	58.4	..	..
70 Mongolia	46.4	45.5	54.0	39.9	20.0	21.0	26.0	39.2	..	..
71 Nicaragua	30.8	24.6	57.0	46.5	16.0	15.8	28.0	37.7	..	-15.8
72 Turkey	42.6	33.8	75.0	39.5	11.0	13.4	14.0	47.1	3.7	-2.3
73 Jordan	23.1	9.9	37.0	10.2	26.0	25.6	37.0	64.2	..	-1.1
74 Peru	31.8	24.1	50.0	35.1	19.0	12.3	32.0	52.6	..	1.2
75 Ecuador	30.4	19.3	55.0	38.5	19.0	19.8	26.0	41.6	2.9	-1.1
76 Iraq	27.0	21.0	50.0	30.4	20.0	22.1	30.0	47.5	..	..
77 United Arab Emirates	49.8	6.2	21.0	4.5	32.0	38.0	47.0	57.3	..	..
78 Thailand	52.5	45.1	82.0	72.4	5.0	5.9	13.0	21.7	1.0	7.2
79 Paraguay	33.0	20.7	55.0	48.6	20.0	20.5	26.0	30.9	..	..
80 Brazil	36.6	27.3	49.0	25.2	20.0	15.8	31.0	59.0	4.0	-1.1
81 Mauritius	39.0	26.0	37.0	18.5	25.0	33.7	38.0	47.8	1.7	-3.1
82 Korea, Dem. Rep.	45.2	45.9	57.0	42.8	23.0	30.3	20.0	26.9	..	..
83 Sri Lanka	36.9	26.8	56.0	42.4	14.0	12.0	30.0	45.6	..	-1.0
84 Albania	48.3	41.1	69.0	55.9	19.0	25.7	12.0	18.4	..	..
High human development	46.3	38.8	24.8	12.9	34.6	26.3	40.8	60.8	2.4	1.4
85 Malaysia	40.5	35.0	59.0	41.6	13.0	19.1	29.0	39.3	2.0	5.4
86 Colombia	32.4	22.0	45.0	1.3	21.0	21.1	34.0	77.6	-0.2	6.2
87 Jamaica	48.4	45.8	37.0	25.3	20.0	11.5	43.0	63.2	-0.2	..
88 Kuwait	39.8	14.2	2.0	1.9	34.0	8.7	64.0	89.4	..	4.1
89 Venezuela	34.4	27.3	30.0	13.6	24.0	17.9	47.0	68.5	3.8	-0.4
90 Romania	50.6	46.3	57.0	30.5	26.0	43.5	18.0	26.0	..	..
91 Mexico	33.8	27.1	50.0	25.8	22.0	14.1	29.0	60.1	1.2	-4.0
92 Cuba	42.0	31.4	33.0	23.8	25.0	28.5	41.0	47.7	..	..
93 Panama	35.6	27.0	46.0	26.2	16.0	9.7	38.0	64.1	0.2	4.4
94 Trinidad and Tobago	38.6	29.9	20.0	9.7	35.0	13.7	45.0	76.6	..	2.4
95 Portugal	45.6	36.5	38.0	20.6	30.0	23.4	32.0	56.0	2.5	1.3
96 Singapore	48.0	32.6	6.0	0.8	27.0	25.6	68.0	73.6	3.6	8.8
97 Korea, Rep.	42.1	33.9	55.0	21.2	15.0	27.3	30.0	51.5	10.0	5.8
98 Poland	51.4	45.5	44.0	28.5	32.0	38.9	25.0	32.6	..	..
99 Argentina	35.8	27.9	18.0	13.0	34.0	33.8	48.0	53.1	1.7	4.4
100 Yugoslavia	45.5	38.7	57.0	28.7	26.0	23.6	17.0	47.7	1.3	-1.9
101 Hungary	49.6	44.7	32.0	20.9	40.0	31.3	29.0	47.8	3.7	1.5
102 Uruguay	38.9	30.9	20.0	15.3	29.0	18.2	51.0	66.5	..	-1.2
103 Costa Rica	34.1	21.7	47.0	27.5	19.0	17.4	34.0	55.1	..	..
104 Bulgaria	49.8	46.3	46.0	16.5	31.0	37.9	23.0	45.6	..	..
105 USSR	51.2	48.1	34.0	20.0	33.0	39.0	33.0	41.0	..	..
106 Czechoslovakia	53.2	46.5	21.0	13.3	47.0	49.4	31.0	37.4	..	..
107 Chile	35.8	28.3	27.0	19.8	29.0	17.2	44.0	63.0	..	-1.5
108 Hong Kong	53.0	33.8	6.0	1.5	53.0	34.1	41.0	64.4	6.1	2.6
109 Greece	38.2	26.5	47.0	26.5	24.0	19.9	29.0	53.6	5.0	-0.3
110 German Dem. Rep.	57.7	45.6	15.0	10.6	49.0	50.0	36.0	39.4	..	..
111 Israel	38.9	33.6	12.0	4.9	35.0	22.8	53.0	72.3	8.8	-10.0
112 USA	48.9	41.5	5.0	3.0	35.0	19.0	60.0	78.0	0.1	1.4
113 Austria	47.3	40.1	19.0	8.4	45.0	28.5	36.0	63.1	3.4	1.3
114 Ireland	39.3	29.2	31.0	12.7	28.0	18.6	41.0	68.7	4.1	8.0
115 Spain	36.3	24.3	34.0	13.8	35.0	21.0	32.0	65.2	4.5	1.9
116 Belgium	41.6	33.8	6.0	2.5	46.0	19.5	48.0	78.0	4.3	-0.1
117 Italy	40.4	31.9	25.0	9.1	42.0	20.4	34.0	70.5	4.1	0.4
118 New Zealand	45.9	34.7	13.0	10.0	36.0	20.1	51.0	69.9	1.2	-1.6
119 Germany, Fed. Rep.	48.3	37.4	11.0	4.8	48.0	31.7	41.0	63.5	3.5	1.0
120 Finland	51.1	46.9	24.0	10.2	35.0	21.9	41.0	67.9	2.6	2.1
121 United Kingdom	48.6	38.7	3.0	2.1	47.0	20.1	50.0	77.8	1.7	3.0
122 Denmark	55.2	44.5	14.0	5.7	37.0	20.0	49.0	74.3	2.5	-0.1
123 France	45.0	39.8	18.0	6.7	39.0	19.8	43.0	73.5	..	..
124 Australia	47.2	38.0	10.0	5.5	38.0	17.3	52.0	77.2	2.9	1.7
125 Norway	50.0	40.9	16.0	6.5	37.0	17.6	48.0	75.9	2.6	1.4
126 Canada	50.3	39.8	10.0	4.9	33.0	18.6	57.0	76.5	4.2	2.8
127 Netherlands	41.2	30.9	9.0	4.5	41.0	17.0	51.0	78.5	2.5	3.6
128 Switzerland	49.2	36.6	9.0	6.4	49.0	29.8	41.0	63.8	..	..
129 Sweden	51.4	44.4	11.0	3.9	43.0	22.0	46.0	74.1	0.4	0.1
130 Japan	50.0	37.9	26.0	8.0	32.0	23.6	42.0	68.4	3.2	1.8
All developing countries	43.5	32.1	71.5	59.0	11.5	13.0	17.1	28.0	0.9	3.1
Least developed countries	38.9	28.2	83.0	71.1	5.8	7.6	11.3	21.2	..	..
Sub-Saharan Africa	40.6	37.8	77.8	70.3	8.6	9.8	13.7	20.0	-1.5	-3.5
Industrial countries	48.5	41.1	22.2	11.5	36.4	27.5	41.5	61.0	2.2	1.4
World	44.6	34.2	56.6	47.9	19.0	16.4	24.5	35.7	1.2	2.7

Note: Summary data for regional and income groups are given in tables 23 and 24.

16 Wealth and poverty

	GNP per capita (US$) 1987	Real GDP per capita (PPP$) 1987	GNP per capita of lowest 40% of households (US$) 1987	Income share of lowest 40% 1975-86	Ratio of highest 20% to lowest 20% 1975-86	Gini coefficient 1967-85	Population below poverty line: Urban (%) 1977-87	Population below poverty line: Rural (%) 1977-87
Low human development	300	970	..	..	..	..	40	55
Excluding India	300	880	..	..	..	..	41	62
1 Niger	260	452	..	..	..	..	..	35
2 Mali	210	543	..	..	..	..	27	48
3 Burkina Faso	190	..	..	..	..	..	..	..
4 Sierra Leone	300	480	..	..	..	0.59	..	65
5 Chad	150	..	..	..	..	..	30	56
6 Guinea	..	..	..	..	..	..	..	..
7 Somalia	290	..	..	..	..	..	40	70
8 Mauritania	440	840	..	..	..	..	..	..
9 Afghanistan	..	..	..	..	..	..	..	..
10 Benin	310	665	..	..	..	..	..	65
11 Burundi	250	450	..	..	..	..	55	85
12 Bhutan	150	..	..	..	..	..	..	..
13 Mozambique	170	..	..	..	..	..	..	..
14 Malawi	160	476	..	..	..	..	25	85
15 Sudan	330	750	..	..	..	..	..	85
16 Central African Rep.	330	591	..	..	..	..	..	91
17 Nepal	160	722	..	..	..	0.53	55	61
18 Senegal	520	1,068	..	..	..	..	..	..
19 Ethiopia	130	454	..	..	..	..	60	65
20 Zaire	150	220	..	..	..	..	..	80
21 Rwanda	300	571	..	..	..	..	30	90
22 Angola	470	..	..	..	..	..	..	..
23 Bangladesh	160	883	70	17	7.2	0.39	86	86
24 Nigeria	370	668	..	..	..	..	..	..
25 Yemen Arab Rep.	590	..	..	..	..	..	..	..
26 Liberia	450	696	..	..	..	..	..	23
27 Togo	290	670	..	..	..	..	42	..
28 Uganda	260	511	..	..	..	..	..	..
29 Haiti	360	775	..	..	..	..	65	80
30 Ghana	390	481	..	..	..	..	59	37
31 Yemen, PDR	420	..	..	..	..	..	..	20
32 Côte d'Ivoire	740	1,123	160	9	25.6	0.55	30	26
33 Congo	870	756	..	..	..	..	..	..
34 Namibia	..	..	..	..	..	..	..	..
35 Tanzania, United Rep.	180	405	..	..	..	..	..	..
36 Pakistan	350	1,585	..	..	..	0.36	32	29
37 India	300	1,053	120	16	7.0	0.42	40	51
38 Madagascar	210	634	..	..	..	..	50	50
39 Papua New Guinea	700	1,843	..	..	..	..	10	75
40 Kampuchea, Dem.	..	..	..	..	..	..	..	..
41 Cameroon	970	1,381	..	..	..	..	15	40
42 Kenya	330	794	70	9	25.0	..	10	55
43 Zambia	250	717	70	11	19.0	..	25	..
44 Morocco	610	1,761	..	..	..	..	28	45
Medium human development	690	2,370	..	..	..	..	16	18
Excluding China	1,250	2,730	..	..	..	..	..	44
45 Egypt	680	1,357	280	16	8.5	0.38	21	25
46 Lao PDR	170	..	..	..	..	..	..	..
47 Gabon	2,700	2,068	..	..	..	..	..	..
48 Oman	5,810	..	..	..	..	..	..	..
49 Bolivia	580	1,380	..	12	..	..	..	85
50 Myanmar	200	752	..	..	..	..	40	40
51 Honduras	810	1,119	..	..	..	0.62	14	55
52 Zimbabwe	580	1,184	..	..	..	..	..	..
53 Lesotho	370	1,585	..	..	..	..	50	55
54 Indonesia	450	1,660	160	14	7.3	0.31	26	44
55 Guatemala	950	1,957	..	..	..	..	66	74
56 Viet Nam	..	..	..	..	..	..	..	..
57 Algeria	2,680	2,633	..	..	..	..	20	..
58 Botswana	1,050	2,496	..	..	..	..	40	55
59 El Salvador	860	1,733	330	15	8.5	0.40	20	32
60 Tunisia	1,180	2,741	..	..	..	0.40	20	15
61 Iran, Islamic Rep.	..	..	..	..	..	0.46	..	..
62 Syrian Arab Rep.	1,640	..	490	..	15.8	..	..	..
63 Dominican Rep.	730	..	..	..	..	..	45	43
64 Saudi Arabia	6,200	8,320	..	..	..	..	..	..
65 Philippines	590	1,878	210	14	10.3	0.45	50	64
66 China	290	2,124	..	..	..	..	..	10

	GNP per capita (US$) 1987	Real GDP per capita (PPP$) 1987	GNP per capita of lowest 40% of households (US$) 1987	Income share of lowest 40% 1975-86	Ratio of highest 20% to lowest 20% 1975-86	Gini coefficient 1967-85	Population below poverty line	
							Urban (%) 1977-87	Rural (%) 1977-87
67 Libyan Arab Jamahiriya	5,460	..	..	..	..	..	..	..
68 South Africa	1,890	4,981	..	..	..	..	..	..
69 Lebanon	..	..	..	..	..	..	..	..
70 Mongolia	..	..	..	..	..	..	..	..
71 Nicaragua	830	2,209	..	..	..	..	21	19
72 Turkey	1,210	3,781	350	11	16.3	0.51	..	..
73 Jordan	1,560	3,161	..	..	..	..	14	17
74 Peru	1,470	3,129	260	7	32.0	0.31	49	..
75 Ecuador	1,040	2,687	..	..	..	..	40	65
76 Iraq	3,020	..	..	..	..	..	..	..
77 United Arab Emirates	15,830	12,191	..	..	..	..	..	..
78 Thailand	850	2,576	320	15	8.8	0.47	15	34
79 Paraguay	990	2,603	..	..	..	..	19	50
80 Brazil	2,020	4,307	350	7	33.7	0.57	..	..
81 Mauritius	1,490	2,617	430	11	15.0	..	12	12
82 Korea, Dem. Rep.	..	..	..	..	..	..	..	..
83 Sri Lanka	400	2,053	160	16	8.3	0.45	..	..
84 Albania	..	..	..	..	..	..	..	..
High human development	9,250	11,860	..	..	..	..	..	..
85 Malaysia	1,810	3,849	510	11	14.4	0.48	13	38
86 Colombia	1,240	3,524	..	..	..	0.45	32	..
87 Jamaica	940	2,506	..	..	..	0.66	..	80
88 Kuwait	14,610	13,843	..	..	..	..	..	..
89 Venezuela	3,230	4,306	830	10	18.2	..	..	..
90 Romania	2,560	..	..	..	..	..	..	..
91 Mexico	1,830	4,624	450	10	19.6	0.50	..	..
92 Cuba	..	..	..	..	..	..	..	..
93 Panama	2,240	4,009	400	7	31.5	0.57	21	30
94 Trinidad and Tobago	4,210	3,664	..	13	..	..	..	39
95 Portugal	2,830	5,597	1,080	15	9.4	..	..	..
96 Singapore	7,940	12,790	..	..	..	0.42	..	..
97 Korea, Rep.	2,690	4,832	1,200	17	6.8	0.36	18	11
98 Poland	2,070	..	..	..	..	..	..	..
99 Argentina	2,390	4,647	840	14	11.3	..	..	..
100 Yugoslavia	2,480	..	1,160	19	5.9	..	..	..
101 Hungary	2,240	..	1,150	20	5.2	..	..	..
102 Uruguay	2,190	5,063	..	..	..	..	22	..
103 Costa Rica	1,610	3,760	..	12	..	0.42	..	..
104 Bulgaria	4,150	..	..	..	..	..	..	..
105 USSR	4,550	..	..	..	..	..	..	..
106 Czechoslovakia	5,820	..	..	..	..	..	..	..
107 Chile	1,310	4,862	..	..	..	0.46	27	..
108 Hong Kong	8,070	13,906	3,270	16	8.7	0.45	..	..
109 Greece	4,020	..	..	..	..	..	..	..
110 German Dem. Rep.	7,180	..	..	..	..	..	..	..
111 Israel	6,800	9,182	3,060	18	6.7	..	..	..
112 USA	18,530	17,615	7,970	17	7.5	..	..	..
113 Austria	11,980	12,386	..	..	..	..	..	..
114 Ireland	6,120	8,566	3,110	20	5.5	..	..	..
115 Spain	6,010	8,989	2,910	19	5.8	..	..	..
116 Belgium	11,480	13,140	6,200	22	4.6	..	..	..
117 Italy	10,350	10,682	4,530	17	7.1	..	..	..
118 New Zealand	7,750	10,541	3,080	16	8.8	..	..	..
119 Germany, Fed. Rep.	14,400	14,730	7,340	20	5.0	..	..	..
120 Finland	14,470	12,795	6,660	18	6.0	..	..	..
121 United Kingdom	10,420	12,270	4,820	18	5.7	..	..	..
122 Denmark	14,930	15,119	6,490	17	7.2	..	..	..
123 France	12,790	13,961	5,440	17	7.7	..	..	..
124 Australia	11,100	11,782	4,270	15	8.7	..	..	..
125 Norway	17,190	15,940	8,120	19	6.4	..	..	..
126 Canada	15,160	16,375	6,480	17	7.5	..	..	..
127 Netherlands	11,860	12,661	6,640	22	4.4	..	..	..
128 Switzerland	21,330	15,403	10,720	20	5.8	..	..	..
129 Sweden	15,550	13,780	7,970	20	5.6	..	..	..
130 Japan	15,760	13,135	8,630	22	4.3	..	..	..
All developing countries	650	1,970	..	..	..	..	27	35
Least developed countries	210	690	..	..	..	..	59	72
Sub-Saharan Africa	440	990	..	..	..	..	34	61
Industrial countries	10,760	14,260	..	..	..	..	..	..
World	3,100	4,110	..	..	..	..	..	..

Note: Summary data for regional and income groups are given in tables 23 and 24.

Urban crowding

		Urban population (%)			Urban population (annual growth rate)		Persons per habitable room	Highest population density	
		1960	1988	2000	1960-88	1988-2000	1970-81	Major city	Population per square kilometre 1980-87
Low human development		17	29	37	4.7	4.6	..	..	..
Excluding India		15	31	39	5.9	5.3	..	..	..
1	Niger	6	18	27	6.9	6.5	..	..	..
2	Mali	11	19	23	4.3	4.9	..	..	..
3	Burkina Faso	5	9	12	4.5	6.0	..	..	..
4	Sierra Leone	13	31	40	5.2	5.0	..	..	..
5	Chad	7	31	44	7.6	5.7	..	..	..
6	Guinea	10	24	33	5.4	5.2	..	Conakry	6,912
7	Somalia	17	35	44	5.8	4.8	..	..	..
8	Mauritania	7	39	54	9.1	5.5	..	..	..
9	Afghanistan	8	21	29	4.8	7.5	..	..	..
10	Benin	10	40	53	7.8	5.8	..	..	..
11	Burundi	2	7	11	6.2	7.7	..	..	..
12	Bhutan	3	5	8	4.4	6.1	..	..	..
13	Mozambique	4	24	41	9.6	7.4	..	..	..
14	Malawi	4	14	21	7.2	7.0	..	..	..
15	Sudan	10	22	26	5.5	4.7	..	..	..
16	Central African Rep.	23	45	55	4.5	4.2	..	..	..
17	Nepal	3	9	14	6.3	6.5	..	..	..
18	Senegal	32	38	44	3.6	4.2	..	..	..
19	Ethiopia	6	12	17	4.7	5.2	..	..	..
20	Zaire	22	38	46	4.8	4.8	..	..	..
21	Rwanda	2	7	11	7.4	7.5	..	..	..
22	Angola	10	27	36	6.0	5.4	..	..	..
23	Bangladesh	5	13	18	6.2	5.6	2.9	Dhaka	9,930
24	Nigeria	14	34	43	6.5	5.6	..	..	..
25	Yemen Arab Rep.	3	23	33	9.5	6.5	2.8	..	..
26	Liberia	19	42	52	6.1	5.1	..	..	..
27	Togo	10	24	33	6.2	5.9	..	..	..
28	Uganda	5	10	14	6.1	6.3	..	..	..
29	Haiti	16	29	37	4.2	4.0	..	Port-au-Prince	6,985
30	Ghana	23	32	38	3.9	4.4	..	..	..
31	Yemen, PDR	28	42	51	3.9	4.8	..	..	..
32	Côte d'Ivoire	19	45	55	7.3	5.6	..	Abidjan	3,030
33	Congo	33	41	50	3.2	4.4	..	..	..
34	Namibia	23	55	66	6.0	4.8	..	..	..
35	Tanzania, United Rep.	5	30	46	10.4	7.6	..	..	..
36	Pakistan	22	31	38	4.3	4.6	3.6	Karachi	3,990
37	India	18	27	34	3.7	4.0	2.8	Calcutta	88,135
38	Madagascar	11	24	32	5.7	5.9	..	..	..
39	Papua New Guinea	3	15	20	9.0	5.0	..	..	..
40	Kampuchea, Dem.	10	11	15	1.7	4.2	..	..	..
41	Cameroon	14	47	60	7.0	4.9	1.2	..	..
42	Kenya	7	22	32	7.9	7.2	..	Nairobi	1,587
43	Zambia	17	53	65	7.6	5.5	2.6	..	..
44	Morocco	29	47	56	4.4	3.7	2.4	Casablanca	12,133
Medium human development		29	42	48	3.7	3.2	..	..	..
Excluding China		34	54	61	4.4	3.4	..	..	..
45	Egypt	38	48	55	3.3	3.3	..	Cairo	29,393
46	Lao PDR	8	18	25	5.0	5.4	..	..	..
47	Gabon	17	44	54	6.4	5.1	..	..	..
48	Oman	4	10	15	7.6	7.1	..	..	..
49	Bolivia	39	50	58	3.4	4.2	..	Cochabamba	6,558
50	Myanmar	19	24	28	3.1	3.3	..	..	..
51	Honduras	23	42	51	5.6	4.7	..	..	..
52	Zimbabwe	13	26	35	5.9	5.4	..	..	..
53	Lesotho	3	19	28	8.8	6.2	..	..	..
54	Indonesia	15	27	36	4.5	3.9	1.5	..	..
55	Guatemala	33	41	47	3.7	4.1	2.2	..	..
56	Viet Nam	15	21	27	3.6	4.2	..	..	..
57	Algeria	30	44	51	4.2	4.0	2.8	Algiers	7,930
58	Botswana	2	22	33	13.1	6.9	..	..	..
59	El Salvador	38	44	49	2.9	3.5	3.2	..	..
60	Tunisia	36	54	59	3.7	2.7	3.1	..	..
61	Iran, Islamic Rep.	34	54	61	5.2	4.0	2.0	Mashhad	21,132
62	Syrian Arab Rep.	37	51	57	4.6	4.5	0.4	..	..
63	Dominican Rep.	30	59	68	5.2	3.2	2.0	..	..
64	Saudi Arabia	30	76	82	7.8	4.5	..	..	..
65	Philippines	30	41	49	3.9	3.7	2.3	Manila	45,839
66	China	19	21	25	2.3	2.7	..	Beijing	4,039

		Urban population (%)			Urban population (annual growth rate)		Persons per habitable room	Highest population density	
		1960	1988	2000	1960-88	1988-2000	1970-81	Major city	Population per square kilometre 1980-87
67	Libyan Arab Jamahiriya	23	68	76	8.3	4.6	1.8	..	..
68	South Africa	47	58	65	3.2	3.1	..	..	..
69	Lebanon	40	82	87	4.2	2.4	..	..	..
70	Mongolia	36	51	55	4.3	3.6	..	..	..
71	Nicaragua	40	59	66	4.7	4.1	..	..	..
72	Turkey	30	47	54	4.1	3.0	..	Ankara	3,438
73	Jordan	43	67	74	4.7	4.9	..	Amman	11,104
74	Peru	46	69	75	4.2	3.0	..	Lima	1,379
75	Ecuador	34	55	65	4.8	4.0	2.3	..	..
76	Iraq	43	73	79	5.4	4.1	..	Baghdad	5,384
77	United Arab Emirates	40	78	78	13.2	2.3	..	..	..
78	Thailand	13	22	29	4.6	4.0	..	Bangkok	3,486
79	Paraguay	36	46	54	4.0	4.0	2.4	..	..
80	Brazil	45	75	83	4.4	2.6	1.0	Recife	6,232
81	Mauritius	33	42	46	2.6	1.8	1.6	Port Louis	3,795
82	Korea, Dem. Rep.	40	66	73	4.5	3.0	..	..	..
83	Sri Lanka	18	21	24	2.5	2.3	2.1	..	..
84	Albania	31	35	40	2.9	2.8	..	..	..
High human development		61	74	77	2.0	1.1	1.0	..	..
85	Malaysia	25	41	50	4.3	3.7	2.6	Kuala Lumpur	3,772
86	Colombia	48	69	75	3.8	2.5	1.8	Medellin	3,106
87	Jamaica	34	51	59	3.0	2.5	..	..	..
88	Kuwait	72	95	97	8.2	3.3	..	..	..
89	Venezuela	67	89	94	4.4	2.7	..	..	..
90	Romania	34	50	53	2.2	1.0	1.4	Bucharest	3,271
91	Mexico	51	71	77	4.2	2.6	2.5	Guadalajara	10,286
92	Cuba	55	74	80	2.4	1.5	1.0	Havana	2,749
93	Panama	41	54	60	3.5	2.8	1.8	Panama	3,890
94	Trinidad and Tobago	23	67	75	5.4	2.4	..	..	..
95	Portugal	23	32	39	1.8	1.9	..	Lisbon	9,893
96	Singapore	100	100	100	1.7	0.9	..	Singapore	4,160
97	Korea, Rep.	28	69	81	5.3	2.3	2.3	Seoul	15,932
98	Poland	48	62	67	1.8	1.1	1.2	Warsaw	3,419
99	Argentina	74	86	89	2.1	1.5	..	Buenos Aires	14,615
100	Yugoslavia	28	49	58	2.9	1.9	1.4	Belgrade	3,022
101	Hungary	40	59	67	1.6	1.0	1.0	Budapest	3,954
102	Uruguay	80	85	87	0.9	0.9	2.1	Montevideo	6,952
103	Costa Rica	37	52	61	4.4	3.5	1.5	..	..
104	Bulgaria	39	69	76	2.6	0.9	1.0	..	..
105	USSR	49	67	71	2.1	1.2	..	Moscow	9,402
106	Czechoslovakia	47	67	74	1.8	1.1	0.9	Prague	2,821
107	Chile	68	85	89	2.7	1.9	1.4	Santiago	9,878
108	Hong Kong	89	93	94	2.4	1.2	..	Hong Kong	5,048
109	Greece	43	62	68	2.0	1.0	0.9	..	..
110	German Dem. Rep.	72	78	81	0.1	0.3	..	Leipzig	3,950
111	Israel	77	91	93	3.3	1.7	1.2	..	..
112	USA	70	74	75	1.3	0.8	0.6	New York City	8,722
113	Austria	50	57	61	0.7	0.5	1.1	Vienna	3,689
114	Ireland	46	58	64	1.8	1.7	0.9	Dublin	4,485
115	Spain	57	77	83	2.0	0.9	..	Barcelona	17,433
116	Belgium	92	97	98	0.4	0.2	0.6	Brussels	4,160
117	Italy	59	68	72	1.0	0.6	..	Naples	10,342
118	New Zealand	76	84	85	1.6	0.8	0.5	..	..
119	Germany, Fed. Rep.	77	86	88	0.7	0.0	0.6	Munich	4,192
120	Finland	38	66	74	2.4	1.2	0.8	Helsinki	2,616
121	United Kingdom	86	92	94	0.6	0.2	..	Birmingham	4,444
122	Denmark	74	86	89	0.9	0.3	0.6	Copenhagen	5,735
123	France	62	74	76	1.3	0.6	0.8	Paris	20,647
124	Australia	81	85	86	1.9	1.1	..	Sydney	3,318
125	Norway	32	74	77	3.6	0.6	0.7	..	..
126	Canada	69	76	79	1.7	1.0	0.5	Montreal	6,357
127	Netherlands	85	88	89	1.0	0.3	..	's Gravenhage	6,386
128	Switzerland	51	59	62	1.2	0.5	0.6	Lausanne	5,512
129	Sweden	73	84	86	0.9	0.2	0.6	Stockholm	3,051
130	Japan	63	77	78	1.7	0.5	0.8	Tokyo	13,973
All developing countries		29	43	50	4.0	3.5	2.4	..	..
Least developed countries		9	22	30	6.2	5.6	..	..	..
Sub-Saharan Africa		19	36	45	6.0	5.3	..	..	..
Industrial countries		62	74	76	1.5	0.8	0.7	..	..
World		42	55	60	3.0	2.4	1.9	..	..

Note: Summary data for regional and income groups are given in tables 23 and 24.

18 Military expenditure imbalances

		Military expenditure (% of GNP) 1960	Military expenditure (% of GNP) 1986	Ratio of military expenditure to health and education expenditure 1986	Arms imports (US$ million) 1987	Ratio of net ODA to military expenditure 1986 Received	Ratio of net ODA to military expenditure 1986 Given	Armed forces as percentage of teachers 1986
	Low human development	1.9	3.7	98	9,160 T	4.6	..	61
	Excluding India	1.8	3.8	117	5,960 T	..	..	96
1	Niger	0.3	0.7	15	10	25.6	..	22
2	Mali	1.7	2.5	64	40	12.0	..	38
3	Burkina Faso	0.6	3.0	88	0	7.7	..	57
4	Sierra Leone	..	1.2	32	0	5.4	..	20
5	Chad	..	6.0	231	100	4.1	..	233
6	Guinea	1.3	3.0	75	50	3.1	..	77
7	Somalia	..	4.4	71	20	7.4	..	525
8	Mauritania	..	4.9	62	0	5.5	..	200
9	Afghanistan	..	..	..	1,300	..	..	209
10	Benin	1.1	2.3	53	0	5.3	..	19
11	Burundi	..	3.5	100	20	4.6	..	71
12	Bhutan	..	..	..	..	..	..	..
13	Mozambique	..	7.0	..	120	1.6	..	73
14	Malawi	..	2.3	38	0	6.6	..	31
15	Sudan	1.5	5.9	140	50	1.8	..	88
16	Central African Rep.	..	1.7	26	0	10.7	..	50
17	Nepal	0.4	1.5	41	0	6.5	..	40
18	Senegal	0.5	2.3	40	5	8.3	..	59
19	Ethiopia	1.6	8.6	165	1,000	1.4	..	494
20	Zaire	..	3.0	250	30	2.9	..	17
21	Rwanda	..	1.9	50	0	6.8	..	33
22	Angola	..	12.0	273	1,600	..	..	165
23	Bangladesh	..	1.5	54	10	5.9	..	26
24	Nigeria	0.2	1.0	56	60	0.0	..	30
25	Yemen Arab Rep.	..	6.8	110	390	0.8	..	185
26	Liberia	1.1	2.2	34	10	4.4	..	55
27	Togo	..	3.2	45	0	7.0	..	36
28	Uganda	0.0	4.2	323	40	0.6	..	30
29	Haiti	2.4	1.5	71	0	5.3	..	27
30	Ghana	1.1	0.9	24	10	6.1	..	14
31	Yemen, PDR	..	22.0	275	300	0.4	..	200
32	Côte d'Ivoire	0.5	1.2	20	0	2.1	..	13
33	Congo	0.3	4.6	66	5	1.4	..	75
34	Namibia	..	..	..	..	..	..	..
35	Tanzania, United Rep.	0.1	3.3	61	110	3.1	..	42
36	Pakistan	5.5	6.7	279	150	0.4	..	154
37	India	1.9	3.5	81	3,200	0.3	..	28
38	Madagascar	0.3	2.4	45	30	4.9	..	37
39	Papua New Guinea	..	1.4	16	0	7.5	..	23
40	Kampuchea, Dem.	..	..	..	350	..	..	111
41	Cameroon	1.7	1.7	49	5	1.4	..	18
42	Kenya	0.5	1.2	20	10	2.8	..	10
43	Zambia	1.1	3.2	49	0	5.5	..	47
44	Morocco	2.0	5.1	74	130	0.6	..	102
	Medium human development	5.5	7.2	133	22,020 T	0.7	..	73
	Excluding China	3.6	7.5	129	21,640 T	..	..	99
45	Egypt	5.5	8.9	153	1,500	0.5	..	168
46	Lao PDR	5.8	..	..	110	..	..	225
47	Gabon	..	3.8	56	0	0.6	..	83
48	Oman	..	27.6	279	30	0.0	..	275
49	Bolivia	2.0	2.4	86	0	3.7	..	50
50	Myanmar	7.0	3.1	100	10	1.8	..	118
51	Honduras	1.2	5.9	78	60	1.4	..	74
52	Zimbabwe	..	5.0	49	80	0.8	..	59
53	Lesotho	..	2.4	47	0	5.9	..	29
54	Indonesia	5.8	2.5	60	250	0.3	..	17
55	Guatemala	0.9	1.3	52	5	1.0	..	114
56	Viet Nam	..	..	..	1,900	..	..	309
57	Algeria	2.1	1.9	23	700	0.0	..	71
58	Botswana	..	2.3	19	0	4.1	..	38
59	El Salvador	1.1	3.7	112	50	2.3	..	183
60	Tunisia	2.2	6.2	81	50	0.4	..	55
61	Iran, Islamic Rep.	4.5	20.0	377	1,500	0.0	..	112
62	Syrian Arab Rep.	7.9	14.7	226	1,900	0.3	..	320
63	Dominican Rep.	5.0	1.4	47	5	1.3	..	58
64	Saudi Arabia	5.7	22.7	155	3,800	..	0.20	46
65	Philippines	1.2	1.7	71	40	1.7	..	30
66	China	12.0	6.0	146	380	0.0	..	51

		Military expenditure (% of GNP)		Ratio of military expenditure to health and education expenditure	Arms imports (US$ million)	Ratio of net ODA to military expenditure 1986		Armed forces as percentage of teachers
		1960	1986	1986	1987	Received	Given	1986
67	Libyan Arab Jamahiriya	1.2	12.0	92	625	..	0.02	104
68	South Africa	0.9	3.9	75	0	..	..	32
69	Lebanon	..	..	..	5	..	..	43
70	Mongolia	4.2	10.5	..	0	..	..	225
71	Nicaragua	1.9	16.0	126	500	0.3	..	326
72	Turkey	5.2	4.9	188	925	0.1	..	271
73	Jordan	16.7	13.8	197	320	1.0	..	245
74	Peru	2.0	6.5	250	430	0.2	..	86
75	Ecuador	2.4	1.6	34	70	0.8	..	46
76	Iraq	8.7	32.0	711	5,600	0.0	..	428
77	United Arab Emirates	..	8.8	275	260	..	0.03	358
78	Thailand	2.6	4.0	74	350	0.3	..	51
79	Paraguay	1.7	1.0	56	0	1.9	..	49
80	Brazil	1.8	0.9	19	100	0.1	..	24
81	Mauritius	0.2	0.2	4	0	18.7	..	10
82	Korea, Dem. Rep.	11.0	10.0	..	420	..	..	..
83	Sri Lanka	1.0	5.7	116	40	1.5	..	12
84	Albania	9.0	4.0	..	0	..	..	138
High human development		6.2	5.4	55	14,760 T	..	0.20	101
85	Malaysia	1.9	6.1	63	60	0.1	..	90
86	Colombia	1.2	1.0	28	10	0.2	..	34
87	Jamaica	..	1.5	18	5	5.2	..	20
88	Kuwait	..	5.8	77	150	..	0.51	44
89	Venezuela	2.5	1.6	17	0	..	0.09	28
90	Romania	2.3	1.6	43	110	..	..	123
91	Mexico	0.7	0.6	13	240	0.2	..	17
92	Cuba	5.1	7.4	79	1,800	..	..	106
93	Panama	0.1	2.0	18	5	0.5	..	50
94	Trinidad and Tobago	..	1.0	11	0	0.3	..	18
95	Portugal	4.2	3.3	33	20	0.1	..	93
96	Singapore	0.4	5.5	86	180	0.0	..	295
97	Korea, Rep.	6.0	5.2	100	550	..	..	272
98	Poland	3.0	3.3	39	725	..	..	131
99	Argentina	2.1	1.5	31	30	0.1	..	50
100	Yugoslavia	6.6	4.0	49	210	0.0	..	139
101	Hungary	1.8	2.4	34	360	..	..	112
102	Uruguay	2.5	1.6	17	0	0.2	..	107
103	Costa Rica	1.2	0.0	0	0	8.2	..	0
104	Bulgaria	3.2	3.6	47	600	..	..	210
105	USSR	11.0	11.5	137	625	..	0.01	147
106	Czechoslovakia	3.8	4.1	53	900	..	..	197
107	Chile	2.8	3.6	49	30	..	..	108
108	Hong Kong	..	..	..	..	..	..	..
109	Greece	4.8	5.7	95	150	..	0.00	263
110	German Dem. Rep.	2.0	4.9	77	220	..	0.02	102
111	Israel	2.9	19.2	204	1,600	0.4	..	191
112	USA	8.8	6.7	..	625	..	0.03	100
113	Austria	1.2	1.3	12	10	..	0.20	63
114	Ireland	1.4	1.9	13	5	..	0.18	39
115	Spain	2.9	2.3	31	875	..	0.04	125
116	Belgium	3.4	3.1	28	140	..	0.19	135
117	Italy	2.7	2.3	27	170	..	0.20	85
118	New Zealand	1.4	2.2	21	30	..	0.15	37
119	Germany, Fed. Rep.	4.0	3.1	29	420	..	0.17	98
120	Finland	1.7	1.7	14	40	..	0.31	117
121	United Kingdom	6.4	5.0	47	380	..	0.07	62
122	Denmark	2.7	2.1	16	100	..	0.51	48
123	France	6.3	3.9	31	220	..	0.21	113
124	Australia	2.4	2.7	26	625	..	0.14	40
125	Norway	3.2	3.2	26	200	..	0.40	74
126	Canada	4.3	2.2	..	150	..	0.21	30
127	Netherlands	3.9	3.1	22	550	..	0.40	94
128	Switzerland	2.4	1.9	16	850	..	0.20	61
129	Sweden	2.8	2.9	19	60	..	0.33	92
130	Japan	1.0	1.0	10	725	..	0.36	25
All developing countries		4.2	5.5	104	34,230 T	1.5	..	68
Least developed countries		2.1	3.8	92	3,680 T	5.3	..	121
Sub-Saharan Africa		0.7	3.3	81	3,360 T	4.7	..	90
Industrial countries		6.3	5.4	55	11,700 T	..	0.20	105
World		6.0	5.4	59	45,930 T	..	..	77

Note: Summary data for regional and income groups are given in tables 23 and 24.

19 Resource flows imbalances

	Official development assistance, 1987				Interest payments on long-term debt (US$ million) 1987	Debt service (as % of exports of goods and services) 1987	Gross international reserves (months of import coverage) 1987	Ratio of exports to imports 1987	Terms of trade (1980=100) 1987
	US$ million		As % of GNP						
	Received	Given	Received	Given					
Low human development	17,130 T	..	3.6	..	5,300 T	24	1.7	0.67	86
Excluding India	15,280 T	..	7.0	..	3,780 T	23	..	..	..
1 Niger	348	..	16.1	..	73	47	..	0.87	86
2 Mali	364	..	18.6	..	13	10	0.5	0.48	86
3 Burkina Faso	283	..	16.2	..	14	..	4.4	0.37	74
4 Sierra Leone	68	..	7.3	..	1	..	1.0	0.91	93
5 Chad	198	..	20.3	..	3	4	1.4	..	..
6 Guinea	214	..	..	..	35	..	..	..	..
7 Somalia	580	..	57.0	..	4	8	0.4	0.21	84
8 Mauritania	178	..	19.0	..	28	18	1.5	0.90	98
9 Afghanistan	260	..	..	..	..	..	5.6	0.39	..
10 Benin	136	..	8.1	..	15	16	0.2	0.40	88
11 Burundi	192	..	15.3	..	15	39	2.8	0.41	75
12 Bhutan	17	..	16.7	..	1	..	..	0.28	..
13 Mozambique	649	..	40.9	..	..	..	..	0.18	..
14 Malawi	280	..	22.8	..	26	23	1.8	0.94	67
15 Sudan	902	..	10.5	..	..	..	0.1	0.69	84
16 Central African Rep.	173	..	16.1	..	9	12	3.2	0.70	84
17 Nepal	345	..	12.7	..	14	..	4.9	0.27	93
18 Senegal	642	..	13.6	..	116	22	0.1	0.55	96
19 Ethiopia	635	..	11.8	..	50	28	2.3	0.35	84
20 Zaire	621	..	10.7	..	119	13	1.8	1.39	74
21 Rwanda	243	..	11.6	..	7	11	4.6	0.34	87
22 Angola	135	..	..	..	..	..	..	..	..
23 Bangladesh	1,637	..	9.3	..	132	24	3.5	0.41	91
24 Nigeria	69	30	0.3	0.13	569	12	2.3	0.94	54
25 Yemen Arab Rep.	349	..	8.2	..	45	25	3.7	0.01	93
26 Liberia	78	..	6.9	..	6	3	0.0	1.85	93
27 Togo	123	..	10.0	..	29	14	..	0.71	86
28 Uganda	276	..	7.2	..	24	20	1.0	0.67	67
29 Haiti	218	..	9.7	..	9	7	0.6	0.69	109
30 Ghana	373	..	7.4	..	58	20	3.0	1.26	85
31 Yemen, PDR	80	..	8.1	..	15	38	2.1	0.28	73
32 Côte d'Ivoire	254	..	2.5	..	597	41	0.1	1.38	86
33 Congo	152	..	7.0	..	45	19	0.1	1.55	64
34 Namibia	17	..	..	..	..	10	..	..	..
35 Tanzania, United Rep.	882	..	25.2	..	38	19	0.3	0.30	90
36 Pakistan	858	..	2.4	..	386	26	2.2	0.72	99
37 India	1,852	129	0.7	0.06	1,517	24	5.9	0.66	114
38 Madagascar	327	..	15.8	..	83	35	3.1	0.80	105
39 Papua New Guinea	322	..	10.6	..	157	37	3.2	0.96	84
40 Kampuchea, Dem.	192	..	..	..	..	..	..	..	..
41 Cameroon	213	..	1.7	..	177	28	0.3	0.79	66
42 Kenya	565	..	7.0	..	244	34	1.4	0.55	80
43 Zambia	429	..	21.1	..	..	..	1.4	1.17	79
44 Morocco	401	..	2.4	..	624	31	1.5	0.66	106
Medium human development	14,600 T	..	0.9	..	18,500 T	24	3.0	0.79	76
Excluding China	13,150 T	..	1.1	..	17,430 T	30	..	..	..
45 Egypt	1,766	..	4.9	..	806	22	2.1	0.48	64
46 Lao PDR	194	..	8.4	..	2	..	..	0.43	..
47 Gabon	82	..	2.3	..	57	5	0.1	1.54	64
48 Oman	16	..	0.2	..	177	..	3.6	2.09	..
49 Bolivia	318	..	7.1	..	62	22	5.2	0.73	51
50 Myanmar	365	..	..	..	69	59	2.7	0.35	65
51 Honduras	258	..	6.4	..	92	26	1.0	0.92	83
52 Zimbabwe	295	..	5.0	..	..	..	2.7	1.29	84
53 Lesotho	108	..	29.4	..	5	4	1.9	..	..
54 Indonesia	1,245	..	1.8	..	2,748	33	3.9	1.19	69
55 Guatemala	241	..	3.4	..	153	26	3.5	0.73	80
56 Viet Nam	2,117	..	..	..	..	..	..	0.56	..
57 Algeria	222	26	0.3	0.04	1,377	49	4.5	1.28	56
58 Botswana	154	..	10.1	..	32	4	17.6	..	..
59 El Salvador	426	..	9.0	..	76	21	3.7	0.65	75
60 Tunisia	282	..	2.9	..	340	29	1.9	0.71	79
61 Iran, Islamic Rep.	70	..	..	..	..	..	..	..	..
62 Syrian Arab Rep.	697	..	2.9	..	112	17	1.3	0.53	78
63 Dominican Rep.	130	..	2.6	..	106	..	2.5	0.40	60
64 Saudi Arabia	22	2,888	0.0	3.40	..	..	7.9	1.13	54
65 Philippines	775	..	2.2	..	1,497	26	2.7	0.79	98
66 China	1,449	266	0.5	0.09	1,069	7	6.7	0.91	87

		Official development assistance, 1987				Interest payments on long-term debt (US$ million) 1987	Debt service (as % of exports of goods and services) 1987	Gross international reserves (months of import coverage) 1987	Ratio of exports to imports 1987	Terms of trade (1980=100) 1987
		US$ million		As % of GNP						
		Received	Given	Received	Given					
67	Libyan Arab Jamahiriya	6	76	0.0	0.30	..	..	15.4	1.24	47
68	South Africa	..	..	..	..	..	..	1.9	1.37	71
69	Lebanon	100	..	..	..	13	..	..	0.31	..
70	Mongolia	3	..	..	..	..	..	..	..	..
71	Nicaragua	141	..	4.4	..	12	..	..	0.33	77
72	Turkey	417	..	0.6	..	1,885	34	2.3	0.72	110
73	Jordan	595	..	12.0	..	183	22	2.6	0.35	106
74	Peru	292	..	0.6	..	203	13	3.2	0.64	69
75	Ecuador	203	..	1.9	..	279	22	2.4	0.90	61
76	Iraq	91	..	..	..	..	..	..	1.22	..
77	United Arab Emirates	1	19	0.5	0.08	..	..	5.7	1.66	54
78	Thailand	506	..	1.1	..	1,057	21	4.1	0.90	81
79	Paraguay	82	..	1.8	..	96	22	4.2	0.79	76
80	Brazil	288	..	0.1	..	5,834	33	3.0	1.58	97
81	Mauritius	65	..	3.7	..	31	7	3.5	0.91	108
82	Korea, Dem. Rep.	79	..	..	..	..	..	..	..	..
83	Sri Lanka	502	..	7.5	..	126	20	1.4	0.67	96
84	Albania	..	..	..	..	..	..	..	..	..
High human development		3,540 T	46,540 T	..	0.31	29,780 T	25	3.6	1.00	99
85	Malaysia	363	..	1.2	..	1,461	20	5.5	1.43	72
86	Colombia	78	..	0.2	..	1,177	36	5.2	1.19	70
87	Jamaica	169	..	5.9	..	231	28	1.1	0.54	100
88	Kuwait	3	316	0.0	1.23	..	..	6.8	1.58	54
89	Venezuela	19	24	0.0	0.06	2,518	32	10.1	1.21	54
90	Romania	..	..	..	..	503	..	1.9	1.10	..
91	Mexico	156	..	0.1	..	7,091	38	6.2	1.64	73
92	Cuba	876	..	..	..	..	..	3.3	..	106
93	Panama	40	..	0.7	..	226	7	0.2	0.29	71
94	Trinidad and Tobago	34	..	0.8	..	121	..	2.8	1.20	61
95	Portugal	65	24	0.2	0.05	1,232	39	9.9	0.68	99
96	Singapore	23	..	0.1	..	305	2	5.0	0.88	102
97	Korea, Rep.	11	..	0.0	..	2,375	28	0.9	1.15	105
98	Poland	..	..	..	..	960	15	1.4	1.13	112
99	Argentina	99	..	0.1	..	3,775	52	3.5	1.09	81
100	Yugoslavia	35	..	0.1	..	1,717	19	1.2	0.91	116
101	Hungary	..	..	..	..	1,130	27	2.9	0.97	89
102	Uruguay	18	..	0.2	..	273	26	12.0	1.04	97
103	Costa Rica	228	..	5.3	..	139	14	3.3	0.84	84
104	Bulgaria	..	..	..	..	..	..	..	..	..
105	USSR	..	4,321	..	0.25	..	..	..	..	..
106	Czechoslovakia	..	..	..	..	..	..	..	..	..
107	Chile	21	..	0.1	..	1,420	26	5.2	1.27	77
108	Hong Kong	19	..	0.0	..	..	..	..	1.00	106
109	Greece	34	28	0.1	0.07	1,260	38	3.6	0.50	93
110	German Dem. Rep.	..	162	..	0.20	..	..	..	..	..
111	Israel	1,251	..	3.6	..	1,864	25	3.9	0.59	89
112	USA	..	8,945	..	0.20	..	..	3.4	0.60	116
113	Austria	..	196	..	0.17	..	..	4.7	0.83	108
114	Ireland	..	51	..	0.28	..	..	3.0	1.17	107
115	Spain	..	176	0.0	0.07	..	..	7.4	0.70	111
116	Belgium	..	689	..	0.49	..	..	2.6	1.00	98
117	Italy	..	2,615	..	0.35	..	..	4.8	0.95	114
118	New Zealand	..	87	..	0.26	..	..	3.5	0.99	98
119	Germany, Fed. Rep.	..	4,391	..	0.39	..	..	5.0	1.29	120
120	Finland	..	433	..	0.50	..	..	3.5	1.01	109
121	United Kingdom	..	1,865	..	0.28	..	..	2.4	0.85	99
122	Denmark	..	859	..	0.88	..	..	..	0.97	..
123	France	..	6,525	..	0.74	..	..	4.0	0.91	104
124	Australia	..	627	..	0.33	..	..	3.5	0.86	72
125	Norway	..	890	..	1.09	..	..	5.0	0.95	72
126	Canada	..	1,885	..	0.47	..	..	1.6	1.00	101
127	Netherlands	..	2,094	..	0.98	..	..	3.8	1.02	93
128	Switzerland	..	547	..	0.31	..	..	10.2	0.90	113
129	Sweden	..	1,337	..	0.88	..	..	2.4	1.09	96
130	Japan	..	7,454	..	0.31	..	..	5.2	1.57	153
All developing countries		33,890 T	..	1.3	..	44,910 T	24	2.7	0.79	83
Least developed countries		10,450 T	..	14.1	..	710 T	25	2.0	0.40	86
Sub-Saharan Africa		10,400 T	..	7.9	..	2,520 T	20	1.5	0.80	84
Industrial countries		..	46,200 T	..	0.31	..	..	3.5	0.96	103
World		..	..	..	..	..	..	3.0	0.87	86

Note: Summary data for regional and income groups are given in tables 23 and 24.

20 Demographic balance sheet

	Population (million)			Annual population growth rate (%)		Dependency ratio		Contraceptive prevalence rate (%)	Fertility rate	Crude birth rate	Crude death rate
	1960	1988	2000 est	1960-88	1988-2000	1960	1985	1985	1988	1988	1988
Low human development	790 T	1,580 T	2,130 T	2.5	2.5	80.9	83.1	23	5.3	39	13
Excluding India	350 T	760 T	1,090 T	2.8	3.0	86.9	95.0	10	6.4	46	16
1 Niger	3.2	6.7	9.8	2.6	3.2	89.3	99.9	1	7.1	51	21
2 Mali	4.6	8.9	13.0	2.3	3.0	85.2	96.0	6	6.7	50	21
3 Burkina Faso	4.5	8.6	12.0	2.3	2.9	81.8	87.5	1	6.5	47	18
4 Sierra Leone	2.2	4.0	5.4	2.0	2.6	77.0	88.7	4	6.5	48	23
5 Chad	3.1	5.4	7.3	2.1	2.6	76.2	84.8	1	5.9	44	19
6 Guinea	3.7	6.6	8.9	2.1	2.6	84.4	85.2	1	6.2	46	22
7 Somalia	2.9	7.1	9.8	3.2	2.7	83.8	94.6	2	6.6	50	20
8 Mauritania	1.0	1.9	2.7	2.4	2.8	81.6	89.6	1	6.5	46	19
9 Afghanistan	11.0	16.0	27.0	1.4	4.5	80.2	80.1	2	6.9	50	22
10 Benin	2.3	4.5	6.6	2.5	3.3	92.3	98.5	6	7.0	50	19
11 Burundi	2.9	5.2	7.3	2.0	2.9	76.8	92.7	9	6.3	46	17
12 Bhutan	0.9	1.5	1.9	1.9	2.3	76.7	76.5	..	5.5	38	17
13 Mozambique	7.5	15.0	20.0	2.5	2.7	79.7	88.2	..	6.4	45	18
14 Malawi	3.5	7.9	12.0	2.9	3.3	93.4	94.5	7	7.0	53	20
15 Sudan	11.0	24.0	34.0	2.7	2.9	89.6	92.2	5	6.4	44	16
16 Central African Rep.	1.6	2.8	3.8	2.0	2.6	74.5	86.5	..	5.9	44	20
17 Nepal	9.4	18.0	24.0	2.4	2.3	73.3	82.7	15	5.9	39	15
18 Senegal	3.0	7.0	9.7	3.0	2.7	83.8	89.8	12	6.4	46	19
19 Ethiopia	24.0	45.0	61.0	2.2	2.6	88.9	101.2	2	6.2	44	23
20 Zaire	16.0	34.0	49.0	2.7	3.2	89.1	94.8	1	6.1	45	14
21 Rwanda	2.7	6.8	10.0	3.3	3.4	90.7	105.5	10	8.3	51	17
22 Angola	4.8	9.5	13.0	2.5	2.8	80.4	90.9	1	6.4	47	20
23 Bangladesh	51.0	110.0	151.0	2.7	2.7	80.8	95.4	25	5.5	42	15
24 Nigeria	42.0	106.0	159.0	3.3	3.5	91.3	102.8	5	7.0	50	15
25 Yemen Arab Rep.	4.0	7.6	11.0	2.3	3.3	83.8	105.2	1	7.0	48	16
26 Liberia	1.0	2.4	3.5	3.0	3.3	81.8	93.2	7	6.5	45	13
27 Togo	1.5	3.3	4.7	2.8	3.2	85.2	92.2	..	6.1	45	14
28 Uganda	6.6	17.0	26.0	3.5	3.6	96.8	102.4	1	6.9	50	15
29 Haiti	3.7	6.3	7.8	1.9	1.9	78.7	78.5	7	4.7	34	13
30 Ghana	6.8	14.0	20.0	2.7	3.1	92.8	92.6	10	6.4	44	13
31 Yemen, PDR	1.2	2.3	3.4	2.4	3.2	91.0	91.8	..	6.7	47	16
32 Côte d'Ivoire	3.8	12.0	19.0	4.1	3.9	90.0	104.5	3	7.4	51	14
33 Congo	1.0	1.9	2.6	2.4	2.8	79.1	88.7	..	6.0	44	16
34 Namibia	0.8	1.8	2.6	2.8	3.2	83.8	93.7	..	6.1	43	12
35 Tanzania, United Rep.	10.0	25.0	40.0	3.4	3.7	95.4	103.6	1	7.1	50	14
36 Pakistan	50.0	115.0	162.0	3.0	2.9	92.3	90.1	11	6.4	46	12
37 India	442.0	820.0	1,043.0	2.2	2.0	76.1	72.1	35	4.3	32	11
38 Madagascar	5.3	11.0	17.0	2.7	3.3	81.3	90.3	..	6.6	45	14
39 Papua New Guinea	1.9	3.8	5.1	2.5	2.5	76.6	78.6	4	5.7	38	12
40 Kampuchea, Dem.	5.4	7.9	10.0	1.3	2.1	82.5	54.3	..	4.7	41	17
41 Cameroon	5.5	11.0	15.0	2.4	2.7	76.9	89.5	2	5.7	42	16
42 Kenya	8.3	23.0	38.0	3.7	4.1	98.8	118.6	17	8.1	53	12
43 Zambia	3.1	7.9	12.0	3.3	3.7	90.8	104.1	1	7.2	50	13
44 Morocco	12.0	24.0	31.0	2.6	2.3	90.1	85.4	36	4.8	35	10
Medium human development	1,130 T	2,060 T	2,500 T	2.1	1.6	80.6	64.6	63	3.2	25	8
Excluding China	470 T	950 T	1,220 T	2.5	2.1	84.6	77.0	45	4.1	32	9
45 Egypt	26.0	51.0	67.0	2.5	2.2	84.3	79.8	32	4.8	35	10
46 Lao PDR	2.2	3.9	5.1	2.1	2.4	80.1	83.6	..	5.7	41	16
47 Gabon	0.5	1.1	1.6	2.9	3.3	63.0	67.1	..	5.0	39	16
48 Oman	0.5	1.4	2.1	3.6	3.4	85.3	87.9	..	7.2	46	13
49 Bolivia	3.4	6.9	9.7	2.5	2.9	85.3	88.5	26	6.0	43	14
50 Myanmar	22.0	40.0	51.0	2.2	2.1	80.3	75.5	5	4.0	31	10
51 Honduras	1.9	4.8	6.8	3.3	2.9	90.3	98.5	35	5.5	39	8
52 Zimbabwe	3.8	9.2	13.0	3.2	3.1	97.6	96.0	40	5.8	41	10
53 Lesotho	0.9	1.7	2.4	2.4	2.9	80.0	86.1	5	5.8	41	12
54 Indonesia	96.0	175.0	208.0	2.2	1.5	77.0	73.0	48	3.2	28	11
55 Guatemala	4.0	8.7	12.0	2.8	2.9	94.9	95.5	23	5.7	41	9
56 Viet Nam	35.0	64.0	83.0	2.2	2.1	75.1	82.1	58	4.0	32	9
57 Algeria	11.0	24.0	33.0	2.9	2.8	90.9	97.3	7	6.0	39	9
58 Botswana	0.5	1.2	1.8	3.3	3.4	103.6	109.0	29	6.2	47	12
59 El Salvador	2.6	5.1	6.7	2.4	2.4	92.5	97.6	48	4.8	37	9
60 Tunisia	4.2	7.8	9.8	2.2	1.9	90.6	76.8	41	4.0	30	7
61 Iran, Islamic Rep.	20.0	53.0	74.0	3.5	2.9	96.7	86.7	23	5.6	41	8
62 Syrian Arab Rep.	4.6	12.0	18.0	3.4	3.5	93.0	103.5	20	6.7	44	7
63 Dominican Rep.	3.2	6.9	8.6	2.7	1.9	98.8	75.2	50	3.7	31	7
64 Saudi Arabia	4.1	13.0	21.0	4.3	3.9	87.4	90.5	..	7.2	42	7
65 Philippines	28.0	59.0	77.0	2.8	2.2	91.0	80.2	45	4.3	33	8
66 China	657.0	1,105.0	1,286.0	1.9	1.3	77.7	53.9	77	2.4	20	7

		Population (million)			Annual population growth rate (%)		Dependency ratio		Contraceptive prevalence rate (%)	Fertility rate	Crude birth rate	Crude death rate
		1960	1988	2000 est	1960-88	1988-2000	1960	1985	1985	1988	1988	1988
67	Libyan Arab Jamahiriya	1.3	4.2	6.5	4.2	3.6	89.7	95.0	..	6.8	44	9
68	South Africa	17.0	34.0	43.0	2.4	2.1	81.2	72.0	..	4.4	31	10
69	Lebanon	1.9	2.8	3.6	1.5	2.0	87.2	74.2	53	3.3	29	8
70	Mongolia	0.9	2.1	3.0	2.9	3.0	84.0	81.4	..	5.4	39	8
71	Nicaragua	1.5	3.6	5.3	3.2	3.1	101.3	97.1	27	5.5	41	8
72	Turkey	28.0	54.0	67.0	2.4	1.8	81.1	68.4	51	3.5	28	8
73	Jordan	1.7	4.0	6.3	3.1	4.0	94.2	103.3	26	7.2	45	7
74	Peru	9.9	21.0	28.0	2.8	2.3	87.8	78.8	46	4.4	34	9
75	Ecuador	4.4	10.0	14.0	3.0	2.6	95.4	83.4	44	4.6	35	8
76	Iraq	6.8	18.0	26.0	3.5	3.4	94.5	98.4	14	6.3	42	8
77	United Arab Emirates	0.1	1.5	2.0	10.5	2.3	88.8	48.1	..	4.8	23	4
78	Thailand	26.0	54.0	64.0	2.6	1.4	90.3	67.0	66	2.5	23	7
79	Paraguay	1.8	4.0	5.5	3.0	2.7	103.9	80.1	49	4.6	34	7
80	Brazil	73.0	144.0	179.0	2.5	1.8	86.9	68.7	66	3.4	28	8
81	Mauritius	0.7	1.1	1.2	1.8	1.2	96.4	51.8	78	1.9	19	6
82	Korea, Dem. Rep.	11.0	22.0	28.0	2.7	2.1	88.8	73.2	..	3.6	29	5
83	Sri Lanka	9.9	17.0	19.0	1.9	1.2	84.1	63.3	62	2.6	23	6
84	Albania	1.6	3.1	3.8	2.4	1.6	86.3	66.0	..	3.0	24	6
High human development		1,080 T	1,470 T	1,590 T	1.0	0.7	62.3	53.9	65	2.1	16	9
85	Malaysia	8.1	17.0	21.0	2.6	1.9	94.9	71.1	51	3.5	28	6
86	Colombia	16.0	31.0	38.0	2.4	1.8	97.5	69.3	65	3.5	29	7
87	Jamaica	1.6	2.4	2.9	1.5	1.4	85.1	74.7	52	2.8	25	5
88	Kuwait	0.3	1.9	2.8	7.2	3.1	58.9	70.4	..	4.8	32	3
89	Venezuela	7.5	19.0	25.0	3.3	2.3	94.4	75.1	49	3.7	30	5
90	Romania	18.0	23.0	24.0	0.8	0.5	53.6	51.8	58	2.1	15	11
91	Mexico	38.0	85.0	107.0	2.9	2.0	94.8	80.2	53	3.5	29	6
92	Cuba	7.0	10.0	11.0	1.3	0.8	64.4	50.7	60	1.7	16	7
93	Panama	1.1	2.3	2.9	2.6	1.8	90.5	72.5	61	3.1	26	5
94	Trinidad and Tobago	0.8	1.2	1.5	1.4	1.5	88.7	62.2	54	2.7	23	6
95	Portugal	8.8	10.0	11.0	0.5	0.3	59.1	55.1	66	1.7	14	10
96	Singapore	1.6	2.6	3.0	1.7	0.9	82.8	42.1	74	1.6	16	6
97	Korea, Rep.	25.0	43.0	48.0	1.9	1.0	82.7	52.1	70	2.0	19	6
98	Poland	30.0	38.0	40.0	0.9	0.5	64.6	53.7	75	2.2	16	10
99	Argentina	21.0	32.0	36.0	1.5	1.2	57.0	64.1	74	2.9	21	9
100	Yugoslavia	18.0	24.0	25.0	0.9	0.5	58.3	48.5	55	1.9	15	9
101	Hungary	10.0	11.0	11.0	0.2	0.0	52.4	51.2	73	1.7	12	13
102	Uruguay	2.5	3.1	3.4	0.7	0.7	56.2	60.3	..	2.6	19	10
103	Costa Rica	1.2	2.9	3.7	3.0	2.2	102.4	68.7	70	3.2	28	4
104	Bulgaria	7.9	9.0	9.1	0.5	0.1	50.5	48.3	76	1.9	13	12
105	USSR	214.0	284.0	308.0	1.0	0.7	59.8	53.4	..	2.4	18	10
106	Czechoslovakia	14.0	16.0	16.0	0.5	0.3	56.4	55.0	..	2.0	14	12
107	Chile	7.6	13.0	15.0	1.9	1.5	79.0	59.5	43	2.7	23	6
108	Hong Kong	3.1	5.7	6.4	2.2	1.1	77.6	44.3	72	1.7	15	6
109	Greece	8.3	10.0	10.0	0.7	0.2	53.2	52.9	..	1.7	12	10
110	German Dem. Rep.	17.0	17.0	17.0	-0.1	0.0	53.3	48.9	..	1.7	13	13
111	Israel	2.1	4.4	5.3	2.7	1.5	69.3	70.5	..	2.9	22	7
112	USA	181.0	245.0	266.0	1.1	0.7	67.4	50.7	68	1.8	15	9
113	Austria	7.0	7.5	7.5	0.2	0.0	51.9	48.6	71	1.5	12	12
114	Ireland	2.8	3.7	4.1	0.9	0.9	73.2	67.3	60	2.5	18	9
115	Spain	30.0	39.0	41.0	0.9	0.4	55.4	53.9	51	1.7	13	9
116	Belgium	9.2	9.9	10.0	0.3	0.1	55.0	48.8	81	1.5	12	11
117	Italy	50.0	57.0	58.0	0.5	0.1	51.7	47.8	78	1.4	11	10
118	New Zealand	2.4	3.3	3.6	1.2	0.7	71.0	53.5	..	1.9	16	8
119	Germany, Fed. Rep.	55.0	61.0	60.0	0.3	-0.1	47.5	43.4	78	1.4	10	12
120	Finland	4.4	4.9	5.1	0.4	0.2	60.3	46.9	77	1.6	12	10
121	United Kingdom	52.0	57.0	58.0	0.3	0.1	53.7	52.3	83	1.8	13	12
122	Denmark	4.6	5.1	5.1	0.4	0.0	55.8	50.6	63	1.5	11	11
123	France	46.0	56.0	58.0	0.7	0.4	61.3	51.9	19	1.8	14	10
124	Australia	10.0	16.0	19.0	1.7	1.1	62.8	50.9	67	1.8	15	7
125	Norway	3.6	4.2	4.3	0.6	0.3	58.7	56.1	71	1.7	12	11
126	Canada	18.0	26.0	29.0	1.3	0.7	69.6	46.9	73	1.6	14	7
127	Netherlands	11.0	15.0	15.0	0.9	0.3	63.9	46.0	72	1.4	12	9
128	Switzerland	5.4	6.5	6.6	0.7	0.1	50.8	46.0	70	1.6	12	10
129	Sweden	7.5	8.3	8.3	0.4	0.0	51.4	54.8	78	1.6	11	12
130	Japan	94.0	122.0	129.0	0.9	0.4	56.1	46.7	64	1.7	12	7
All developing countries		2,060 T	3,900 T	4,960 T	2.3	2.0	80.9	72.3	47	4.0	31	10
Least developed countries		210 T	410 T	580 T	2.5	2.8	83.7	92.6	10	6.0	44	17
Sub-Saharan Africa		210 T	470 T	680 T	2.8	3.1	87.9	96.8	6	6.5	46	16
Industrial countries		940 T	1,200 T	1,270 T	0.8	0.5	59.1	50.8	66	1.9	15	10
World		3,000 T	5,100 T	6,230 T	1.8	1.6	74.1	67.3	50	3.5	27	10

Note: Summary data for regional and income groups are given in tables 23 and 24.

21 Natural resources balance sheet

	Land area (millions hectares)	Arable land: As % of total land 1984-86	Arable land: Annual percentage increase 1965-85	Population density (per 1,000 hectares) 1988	Livestock: Per capita 1984-86	Livestock: Annual percentage increase 1975-85	Fuelwood production: Per capita 1984-86	Fuelwood production: Annual percentage increase 1975-85	Annual deforestation (1,000 hectares) 1980-88	Internal renewable water resources per capita (1,000 m³) 1988
Low human development	2,660 T	36	0.21	2,317	0.73	1.8	0.46	2.5	121	3.4
Excluding India	2,360 T	33	0.22	1,863	0.91	1.7	0.65	3.0	..	5.5
1 Niger	127	10	0.45	54	2.57	3.7	0.60	2.7	67	..
2 Mali	122	26	0.07	74	1.98	1.2	0.57	2.6	36	..
3 Burkina Faso	27	46	0.21	282	1.13	2.8	0.82	2.3	80	..
4 Sierra Leone	7	56	0.50	543	0.24	2.6	2.08	1.7	6	..
5 Chad	126	38	0.03	44	2.18	2.5	0.61	2.3	80	..
6 Guinea	25	19	0.01	273	0.46	2.0	0.60	2.3	86	..
7 Somalia	63	48	0.02	81	5.53	0.7	0.68	3.6	13	2.3
8 Mauritania	103	38	-0.01	21	4.96	0.7	..	3.4	13	0.2
9 Afghanistan	65	59	0.02	311	1.98	0.1	0.34	1.0	..	2.5
10 Benin	11	20	0.97	414	0.91	3.1	1.03	2.9	67	5.7
11 Burundi	3	87	1.61	2,062	0.35	0.9	0.76	2.3	1	..
12 Bhutan	5	7	0.82	327	0.41	1.8	2.16	1.3	1	..
13 Mozambique	78	60	0.05	198	0.14	-0.6	1.04	4.4	120	..
14 Malawi	9	45	0.50	843	0.28	2.3	0.87	3.0	150	..
15 Sudan	238	29	0.10	102	2.81	3.7	0.81	3.0	104	1.2
16 Central African Rep.	62	8	0.15	46	1.40	8.0	1.14	2.8	55	..
17 Nepal	14	31	0.87	1,320	0.92	0.3	0.90	2.4	84	9.4
18 Senegal	19	57	0.37	374	0.94	1.3	0.54	2.9	50	..
19 Ethiopia	110	54	0.06	442	1.77	0.1	0.85	2.4	88	2.3
20 Zaire	227	7	0.30	149	0.19	1.3	0.91	3.0	347	..
21 Rwanda	2	58	1.05	2,781	0.34	2.9	0.91	1.1	5	..
22 Angola	125	26	0.03	78	0.58	1.6	0.45	3.0	84	..
23 Bangladesh	13	73	0.07	8,369	0.36	0.6	0.26	2.8	8	12.1
24 Nigeria	91	57	0.33	1,201	0.56	1.3	0.92	3.5	400	..
25 Yemen Arab Rep.	20	45	0.43	394	0.82	-5.8	..	..	..	0.1
26 Liberia	10	6	-0.01	259	0.30	3.3	1.80	3.6	46	..
27 Togo	5	30	1.07	615	0.69	0.3	0.20	2.7	12	3.4
28 Uganda	20	57	0.85	891	0.66	3.1	0.70	3.3	50	..
29 Haiti	3	51	0.43	2,653	0.65	1.3	0.96	2.5	2	1.5
30 Ghana	23	27	0.01	675	0.45	2.0	0.64	3.3	72	3.4
31 Yemen, PDR	33	28	0.05	72	1.25	1.5	0.13	2.6	..	0.6
32 Côte d'Ivoire	32	22	1.29	355	0.41	4.3	0.78	3.8	510	6.6
33 Congo	34	31	0.03	57	0.21	4.8	0.91	2.6	22	..
34 Namibia	..	..	..	..	..	..	..	..	..	..
35 Tanzania, United Rep.	89	45	0.23	294	1.10	2.5	0.95	3.5	130	..
36 Pakistan	77	33	0.24	1,424	0.86	3.1	0.19	3.0	9	2.7
37 India	297	61	0.11	2,736	0.57	2.0	0.29	2.0	147	2.3
38 Madagascar	58	64	0.14	193	1.33	2.5	0.60	2.8	156	3.6
39 Papua New Guinea	45	1	1.12	86	0.46	1.5	1.58	1.5	23	..
40 Kampuchea, Dem.	18	21	0.21	457	0.46	4.1	0.65	0.3	30	10.9
41 Cameroon	47	33	0.30	235	0.98	2.6	0.93	2.7	110	18.8
42 Kenya	57	11	0.33	428	1.17	5.0	1.52	4.1	39	0.6
43 Zambia	74	54	0.04	103	0.47	3.9	1.34	2.6	80	..
44 Morocco	45	53	0.92	540	0.92	-2.8	0.06	5.0	13	1.3
Medium human development	4,020 T	34	0.42	1,026	0.66	1.9	0.35	1.8	..	6.9
Excluding China	3,090 T	32	0.56	834	0.79	2.8	0.57	1.5	..	12.6
45 Egypt	100	2	-0.34	516	0.26	3.3	0.04	2.6	..	0.0
46 Lao PDR	23	7	0.19	197	0.84	6.9	1.06	1.9	130	59.5
47 Gabon	26	20	0.21	48	0.30	2.4	2.56	1.4	15	..
48 Oman	21	5	0.15	67	0.89	15.5	..	..	..	1.4
49 Bolivia	108	28	0.39	66	2.94	2.3	0.18	2.7	117	..
50 Myanmar	66	16	-0.10	610	0.44	3.9	0.42	2.0	105	27.0
51 Honduras	11	46	0.30	443	0.82	3.6	1.04	3.5	90	20.6
52 Zimbabwe	39	20	0.47	262	0.88	-2.0	0.71	3.5	80	..
53 Lesotho	3	76	-0.57	556	2.09	0.9	0.34	2.5	..	..
54 Indonesia	181	18	0.49	985	0.20	4.3	0.76	2.1	620	14.2
55 Guatemala	11	29	0.61	824	0.50	3.4	0.84	2.8	90	13.0
56 Viet Nam	33	22	0.79	1,992	0.29	3.0	0.36	2.2	65	..
57 Algeria	238	16	-0.53	104	0.93	4.2	0.08	3.1	40	0.8
58 Botswana	57	80	0.30	23	3.71	1.8	1.03	3.9	20	0.8
59 El Salvador	2	65	0.28	3,034	0.31	-0.8	0.98	3.0	4	3.0
60 Tunisia	16	50	0.67	497	1.08	-0.3	0.37	2.3	5	0.5
61 Iran, Islamic Rep.	164	36	-0.08	305	1.23	..	0.05	0.4	20	2.4
62 Syrian Arab Rep.	18	76	-0.05	662	1.35	6.7	..	9.6	..	0.6
63 Dominican Rep.	5	49	-1.25	1,411	0.74	5.1	0.15	8.3	4	2.9
64 Saudi Arabia	215	40	0.03	63	0.60	6.0	..	..	..	0.2
65 Philippines	30	38	0.62	2,001	0.27	1.4	0.53	2.6	92	5.4
66 China	933	41	-0.04	1,191	0.56	1.2	0.16	2.0	..	2.5

	Land area (millions hectares)	Arable land		Population density (per 1,000 hectares) 1988	Livestock		Fuelwood production		Annual defore-station (1,000 hectares) 1980-88	Internal renewable water resources per capita (1,000 m³) 1988
		As % of total land 1984-86	Annual percentage increase 1965-85		Per capita 1984-86	Annual percentage increase 1975-85	Per capita 1984-86	Annual percentage increase 1975-85		
67 Libyan Arab Jamahiriya	176	9	1.29	24	1.82	1.7	0.14	1.0	..	0.2
68 South Africa	122	77	-0.19	294	1.59	..	0.22	0.2	..	1.4
69 Lebanon	1	30	0.14	2,833	0.25	1.7	0.17	-0.3	..	1.7
70 Mongolia	157	80	-0.58	14	12.02	-0.4	0.71	..	..	11.6
71 Nicaragua	12	54	0.95	315	1.02	-0.3	0.83	3.1	121	46.7
72 Turkey	77	47	-0.11	695	1.53	..	0.23	-10.5	..	3.7
73 Jordan	10	8	3.84	424	0.46	2.9	..	2.3	..	0.2
74 Peru	128	23	-0.08	170	1.14	-0.9	0.33	1.9	270	1.8
75 Ecuador	28	26	2.80	379	1.17	3.8	0.64	4.1	340	29.9
76 Iraq	43	22	0.31	419	0.83	-3.4	0.01	3.2	..	1.9
77 United Arab Emirates	8	3	0.27	183	0.83	10.3	..	..	..	0.2
78 Thailand	51	39	2.19	1,071	0.30	1.3	0.62	2.2	379	2.0
79 Paraguay	40	44	1.07	104	2.46	2.8	1.32	2.6	212	..
80 Brazil	846	28	1.48	174	1.46	2.4	1.24	2.3	2,323	35.2
81 Mauritius	0	62	0.66	6,070	0.14	1.8	0.01	-4.4	..	2.0
82 Korea, Dem. Rep.	12	20	0.91	1,862	0.22	4.7	0.19	1.8	..	..
83 Sri Lanka	6	41	1.07	2,659	0.21	0.9	0.49	1.8	58	2.5
84 Albania	3	42	0.39	1,211	0.96	1.4	0.54	..	..	3.0
High human development	6,170 T	36	0.16	1,168	1.12	0.9	0.26	3.9	..	10.5
85 Malaysia	33	13	0.82	516	0.22	3.6	0.49	2.3	255	26.9
86 Colombia	104	34	0.10	300	1.12	0.8	0.51	2.2	890	34.3
87 Jamaica	1	43	-0.66	2,294	0.43	1.5	0.01	6.4	2	3.3
88 Kuwait	2	8	0.10	1,205	0.36	11.2	..	..	..	0.0
89 Venezuela	88	24	0.40	218	1.02	3.0	0.04	3.2	245	44.5
90 Romania	23	65	0.07	1,027	1.85	3.6	0.20	-1.6	..	1.6
91 Mexico	192	51	0.05	452	1.00	1.3	0.17	2.7	615	4.1
92 Cuba	11	52	1.64	940	0.99	2.2	0.28	3.3	2	3.3
93 Panama	8	23	0.45	312	0.84	0.8	0.78	1.6	36	60.8
94 Trinidad and Tobago	1	30	0.40	2,462	0.19	2.4	0.02	2.6	1	..
95 Portugal	9	42	-1.00	1,143	1.03	3.1	0.06	1.0	..	3.3
96 Singapore	0	11	-4.07	46,895	0.30	-3.6	..	..	..	0.2
97 Korea, Rep.	10	23	0.25	4,494	0.15	5.5	0.15	-1.4	..	1.4
98 Poland	30	62	-0.29	1,257	0.94	-0.7	0.10	7.6	..	1.3
99 Argentina	274	65	0.13	118	3.13	-0.9	0.21	1.8	1,550	21.4
100 Yugoslavia	26	55	-0.20	930	0.95	0.1	0.18	0.9	..	6.3
101 Hungary	9	71	-0.30	1,155	1.30	1.7	0.27	1.7	..	0.6
102 Uruguay	17	87	-0.01	179	10.87	2.5	0.78	4.9	..	..
103 Costa Rica	5	56	3.22	566	1.06	2.9	0.97	2.8	65	33.1
104 Bulgaria	11	56	0.73	833	1.89	0.9	0.20	5.4	..	2.0
105 USSR	2,227	27	0.03	130	1.28	0.7	0.31	0.3	..	..
106 Czechoslovakia	13	54	-0.16	1,258	0.85	0.9	0.09	-2.7	..	1.8
107 Chile	75	23	0.90	171	0.96	0.4	0.50	1.6	50	..
108 Hong Kong	..	..	..	..	..	..	..	..	..	..
109 Greece	13	71	0.17	768	1.71	1.2	0.19	-0.4	..	4.5
110 German Dem. Rep.	11	59	-0.15	1,597	1.29	1.4	0.04	2.3	..	1.0
111 Israel	2	61	0.08	2,236	0.20	1.6	..	..	..	0.4
112 USA	917	47	-0.04	269	0.78	-1.0	0.43	18.2	..	10.1
113 Austria	8	43	-0.60	907	0.92	0.9	0.19	3.5	..	7.5
114 Ireland	7	84	0.24	551	2.67	-2.1	0.01	19.1	..	13.2
115 Spain	50	62	-0.23	791	0.96	1.6	0.06	-1.3	..	2.8
116 Belgium	3	46	-0.91	3,029	0.88	0.9	0.05	9.5	..	0.8
117 Italy	29	59	-0.81	1,955	0.51	1.0	0.09	4.4	..	3.1
118 New Zealand	27	54	0.42	128	24.22	1.9	0.02	-14.9	..	115.6
119 Germany, Fed. Rep.	24	49	-0.38	2,473	0.67	1.3	0.06	2.3	..	1.3
120 Finland	31	8	-0.28	162	0.62	..	0.64	-6.5	..	22.2
121 United Kingdom	24	77	0.39	2,326	0.79	-1.3	..	-2.8	..	2.1
122 Denmark	4	67	-0.27	1,209	2.29	0.9	0.07	17.0	..	2.2
123 France	55	57	-0.42	1,014	0.84	-0.2	0.19	-0.3	..	3.1
124 Australia	762	64	0.07	22	11.06	-0.4	0.18	7.8	..	20.8
125 Norway	31	3	-0.18	135	1.02	2.6	0.19	4.5	..	97.1
126 Canada	922	8	0.82	29	0.93	2.6	0.24	5.2	..	109.5
127 Netherlands	3	59	-0.56	4,335	1.26	4.1	0.01	11.5	..	0.7
128 Switzerland	4	51	-0.36	1,606	0.68	-0.2	0.13	1.2	..	6.7
129 Sweden	41	9	-0.32	202	0.58	0.1	0.53	5.8	..	21.2
130 Japan	38	14	0.98	3,275	0.13	3.1	..	-5.1	..	4.4
All developing countries	7,490 T	36	0.33	1,583	0.73	1.9	0.39	2.1	582	6.2
Least developed countries	1,530 T	37	0.15	2,635	1.07	1.5	0.63	2.7	74	10.5
Sub-Saharan Africa	2,030 T	37	0.18	595	0.99	1.8	0.84	2.9	125	3.5
Industrial countries	5,360 T	34	0.14	1,088	1.11	0.7	0.26	4.3	..	9.3
World	12,850 T	35	0.25	1,467	0.82	1.6	0.36	2.6	..	6.8

Note: Summary data for regional and income groups are given in tables 23 and 24.

22 National income accounts

	GDP (US$ billion) 1987	GNP (US$ billion) 1987	GNP per capita (US$) 1987	Annual GNP per capita growth rate (%) 1965-80	Annual GNP per capita growth rate (%) 1980-87	Annual rate of inflation (%) 1980-87	Overall budget surplus or deficit (as % of GNP) 1987	Gross domestic saving (as % of GDP) 1987
Low human development	410 T	450 T	300	1.4	1.3	8.2	-7.8	17
Excluding India	190 T	210 T	300	1.2	-1.0	..	-7.3	11
1 Niger	2.2	1.9	260	-2.5	-4.9	4.1	..	5
2 Mali	2.0	1.6	210	2.1	0.5	4.2	-10.0	0
3 Burkina Faso	1.7	1.4	190	1.7	2.5	4.4	1.6	1
4 Sierra Leone	0.9	1.2	300	0.7	-2.0	50.0	-8.9	10
5 Chad	1.0	0.8	150	-1.9	..	5.3	-1.3	-12
6 Guinea	..	..	..	1.3	-0.1	..	..	..
7 Somalia	1.9	1.7	290	-0.1	-2.5	37.8	..	1
8 Mauritania	0.8	0.8	440	-0.1	-1.6	9.8	..	14
9 Afghanistan	..	..	..	0.6	..	..	..	..
10 Benin	1.6	1.3	310	-0.3	-0.8	8.2	..	4
11 Burundi	1.2	1.2	250	2.4	-0.1	7.5	..	8
12 Bhutan	0.3	0.2	150	0.6	..	..	..	..
13 Mozambique	1.5	2.1	170	0.6	-8.2	26.9	..	-10
14 Malawi	1.1	1.2	160	3.2	0.0	12.4	-10.3	12
15 Sudan	8.2	7.6	330	0.8	-4.3	31.7	..	6
16 Central African Rep.	1.0	0.9	330	0.8	-0.7	7.9	..	-2
17 Nepal	2.6	2.8	160	0.0	..	8.8	-7.5	11
18 Senegal	4.7	3.5	520	-0.5	0.1	9.1	..	6
19 Ethiopia	4.8	5.5	130	0.4	-1.6	2.6	..	3
20 Zaire	5.8	5.3	150	-1.3	-2.5	53.5	..	10
21 Rwanda	2.1	2.0	300	1.6	-1.0	4.5	..	5
22 Angola	..	..	470	0.6	0.1	..	..	..
23 Bangladesh	17.6	17.4	160	-0.3	0.8	11.1	-1.4	2
24 Nigeria	24.4	39.5	370	4.2	-4.7	10.1	-10.3	20
25 Yemen Arab Rep.	4.3	4.9	590	6.5	2.0	11.4	-19.9	-12
26 Liberia	1.0	1.0	450	0.5	-5.2	1.5	-7.9	18
27 Togo	1.2	1.0	290	1.7	-3.4	6.6	-5.0	6
28 Uganda	3.6	4.1	260	-2.2	-2.4	95.2	-4.4	5
29 Haiti	2.3	2.2	360	0.9	-2.1	7.9	..	5
30 Ghana	5.1	5.3	390	-0.8	-2.0	48.3	0.6	4
31 Yemen, PDR	0.8	1.0	420	0.6	-6.1	5.0	..	..
32 Côte d'Ivoire	7.7	8.3	740	2.8	-3.0	4.4	..	19
33 Congo	2.2	1.8	870	2.7	1.7	1.8	..	21
34 Namibia	..	..	..	0.6	..	..	..	..
35 Tanzania, United Rep.	3.1	5.2	180	0.8	-1.7	24.9	-4.9	-6
36 Pakistan	31.7	36.2	350	1.8	3.3	7.3	-8.2	11
37 India	220.8	241.3	300	1.5	3.2	7.7	-8.1	22
38 Madagascar	2.1	2.2	210	-0.4	-3.7	17.4	..	7
39 Papua New Guinea	3.0	2.6	700	0.6	0.1	4.4	-3.3	17
40 Kampuchea, Dem.	..	..	..	0.6	..	..	..	..
41 Cameroon	12.7	10.4	970	2.4	4.5	8.1	-3.5	15
42 Kenya	6.9	7.5	330	3.1	-0.9	10.3	-4.6	20
43 Zambia	2.0	1.7	250	-1.2	-5.6	28.7	-15.8	20
44 Morocco	16.8	14.2	610	2.7	0.3	7.3	-9.3	14
Medium human development	1,230 T	1,280 T	690	3.9	5.5	11.4	..	26
Excluding China	930 T	960 T	1,250	3.7	0.7	..	..	22
45 Egypt	34.5	36.0	680	2.8	2.9	9.2	-6.6	8
46 Lao PDR	0.7	0.6	170	0.6	..	46.5	..	..
47 Gabon	3.5	2.9	2,700	5.6	-3.5	2.6	0.1	34
48 Oman	8.2	7.8	5,810	9.0	8.6	-6.5	-5.2	..
49 Bolivia	4.5	4.2	580	1.7	-4.9	601.8	..	2
50 Myanmar	..	..	200	1.6	1.2	..	-0.8	..
51 Honduras	3.5	3.6	810	1.1	-2.0	4.9	..	13
52 Zimbabwe	5.2	5.3	580	1.7	-1.3	12.4	-10.8	22
53 Lesotho	0.3	0.6	370	6.8	-0.9	12.3	-2.6	-73
54 Indonesia	69.7	76.8	450	5.2	1.7	8.5	-0.9	29
55 Guatemala	7.0	6.8	950	3.0	-3.6	12.7	..	7
56 Viet Nam	..	..	..	0.6	..	..	..	..
57 Algeria	64.6	63.6	2,680	4.2	0.6	5.6	..	29
58 Botswana	1.5	1.2	1,050	9.9	8.0	8.4	28.2	..
59 El Salvador	4.8	4.2	860	1.5	-2.0	16.5	0.6	8
60 Tunisia	8.5	9.0	1,180	4.7	0.7	8.2	..	20
61 Iran, Islamic Rep.	..	..	..	2.9	3.5	..	-3.9	..
62 Syrian Arab Rep.	24.0	20.4	1,640	5.1	-3.2	11.0	-10.9	10
63 Dominican Rep.	4.9	4.9	730	3.8	-1.5	16.3	-2.0	..
64 Saudi Arabia	71.5	90.5	6,200	0.6	-11.8	-2.8	..	17
65 Philippines	34.6	34.6	590	3.2	-3.3	16.7	-5.0	16
66 China	293.4	319.8	290	4.1	9.1	4.2	..	38

		GDP (US$ billion) 1987	GNP (US$ billion) 1987	GNP per capita (US$) 1987	Annual GNP per capita growth rate (%) 1965-80	Annual GNP per capita growth rate (%) 1980-87	Annual rate of inflation (%) 1980-87	Overall budget surplus or deficit (as % of GNP) 1987	Gross domestic saving (as % of GDP) 1987
67	Libyan Arab Jamahiriya	..	22.3	5,460	0.6	-10.5	0.1	..	..
68	South Africa	74.3	62.9	1,890	3.2	-1.3	13.8	-4.4	28
69	Lebanon	..	..	..	0.6	..	..	..	..
70	Mongolia	..	..	..	0.6	..	..	..	..
71	Nicaragua	3.2	3.0	830	-0.7	-4.7	86.6	-16.3	..
72	Turkey	60.8	63.6	1,210	3.6	3.0	37.4	-4.2	23
73	Jordan	4.3	4.4	1,560	5.8	-0.7	2.8	-8.4	-3
74	Peru	45.2	29.7	1,470	0.8	-1.0	101.5	0.2	23
75	Ecuador	10.6	10.3	1,040	5.4	-2.0	29.5	2.1	17
76	Iraq	..	..	3,020	0.6	..	..	..	..
77	United Arab Emirates	23.7	22.8	15,830	0.6	-9.3	-0.3	..	41
78	Thailand	48.2	44.8	850	4.4	3.4	2.8	-2.3	26
79	Paraguay	4.6	3.9	990	4.1	-2.1	21.0	1.5	18
80	Brazil	299.2	314.6	2,020	6.3	1.0	166.3	-13.3	23
81	Mauritius	1.5	1.5	1,490	3.7	4.4	8.1	0.2	29
82	Korea, Dem. Rep.	..	..	..	0.6	..	..	..	..
83	Sri Lanka	6.0	6.6	400	2.8	3.0	11.8	-8.9	13
84	Albania	..	..	..	0.6	..	..	..	..
High human development		12,880 T	11,760 T	9,250	2.6	1.7	7.5	-3.8	21
85	Malaysia	31.2	29.6	1,810	4.7	1.1	1.1	-8.2	37
86	Colombia	31.9	36.0	1,240	3.7	0.9	23.7	-0.7	26
87	Jamaica	2.9	2.3	940	-0.1	-2.5	19.4	..	23
88	Kuwait	17.9	27.3	14,610	0.6	-3.2	-4.6	23.5	..
89	Venezuela	49.6	48.2	3,230	2.3	-3.1	11.4	-2.1	25
90	Romania	..	..	2,560	0.6	..	..	..	..
91	Mexico	141.9	149.4	1,830	3.6	-1.6	68.9	-9.5	17
92	Cuba	..	..	..	0.6	..	..	..	..
93	Panama	5.5	5.1	2,240	2.8	0.3	3.3	-4.2	..
94	Trinidad and Tobago	4.3	5.1	4,210	3.1	-6.5	6.2	..	18
95	Portugal	34.3	29.6	2,830	4.6	1.4	20.8	..	18
96	Singapore	19.9	20.7	7,940	8.3	5.7	1.3	1.4	40
97	Korea, Rep.	121.3	112.9	2,690	7.3	7.3	5.0	0.5	38
98	Poland	..	72.4	2,070	0.6	1.4	29.2	-1.7	30
99	Argentina	71.5	74.5	2,390	1.7	-1.8	298.7	..	10
100	Yugoslavia	60.0	58.0	2,480	5.2	0.0	57.2	0.0	40
101	Hungary	26.1	23.8	2,240	5.1	1.8	5.7	-3.6	26
102	Uruguay	6.4	6.6	2,190	2.5	-2.3	54.5	-0.7	11
103	Costa Rica	4.3	4.3	1,610	3.3	-0.5	28.6	-4.8	18
104	Bulgaria	..	..	4,150	0.6	..	..	..	..
105	USSR	..	..	4,550	0.6	..	..	..	..
106	Czechoslovakia	..	..	5,820	0.6	..	..	..	..
107	Chile	19.0	16.5	1,310	0.0	-1.1	20.6	0.1	21
108	Hong Kong	36.5	45.3	8,070	6.2	5.3	6.7	..	31
109	Greece	40.9	43.6	4,020	4.8	0.0	19.7	-14.4	8
110	German Dem. Rep.	..	..	7,180	0.6	..	..	..	..
111	Israel	35.0	29.8	6,800	3.7	1.5	159.0	0.8	11
112	USA	4,497.2	4,486.2	18,530	1.8	2.0	4.3	-3.3	13
113	Austria	117.7	90.5	11,980	4.0	1.6	4.3	-5.3	25
114	Ireland	21.9	21.8	6,120	2.8	..	10.2	-13.0	27
115	Spain	288.0	233.4	6,010	4.1	1.6	10.7	-5.2	22
116	Belgium	142.3	112.0	11,480	3.6	1.3	5.1	-10.6	19
117	Italy	748.6	597.0	10,350	3.2	1.8	11.5	-16.5	21
118	New Zealand	31.9	27.1	7,750	1.7	1.3	11.5	0.6	29
119	Germany, Fed. Rep.	1,117.8	879.6	14,400	3.0	1.8	2.9	-1.1	25
120	Finland	77.9	71.1	14,470	3.6	2.5	7.2	-1.0	22
121	United Kingdom	575.7	592.9	10,420	2.0	2.6	5.7	-1.8	18
122	Denmark	85.5	76.6	14,930	2.2	2.5	6.8	-0.6	21
123	France	873.4	715.0	12,790	3.7	0.9	7.7	-0.8	20
124	Australia	183.3	176.3	11,100	2.2	1.4	7.8	-1.2	22
125	Norway	83.1	71.4	17,190	3.6	3.7	6.1	3.9	27
126	Canada	373.7	390.1	15,160	3.3	2.1	5.0	-4.1	22
127	Netherlands	214.4	173.4	11,860	2.7	6.8	2.3	-3.2	23
128	Switzerland	170.9	138.2	21,330	1.5	1.6	3.9	..	31
129	Sweden	137.7	131.1	15,550	2.0	1.9	7.9	1.9	21
130	Japan	2,376.4	1,925.6	15,760	5.1	3.2	1.4	-4.9	34
All developing countries		2,210 T	2,320 T	650	2.9	3.4	9.1	-6.1	24
Least developed countries		70 T	70 T	210	0.5	-0.8	8.6	-4.0	1
Sub-Saharan Africa		190 T	190 T	440	1.6	-2.4	8.8	-5.7	18
Industrial countries		12,310 T	11,170 T	10,760	2.3	2.0	7.2	-3.9	21
World		14,520 T	13,490 T	3,100	2.7	3.2	8.3	-4.2	21

Notes: Summary data for regional and income groups are given in tables 23 and 24.

Regional aggregates of human development indicators

	Sub-Saharan Africa	Middle East and North Africa	Asia and Oceania	South Asia	East and Southeast Asia	Latin America and the Caribbean	Least developed countries	All developing countries	Industrial countries	World
Table 2: Profile of human development										
Life expectancy at birth	51	62	64	58	68	67	50	62	74	65
Access to health services	45	76	66	56	75	61	46	61	..	..
Access to safe water	37	69	52	54	48	73	35	55	..	..
Access to sanitation	..	..	22	11	51	60	20	32	..	..
Daily calorie supply	91	120	107	100	112	115	89	107	132	113
Adult literacy rate	48	54	59	41	71	83	37	60	..	..
GNP per capita	440	1,780	390	290	470	1,790	210	650	10,760	3,100
Real GDP per capita	990	2,850	1,740	1,100	2,200	3,980	690	1,970	14,260	4,110
Table 3: Profile of human deprivation (in millions of people)										
No health services	..	..	..	..	..	..	225 T	1,500 T	..	..
No safe water	..	..	..	..	..	..	250 T	1,750 T	..	..
No sanitation	..	..	..	..	..	..	350 T	2,800 T	..	..
Malnourished children	..	..	..	..	..	..	35 T	150 T	..	..
Illiterate adults	..	..	..	..	..	..	150 T	870 T	..	..
Out-of-school children	..	..	..	..	..	..	70 T	240 T	..	..
Below poverty line	..	..	..	..	..	..	250 T	1,150 T	..	..
Table 4: Trends in human development										
Life expectancy 1960	40	47	46	44	47	56	39	46	69	53
1975	46	55	57	51	62	62	45	57	71	61
1987	51	62	64	58	68	67	50	62	74	65
Under-five mortality rate 1960	284	273	241	278	199	160	288	243	46	218
1988	183	115	111	156	61	79	205	121	18	108
Access to safe water 1975	24	..	31	32	..	58	31	35	..	..
1985-87	37	69	52	54	48	73	35	55	..	..
Daily calorie supply 1964-66	92	92	87	88	87	102	87	90	124	100
1984-86	91	120	107	100	112	115	89	107	132	113
Adult literacy rate 1970	26	34	41	31	67	72	25	43	..	..
1985	48	54	59	41	71	83	37	60	..	..
GNP per capita 1976	350	940	330	210	410	1,090	140	450	4,850	1,800
1987	440	1,780	390	290	470	1,790	210	650	10,760	3,100
Table 5: Human capital formation										
Literacy rate, 1970 male	34	48	53	44	76	75	33	53	..	..
female	17	19	29	19	57	69	16	33	..	..
total	26	34	41	31	67	72	25	43	..	..
Literacy rate, 1985 male	59	66	71	54	83	85	47	71	..	..
female	38	41	47	28	62	81	27	50	..	..
total	48	54	59	41	71	83	37	60	..	..
Scientists and technicians per 1,000 people	2.6	25.8	6.3	3.5	8.3	37.8	1.4	9.7	140.5	44.4
Table 6: South-North human gaps (Index: North = 100)										
Life expectancy	69	87	84	74	86	90	66	80	100	..
Adult literacy rate	44	59	79	35	85	93	35	66	100	..
Female literacy rate	34	42	71	24	86	92	24	56	100	..
Under-five mortality rate	10	20	21	10	31	27	9	16	100	..
Maternal mortality rate	6	10	17	4	34	34	4	10	100	..
Combined primary and secondary enrolment ratio	45	80	73	50	78	86	42	71	100	..
Doctors per 1,000 people	3	30	16	8	29	37	3	16	100	..
Scientists and technicians per 1,000 people	2	5	3	2	5	8	1	4	100	..
GNP per capita	3	15	4	2	7	12	2	6	100	..
Real GDP per capita	5	21	13	7	16	23	4	12	100	..
Table 7: South-North real expenditure (Index of real per capita expenditure: North = 100)										
All food	17	..	..	..	..	59	..	41	100	..
Meat	15	..	..	..	..	48	..	24	100	..
Dairy, oils, fats	9	..	..	..	..	36	..	22	100	..
Cereals and bread	52	..	..	..	..	80	..	87	100	..
Health services	3	..	..	..	..	21	..	8	100	..
Pharmaceuticals	3	..	..	..	..	40	..	21	100	..
Education services	21	..	..	..	..	57	..	36	100	..
Books	3	..	..	..	..	8	..	5	100	..
Table 8: Rural-urban gaps										
Rural population	70	51	74	73	75	29	82	67	28	58
Rural access to health	38	58	..	..	..	42	36	45	..	..
Rural access to water	24	47	45	46	42	46	23	41	..	..
Rural access to sanitation	17	..	13	3	..	15	15	14	..	..
Urban access to health	72	97	..	..	..	82	..	88	..	..
Urban access to water	74	94	75	76	75	84	52	79	..	..
Urban access to sanitation	..	..	44	33	66	79	49	61	..	..
Rural-urban disparity: health	53	60	..	..	..	51	..	51	..	..
Rural-urban disparity: water	32	50	60	61	56	54	59	52	..	..
Rural-urban disparity: sanitation	..	..	30	9	..	19	31	23	..	..

	Sub-Saharan Africa	Middle East and North Africa	Asia and Oceania	South Asia	East and Southeast Asia	Latin America and the Caribbean	Least developed countries	All developing countries	Industrial countries	World
Table 9: Female-male gaps (Female as % of male: male value=100)										
Life expectancy	107.2	105.0	103.0	100.2	105.0	108.6	104.0	104.2	110.4	105.7
Literacy rate 1970	42	36	50	41	74	91	37	54	98	69
Literacy rate 1985	61	58	64	51	73	95	51	67	..	..
Primary enrolment 1960	52	56	55	46	78	95	44	61	99	74
Primary enrolment 1986-88	77	81	82	72	90	98	72	83	100	86
Secondary enrolment	58	68	66	53	77	113	53	71	104	77
Labour force	62	25	54	29	71	36	46	51	71	56
Parliament	6.3	3.2	13.7	7.7	18.1	7.2	7.6	11.8	16.0	12.9
Table 10: Child survival and development										
Births attended by health personnel	36	46	37	31	55	77	23	42	99	51
Low birthweight babies	16	7	19	28	9	11	21	17	6	15
Infant mortality rate	110	78	74	101	43	55	124	79	15	71
Breastfeeding at one year	71	62	..	76	..	36	75	64	..	..
One-year-olds immunised	52	72	71	59	85	70	37	68	75	69
Under-fives underweight	29	..	45	45	43	14	46	38	..	..
Wasted children	16	..	16	18	..	3	16	13	..	..
Stunted children	38	..	50	50	..	31	54	42	..	..
Under-five mortality rate	183	115	111	156	61	79	205	121	18	108
Table 11: Health profile										
Population with access to health services	45	76	66	56	75	61	46	61	..	..
Maternal mortality rate	530	280	270	410	120	110	520	290	24	250
Thousands of people per doctor	24.6	3.7	2.9	3.5	2.4	1.2	23.3	4.8	0.5	4.0
Thousands of people per nurse	2.2	1.0	2.0	2.7	1.5	1.0	4.7	1.9	0.1	1.6
Nurses per doctor	12.4	4.2	1.8	1.5	1.9	1.4	8.4	2.8	4.2	3.1
Health expenditure as % of GNP 1960	0.7	0.9	0.7	0.6	0.8	1.2	0.7	0.9	2.2	2.0
Health expenditure as % of GNP 1986	0.8	2.0	1.1	1.0	1.1	1.6	0.9	1.4	4.7	4.2
Private expenditure on health per capita	2.6	..	..	..	..	16.3	..	7.1	90.9	33.0
Table 12: Food security										
Food production per capita index	93	112	116	107	122	101	96	111	101	109
Daily calorie supply per capita	2,160	3,000	2,460	2,270	2,600	2,700	2,070	2,480	3,390	2,650
Food import dependency ratio 1979-81	11.1	35.3	6.0	4.5	7.0	17.1	9.1	9.6	23.1	12.9
Food import dependency ratio 1984-86	10.8	38.2	5.6	6.0	5.2	14.9	12.8	9.2	21.4	12.2
Food aid in cereals 1981-82	2,070 T	2,860 T	2,770 T	2,000 T	770 T	710 T	2,800 T	8,410 T	..	..
Food aid in cereals 1987-88	2,850 T	3,380 T	4,090 T	2,760 T	1,330 T	2,440 T	4,410 T	12,770 T	..	..
Table 13: Education profile										
Gross primary enrolment ratio male	75	101	116	101	131	110	69	110	103	109
Gross primary enrolment ratio female	61	81	96	73	119	108	50	92	103	93
Net primary enrolment ratio male	48	85	95	78	..	75	53	87	97	89
Net primary enrolment ratio female	37	72	87	59	..	75	38	79	97	82
Gross secondary enrolment ratio male	21	57	47	44	52	47	19	45	90	51
Gross secondary enrolment ratio female	13	39	32	24	41	51	9	33	93	40
Gross tertiary enrolment ratio male	2.1	15.7	7.9	10.9	5.4	21.3	3.9	8.9	32.0	15.1
Gross tertiary enrolment ratio female	0.7	9.1	4.0	4.5	3.6	18.6	1.0	5.1	33.3	12.7
Radios per 1,000 people	139	230	148	83	195	327	85	172	1,013	369
Televisions per 1,000 people	14	106	22	10	30	145	6	40	477	134
Table 14: Education imbalances										
Dropout rate	41	27	36	55	31	55	60	39	11	35
Primary pupil to teacher ratio	43	29	35	45	26	28	48	35	19	33
Combined primary to secondary enrolment ratio	49	71	78	79	78	85	40	75	97	78
Secondary technical enrolment	6.3	12.0	4.6	1.6	7.6	17.9	4.2	6.9	23.0	9.0
Education expenditure as % of GNP 1960	2.3	3.7	2.1	2.1	2.1	2.0	1.5	2.4	3.7	3.5
Education expenditure as % of GNP 1986	3.6	5.9	3.4	3.3	3.5	3.5	3.3	3.9	5.2	5.0
Primary education expenditure	44.3	48.0	38.4	43.2	..	40.9	46.2	41.1	34.1	35.0
Educated unemployed	..	..	..	..	..	..	..	6.1	13.3	8.0
Table 15: Employment										
Labour force	40.6	30.9	46.4	35.6	54.2	35.1	38.9	43.5	48.5	44.6
Women in labour force	37.8	18.7	33.2	22.0	41.2	26.3	28.2	32.1	41.1	34.2
Labour force in agriculture 1965	77.8	62.5	75.1	72.0	77.2	45.1	83.0	71.5	22.2	56.6
Labour force in agriculture 1985-87	70.3	38.9	64.1	59.6	67.3	25.3	71.1	59.0	11.5	47.9
Labour force in industry 1965	8.6	14.0	10.2	12.3	8.9	21.8	5.8	11.5	36.4	19.0
Labour force in industry 1985-87	9.8	15.3	12.7	11.9	13.3	17.4	7.6	13.0	27.5	16.4
Labour force in services 1965	13.7	23.5	14.7	15.8	14.1	33.3	11.3	17.1	41.5	24.5
Labour force in services 1985-87	20.0	45.7	23.2	28.5	19.4	57.3	21.2	28.0	61.0	35.7
Earnings per employee growth rate 1970-80	-1.5	..	0.8	..	3.4	2.2	..	0.9	2.2	1.2
Earnings per employee growth rate 1980-86	-3.5	..	5.5	4.9	..	-0.6	..	3.1	1.4	2.7

(table 23 continues)

 Regional aggregates of human development indicators (continued)

	Sub-Saharan Africa	Middle East and North Africa	Asia and Oceania	South Asia	East and Southeast Asia	Latin America and the Caribbean	Least developed countries	All developing countries	Industrial countries	World
Table 16: Wealth and poverty										
GNP per capita	440	1,780	390	290	470	1,790	210	650	10,760	3,100
GDP per capita (PPP$)	990	2,850	1,740	1,100	2,200	3,980	690	1,970	14,260	4,110
GNP per capita of lowest 40%	..	..	..	..	..	..	..	..	..	..
Income share of lowest 40%	..	..	..	..	..	..	..	..	..	..
Highest 20% to lowest 20%	..	..	..	..	..	..	..	..	..	..
Gini coefficient	..	..	..	..	..	..	..	..	..	..
Population below poverty line										
Urban	34	..	25	41	..	36	59	27	..	..
Rural	61	..	32	53	18	61	72	35	..	..
Table 17: Urban crowding										
Urban population 1960	19	33	21	19	22	50	9	29	62	42
1988	36	54	30	29	30	73	22	43	74	55
2000	45	60	36	36	36	79	30	50	76	60
Urban population growth rate 1960-88	6.0	4.8	3.5	4.1	3.1	3.9	6.2	4.0	1.5	3.0
1988-2000	5.3	3.7	3.5	4.2	3.0	2.6	5.6	3.5	0.8	2.4
Persons per habitable room	..	..	..	..	..	1.6	..	2.4	0.7	1.9
Highest population density of a major city	..	..	..	..	..	..	..	..	..	..
Table 18: Military expenditure imbalances										
Military expenditure as % of GNP 1960	0.7	4.9	6.3	2.8	8.2	1.8	2.1	4.2	6.3	6.0
1986	3.3	12.6	5.9	7.2	5.2	1.5	3.8	5.5	5.4	5.4
Ratio of military expenditure to health and education expenditure	81	189	130	161	112	33	92	104	55	59
Arms imports	3,360 T	16,730 T	10,800 T	6,200 T	4,600 T	3,340 T	3,680 T	34,230 T	11,700 T	45,930 T
Ratio of ODA received to military expenditure	4.7	0.4	0.4	1.0	0.3	0.8	5.3	1.5	..	..
Ratio of ODA given to military expenditure	..	..	..	..	..	..	..	..	0.20	..
Armed forces as % of teachers	90	183	58	47	66	42	121	68	105	77
Table 19: Resource flows imbalances										
ODA received	10,400 T	5,950 T	13,200 T	5,540 T	7,660 T	4,340 T	10,450 T	33,890 T	..	..
ODA given	..	..	..	..	..	..	..	..	46,200 T	..
ODA received as % of GNP	7.9	1.4	1.0	1.7	0.7	0.4	14.1	1.3	..	..
ODA given as % of GNP	..	..	..	..	..	..	..	..	0.31	..
Interest payments on long-term debt	2,520 T	5,580 T	12,920 T	2,180 T	10,740 T	23,890 T	710 T	44,910 T	..	..
Debt service as % of exports of goods and services	20	..	18	24	18	34	25	24	..	..
Gross international reserves	1.5	2.6	3.9	4.2	3.9	3.3	2.0	2.7	3.5	3.0
Ratio of exports to imports	0.80	0.71	0.72	0.41	0.91	0.84	0.40	0.79	0.96	0.87
Terms of trade	84	76	93	96	86	77	86	83	103	86
Table 20: Demographic balance sheet										
Population 1960	210 T	120 T	1,520 T	590 T	920 T	210 T	210 T	2,060 T	940 T	3,000 T
1988	470 T	250 T	2,760 T	1,150 T	1,600 T	420 T	410 T	3,900 T	1,200 T	5,100 T
2000 (estimate)	680 T	340 T	3,400 T	1,500 T	1,900 T	530 T	580 T	4,960 T	1,270 T	6,230 T
Population growth rate 1960-88	2.8	2.7	2.1	2.4	2.0	2.5	2.5	2.3	0.8	1.8
1988-2000	3.1	2.6	1.8	2.2	1.4	1.9	2.8	2.0	0.5	1.6
Dependency ratio 1960	87.9	86.7	78.8	78.7	78.8	85.9	83.7	80.9	59.1	74.1
1985	96.8	84.8	66.8	77.0	59.6	73.3	92.6	72.3	50.8	67.3
Contaceptive prevalence rate	6	29	53	31	69	57	10	47	66	50
Fertility rate 6.5	5.2	3.5	4.7	2.7	3.6	6.0	4.0	1.9	3.5	
Crude birth rate 46	37	28	35	23	29	44	31	15	27	
Crude death rate 16	10	9	12	7	7	17	10	10	10	

	Sub-Saharan Africa	Middle East and North Africa	Asia and Oceania	South Asia	East and Southeast Asia	Latin America and the Caribbean	Least developed countries	All developing countries	Industrial countries	World
Table 21. Natural resources balance sheet										
Land area	2,030 T	1,260 T	2,230 T	640 T	1,590 T	1,970 T	1,530 T	7,490 T	5,360 T	12,850 T
Arable land as % of total	37	26	42	50	38	37	37	36	34	35
Annual percentage increase of arable land	0.18	0.16	0.12	0.10	0.13	0.82	0.15	0.33	0.14	0.25
Population density	595	463	2,038	2,971	1,368	388	2,635	1,583	1,088	1,467
Livestock per capita	0.99	1.09	0.54	0.63	0.48	1.42	1.07	0.73	1.11	0.82
Annual percentage increase of livestock	1.8	2.2	1.9	1.9	1.9	1.6	1.5	1.9	0.7	1.6
Production of fuelwood per capita	0.84	0.19	0.28	0.28	0.28	0.66	0.63	0.39	0.26	0.36
Annual percentage increase of fuelwood	2.9	0.2	2.0	2.1	2.0	2.6	2.7	2.1	4.3	2.6
Average annual deforestation	125	..	195	86	326	1,381	74	582	..	..
Internal renewable water resources per capita	3.5	1.3	4.3	3.4	5.1	22.9	10.5	6.2	9.3	6.8
Table 22: National income accounts										
GDP	190 T	350 T	940 T	280 T	660 T	730 T	70 T	2,210 T	12,310 T	14,520 T
GNP	190 T	400 T	990 T	300 T	690 T	740 T	70 T	2,320 T	11,170 T	13,490 T
GNP per capita	440	1,780	390	290	470	1,790	210	650	10,760	3,100
Annual GNP per capita growth rate 1965-80	1.6	2.9	2.9	1.4	4.0	3.8	0.5	2.9	2.3	2.7
1980-87	-2.4	-0.1	5.4	3.0	7.2	-0.7	-0.8	3.4	2.0	3.2
Annual rate of inflation	8.8	5.6	7.3	8.8	4.7	21.0	8.6	9.1	7.2	8.3
Overall budget or surplus deficit	-5.7	..	-4.5	-6.9	-1.6	-9.1	-4.0	-6.1	-3.9	-4.2
Gross domestic saving	18.0	19.7	29.9	19.3	34.6	19.9	1.1	24.1	20.5	21.1

HDI and income aggregates of human development indicators

		Low human development		Medium human development			Low-income			
		All countries	Excluding India	All countries	Excluding China	High human development	All countries	Excluding China and India	Middle-income	High-income
TABLE 2: Profile of human development										
Life expectancy at birth		55	52	67	63	73	61	54	67	76
Access to health services		47	..	75	..	..	56	..	69	..
Access to safe water		48	39	59	..	..	47	38	70	..
Access to sanitation		14	22	50	..	..	17	26	63	..
Daily calorie supply		95	91	113	115	131	103	96	122	132
Adult literacy rate		41	40	71	73	..	55	50	75	..
GNP per capita		300	300	690	1,250	9,250	300	300	2,380	14,260
Real GDP per capita		970	880	2,370	2,730	11,860	1,470	1,020	3,420	14,260
TABLE 3: Profile of human deprivation (in millions of people)										
No health services		775 T	..	475 T	..	..	..	..	..	..
No safe water		750 T	..	750 T	..	..	..	..	..	..
No sanitation		1,300 T	..	1,100 T	..	..	..	..	..	..
Malnourished children		90 T	..	55 T	..	..	..	..	..	..
Illiterate adults		500 T	..	370 T	..	..	..	..	..	..
Out-of-school children		140 T	..	100 T	..	..	..	..	..	..
Below poverty line		700 T	..	450 T	..	..	..	..	..	..
TABLE 4: Trends in human development										
Life expectancy	1960	42	40	48	48	68	44	41	59	70
	1975	49	46	61	56	71	56	47	63	73
	1987	55	52	67	63	73	61	54	67	76
Under-five mortality rate	1960	285	287	209	214	67	259	275	170	49
	1988	170	186	72	94	27	133	172	75	16
Access to safe water	1975	31	30	33	..	..	28	24	55	..
	1985-87	48	39	59	..	..	47	38	70	..
Daily calorie supply	1964-66	89	88	88	92	121	87	87	108	123
	1984-86	95	91	113	115	131	103	96	122	132
Adult literacy rate	1970	29	23	57	..	..	34	34	76	..
	1985	41	40	71	73	..	55	50	75	..
GNP per capita	1976	180	220	540	740	4,350	280	200	1,600	6,030
	1987	300	300	690	1,250	9,250	300	300	2,380	14,260
TABLE 5: Human capital formation										
Literacy rate, 1970	male	40	32	67	..	..	45	43	80	..
	female	17	13	48	..	..	22	23	71	..
	total	29	23	57	..	..	34	34	76	..
Literacy rate, 1985	male	54	51	81	80	..	68	61	81	..
	female	29	29	61	66	..	43	41	70	..
	total	41	40	71	73	..	55	50	75	..
Scientists and technicians per 1,000 people		2.9	1.9	11.8	19.6	135.5	5.5	5.4	74.7	140.8
TABLE 6: South-North human gaps (Index: North=100)										
Life expectancy		69	..	87	..	100	69	..	89	102
Adult literacy rate		41	..	79	..	99	44	..	81	..
Female literacy rate		30	..	74	..	96	32	..	75	..
Under-five mortality rate		10	..	21	..	100 +	10	..	27	100 +
Maternal mortality rate		4	..	22	..	100 +	5	..	30	100 +
Combined primary and secondary enrolment ratio		43	..	80	..	96	46	..	85	98
Doctors per 1,000 people		3	..	27	..	90	4	..	31	102
Scientists and technicians per 1,000 people		2	..	5	..	89	2	..	8	96
GNP per capita		3	..	9	..	57	3	..	14	111
Real GDP per capita		5	..	16	..	79	5	..	18	90
TABLE 7: South-North real expenditure (Index of real per capita expenditure: North=100)										
All food		19	..	44	..	95	19	..	54	100
Meat		16	..	31	..	94	8	..	44	106
Dairy, oils, fats		10	..	25	..	72	6	..	31	97
Cereals and bread		63	..	103	..	104	66	..	95	101
Health services		3	..	16	..	66	3	..	19	103
Pharmaceuticals		3	..	26	..	88	2	..	39	117
Education services		21	..	48	..	94	19	..	52	105
Books		3	..	7	..	62	2	..	8	107
TABLE 8: Rural-urban gaps										
Rural population		73	74	68	55	28	76	75	40	23
Rural access to health		37	..	57	..	..	40	..	56	..
Rural access to water		38	23	45	..	..	38	27	51	..
Rural access to sanitation		6	11	33	..	..	10	18	33	..
Urban access to health		81	..	95	..	..	85	..	90	..
Urban access to water		74	73	80	77	..	75	65	84	..
Urban access to sanitation		39	56	72	..	..	37	45	83	..
Rural-urban disparity: health		46	..	59	..	..	47	..	62	..
Rural-urban disparity: water		53	38	56	..	..	52	41	60	..
Rural-urban disparity: sanitation		15	20	46	..	..	27	40	40	..

		Low human development		Medium human development		High human development	Low-income		Middle-income	High-income
		All countries	Excluding India	All countries	Excluding China		All countries	Excluding China and India		
TABLE 9: Female-male gaps (Female as % of male: male value=100)										
Life expectancy		102.1	104.2	105.2	106.3	110.1	103.3	104.5	108.8	109.0
Literacy rate	1970	40	36	69	..	97	44	46	83	98
	1985	52	54	74	82	97	61	62	84	..
Primary enrolment	1960	46	40	77	..	99	50	50	89	100
	1986-88	71	70	90	92	99	81	79	93	99
Secondary enrolment		53	52	81	94	103	64	58	96	102
Labour force		39	43	62	45	66	55	48	54	61
In parliament		7.6	7.0	15.1	7.4	14.8	13.9	9.3	13.6	8.0
TABLE 10: Child survival and development										
Births attended by health personnel		30	27	61	..	94	33	33	71	98
Low birthweight babies		25	20	9	12	8	19	20	9	6
Infant mortality rate		107	114	51	66	20	86	107	52	12
Breastfeeding at one year		72	..	60	..	..	74	..	49	..
One-year-olds immunised		55	49	81	70	76	66	51	75	69
Under-fives underweight		42	43	32	..	..	44	46	21	..
Wasted children		16	..	10	..	..	17	..	6	..
Stunted children		44	..	43	..	..	48	..	36	..
Under-five mortality rate		170	186	72	94	27	133	172	75	16
TABLE 11: Health profile										
Population with access to health services		47	..	75	..	..	56	..	69	..
Maternal mortality rate		460	570	130	210	37	330	510	130	10
Thousands of people per doctor		8.8	16.8	2.3	3.9	0.6	5.5	13.6	2.4	0.5
Thousands of people per nurse		2.6	3.9	1.5	1.1	0.3	2.1	3.0	1.0	0.1
Nurses per doctor		4.3	7.9	1.9	3.6	3.6	2.8	6.9	2.8	4.5
Health expenditure as % GNP	1960	0.6	0.7	0.8	0.7	2.2	0.8	0.5	2.5	1.9
	1986	0.8	0.7	1.5	1.5	4.6	1.0	0.7	2.6	5.1
Private expenditure on health per capita		4.2	7.5	11.8	..	82.5	4.1	6.5	21.7	95.8
TABLE 12: Food security										
Food production per capita index		105	97	118	108	100	114	102	103	99
Daily calorie supply per capita		2,190	2,130	2,650	2,680	3,280	2,380	2,220	2,860	3,380
Food import dependency ratio	1979-81	6.9	12.2	8.8	15.0	24.4	5.5	9.9	18.7	29.6
	1984-86	7.7	12.9	7.7	14.4	22.9	4.8	9.2	18.6	27.7
Food aid in cereals	1981-82	4,630 T	..	3,190 T	..	..	4,520 T	..	4,310 T	..
	1987-88	6,390 T	..	5,790 T	..	..	6,950 T	..	5,820 T	..
TABLE 13: Education profile										
Gross primary enrolment ratio	male	92	70	127	111	106	110	81	109	103
	female	67	51	114	103	106	89	66	101	102
Net primary enrolment ratio	male	55	..	98	95	96	87	69	90	96
	female	38	..	91	90	96	78	58	85	95
Gross secondary enrolment ratio	male	37	23	51	53	81	42	25	56	93
	female	20	12	41	46	83	27	15	51	94
Gross tertiary enrolment ratio	male	9.3	4.8	6.9	16.4	30.2	6.3	5.5	19.6	38.6
	female	3.8	1.7	4.4	11.3	30.9	2.7	2.2	18.7	38.2
Radios per 1,000 people		92	108	197	213	910	129	112	382	1,229
Televisions per 1,000 people		8	10	48	85	413	14	15	166	592
TABLE 14: Education imbalances										
Dropout rate		50	..	35	37	15	40	46	32	4
Primary pupil to teacher ratio		45	44	27	29	21	37	41	26	20
Combined primary to secondary enrolment ratio		68	43	79	76	95	73	49	84	96
Secondary technical enrolment		3.0	4.7	9.4	11.7	21.4	4.3	4.6	18.0	18.9
Education expenditure as % of GNP	1960	2.1	1.8	2.5	2.7	3.7	2.0	1.9	3.8	3.6
	1986	3.2	3.1	4.0	4.3	5.1	3.0	2.9	4.6	5.3
Primary education expenditure		43.2	43.0	43.8	49.7	34.1	36.7	42.6	41.2	32.7
Educated unemployed		..	..	..	..	12.0	..	..	7.3	12.8
TABLE 15: Employment										
Labour force		37.3	36.5	49.1	37.3	46.3	46.1	38.2	40.1	47.1
Women in labour force		26.5	27.6	36.8	29.2	38.8	33.7	29.9	33.2	37.5
Labour force in agriculture	1965	75.3	78.2	72.8	61.5	24.8	77.2	76.1	47.1	14.2
	1985-87	64.7	67.0	59.8	43.7	12.9	67.4	64.4	30.4	6.4
Labour force in industry	1965	10.3	8.2	10.8	14.7	34.6	9.3	8.5	24.1	38.4
	1985-87	10.3	9.8	14.1	14.7	26.3	11.5	9.6	23.8	21.9
Labour force in services	1965	14.4	13.6	16.5	24.1	40.8	13.6	15.7	28.9	47.6
	1985-87	25.0	23.3	26.1	41.6	60.8	21.1	26.0	45.8	71.7
Earnings per employee growth rate	1970-80	..	..	2.8	..	2.4	0.2	0.6	2.3	2.2
	1980-86	..	..	2.5	..	1.4	4.4	2.7	0.4	1.6

(table 24 continues)

	Low human development: All countries	Low human development: Excluding India	Medium human development: All countries	Medium human development: Excluding China	High human development	Low-income: All countries	Low-income: Excluding China and India	Middle-income	High-income
Table 16: Wealth and poverty									
GNP per capita	300	300	690	1,250	9,250	300	300	2,380	14,260
GDP per capita (PPP$)	970	880	2,370	2,730	11,860	1,470	1,020	3,420	14,270
GNP per capita of lowest 40%	..	..	..	..	..	..	..	..	..
Income share of lowest 40%	..	..	..	..	..	..	..	..	..
Ratio of highest 20% to lowest 20%	..	..	..	..	..	..	..	..	..
Gini coefficient	..	..	..	..	..	..	..	..	..
Urban population below poverty line	40	41	16	..	..	26	38	29	..
Rural population below poverty line	55	62	18	44	..	35	57	41	..
TABLE 17: Urban crowding									
Urban population 1960	17	15	29	34	61	17	14	43	69
1988	29	31	42	54	74	25	28	64	78
2000	37	39	48	61	77	32	36	70	80
Urban population growth rate 1960-88	4.7	5.9	3.7	4.4	2.0	3.8	5.3	3.6	1.4
1988-2000	4.6	5.3	3.2	3.4	1.1	3.9	4.9	2.4	0.7
Persons per habitable room	..	..	..	..	1.0	..	..	1.8	0.6
Highest population density of a major city	..	..	..	..	..	..	..	..	..
TABLE 18: Military expenditure imbalances									
Military expenditure as % of GNP 1960	1.9	1.8	5.5	3.6	6.2	6.5	3.5	7.5	5.4
1986	3.7	3.8	7.2	7.5	5.4	4.5	3.5	8.4	4.4
Ratio of military expenditure to health and education expenditure	98	117	133	129	55	116	110	114	29
Arms imports	9,160 T	5,960 T	22,020 T	21,640 T	14,760 T	11,310 T	7,730 T	22,230 T	12,390 T
Ratio of ODA received to military expenditure	4.6	..	0.7	..	..	4.6	..	0.6	..
Ratio of ODA given to military expenditure	..	..	..	..	0.20	..	..	..	0.20
Armed forces as % of teachers	61	96	73	99	101	60	97	110	82
Table 19: Resource flows imbalances									
ODA received	17,130 T	15,280 T	14,600 T	13,150 T	3,540 T	20,760 T	17,460 T	13,200 T	..
ODA given	..	..	..	..	46,540 T	..	..	..	45,050 T
ODA received as % of GNP	3.6	7.0	0.9	1.1	..	2.2	6.0	0.8	..
ODA given as % of GNP	..	..	..	..	0.31	..	..	..	0.35
Interest payments on long-term debt	5,300 T	3,780 T	18,500 T	17,430 T	29,780 T	7,560 T	4,970 T	43,850 T	..
Debt service as % of goods and services	24	23	24	30	25	17	25	29	..
Gross international reserves	1.7	..	3.0	..	3.6	2.0	..	3.0	4.0
Ratio of exports to imports	0.67	..	0.79	..	1.00	0.66	..	0.91	0.99
Terms of trade	86	..	76	..	99	86	..	81	102
TABLE 20: Demographic balance sheet									
Population 1960	790 T	350 T	1,130 T	470 T	1,080 T	1,580 T	480 T	800 T	620 T
1988	1,580 T	760 T	2,060 T	950 T	1,470 T	2,920 T	990 T	1,390 T	790 T
2000 (estimate)	2,130 T	1,090 T	2,500 T	1,220 T	1,590 T	3,690 T	1,360 T	1,690 T	850 T
Population growth rate 1960-88	2.5	2.8	2.1	2.5	1.0	2.2	2.6	1.9	0.8
1988-2000	2.5	3.0	1.6	2.1	0.7	1.9	2.6	1.6	0.5
Dependency ratio 1960	80.9	86.9	80.6	84.6	62.3	79.1	83.8	75.4	59.4
1985	83.1	95.0	64.6	77.0	53.9	71.0	89.1	69.2	50.2
Contaceptive prevalence rate	23	10	63	45	65	46	21	50	66
Fertility rate 5.3	6.4	3.2	4.1	2.1	4.0	5.5	3.5	1.8	
Crude birth rate 39	46	25	32	16	31	41	27	14	
Crude death rate 13	16	8	9	9	10	14	9	9	

	Low human development		Medium human development		High human development	Low-income		Middle-income	High-income
	All countries	Excluding India	All countries	Excluding China		All countries	Excluding China and India		
TABLE 21. Natural resources balance sheet									
Land area	2,660 T	2,360 T	4,020 T	3,090 T	6,170 T	3,660 T	2,430 T	5,970 T	3,220 T
Arable land as % of total	36	33	34	32	36	36	31	33	39
Annual percentage increase of arable land	0.21	0.22	0.42	0.56	0.16	0.13	0.20	0.34	0.22
Population density	2,317	1,863	1,026	834	1,168	1,822	1,770	660	1,568
Livestock per capita	0.73	0.91	0.66	0.79	1.12	0.62	0.72	1.13	1.01
Annual percentage increase of livestock	1.8	1.7	1.9	2.8	0.9	1.8	2.4	1.5	0.7
Production of fuelwood per capita	0.46	0.65	0.35	0.57	0.26	0.36	0.65	0.40	0.25
Annual percentage increase of fuelwood	2.5	3.0	1.8	1.5	3.9	2.2	2.7	1.4	6.2
Average annual deforestation	121	..	..	..	..	155	..	980	..
Internal renewable water resources per capita	3.4	5.5	6.9	12.6	10.5	4.2	9.4	10.8	10.3
TABLE 22: National income accounts									
GDP	410 T	190 T	1,230 T	930 T	12,880 T	730 T	220 T	1,460 T	12,330 T
GNP	450 T	210 T	1,280 T	960 T	11,760 T	810 T	250 T	1,530 T	11,150 T
GNP per capita	300	300	690	1,250	9,250	290	300	2,380	14,260
Annual GNP per capita growth rate 1965-80	1.4	1.2	3.9	3.7	2.6	2.6	1.8	2.9	2.9
1980-87	1.3	-1.0	5.5	0.7	1.7	4.4	-0.4	0.6	1.9
Annual rate of inflation	8.2	..	11.4	..	7.5	9.3	..	11.4	5.7
Overall budget or surplus deficit	-7.8	-7.3	..	..	-3.8	-6.5	-4.7	-6.4	-3.8
Gross domestic saving	16.7	10.6	26.5	22.5	20.7	26.5	15.9	22.9	20.5

25 Profile of human development for countries with fewer than one million people

		Total population 1987	Life expectancy (years) 1988	Adult literacy rate (%) 1985	Daily calorie supply (as % of requirements) 1984-86	Population with access to health services (%) 1985-87	Population with access to safe water (%) 1985-87	GNP per capita (US$) 1987	Human development index
1	Djibouti	400,000	47	12	..	37	44	480	0.073
2	Guinea-Bissau	925,000	39	32	105	..	21	160	0.074
3	Gambia	800,000	43	26	103	..	60	220	0.094
4	Equatorial Guinea	400,000	46	*42*	..	9	..	180	0.229
5	Swaziland	700,000	55	38	105	..	34	700	0.329
6	Solomon Islands	300,000	66	15	84	..	50	420	0.349
7	Comoros	500,000	56	48	82	82	28	370	0.399
8	Vanuatu	200,000	63	53	..	..	63	880	0.521
9	Cape Verde	400,000	65	50	125	..	69	500	0.534
10	Sao Tome and Principe	100,000	65	58	103	80	82	280	0.576
11	Qatar	300,000	69	51	..	..	95	12,510	0.594
12	Maldives	200,000	59	93	81	97	24	300	0.692
13	Bahrain	500,000	71	72	..	100	100	8,510	0.737
14	Brunei Darussalam	200,000	74	*70*	..	..	..	15,390	0.770
15	Saint Vincent	100,000	69	*83*	99	..	95	1,000	0.775
16	Suriname	400,000	67	90	..	..	83	2,270	0.788
17	Saint Lucia	100,000	70	*83*	102	..	67	1,400	0.789
18	Saint Kitts and Nevis	47,000	68	*90*	103	..	100	1,700	0.801
19	Belize	100,000	67	93	114	87	63	1,240	0.805
20	Western Samoa	200,000	65	98	..	..	80	550	0.806
21	Fiji	700,000	70	86	108	..	77	1,570	0.806
22	Guyana	989,000	66	96	111	89	77	500	0.808
23	Seychelles	100,000	70	88	..	100	83	3,120	0.817
24	Grenada	100,000	69	96	94	100	85	1,340	0.849
25	Bahamas	300,000	70	*99*	..	..	..	10,280	0.880
26	Antigua and Barbuda	100,000	73	95	84	..	100	2,540	0.898
27	Malta	300,000	73	*95*	..	..	100	4,190	0.898
28	Dominica	100,000	74	*94*	100	..	77	1,440	0.906
29	Barbados	300,000	75	*95*	128	..	99	5,350	0.925
30	Cyprus	700,000	76	*93*	..	100	100	5,200	0.928
31	Luxembourg	400,000	74	99	..	..	..	18,550	0.934
32	Iceland	200,000	77	99	..	..	..	16,330	0.975

Selected definitions

Birth rate (crude) The annual number of births per 1,000 population.

Births attended The percentage of births attended by physicians, nurses, midwives, trained primary health care workers or trained traditional birth attendants.

Budget surplus or deficit Current capital revenue and grants received, less total expenditure and lending minus repayments.

Calorie supply See *Daily calorie supply*.

Child malnutrition See *Underweight.*

Child mortality See *Under-five mortality.*

Contraceptive prevalence rate The percentage of married women of childbearing age who are using, or whose husbands are using, any form of contraception: that is, modern or traditional methods.

Daily calorie requirement per capita The average number of calories needed to sustain a person at normal levels of activity and health, taking into account the distribution by age, sex, body weight and environmental temperature.

Daily calorie supply per capita The calorie equivalent of the net food supplies in a country, divided by the population, per day.

Death rate (crude) The annual number of deaths per 1,000 population.

Debt service The sum of repayments of principle (amortisation) and payments of interest made in foreign currencies, goods or services on external, public, publicly guaranteed and private nonguaranteed debt.

Dependency ratio The ratio of the population defined as dependent, under 15 and over 64 years, to the working-age population, aged 15-64.

Domestic savings (gross) The gross domestic product less government and private consumption.

Dropout rate The proportion of the children entering the first grade of primary school who do not successfully complete that level in due course.

Earnings per employee Earnings in constant prices derived by deflating nominal earnings per employee by the country's consumer price index.

Educated unemployed as a percentage of total unemployed The number of professional, technical and related workers (International Standard Classification of Occupations (ISCO) major group 0/1) and administrative and managerial workers (ISCO major group 0/2) reported as unemployed as a proportion of the total number of workers reported as unemployed. The number of persons unemployed includes those without work and those currently available for work as well as those seeking work.

Education expenditures Expenditures on the provision, management, inspection and support of preprimary, primary and secondary schools; universities and colleges; vocational, technical and other training institutions; and general administration and subsidiary services.

Employees Regular employees, working proprietors, active business partners, and unpaid family workers, but excluding homeworkers.

Enrolment ratios (gross and net) The *gross* enrolment ratio is the number enrolled in a level of education, whether or not they belong in the relevant age group for that level, expressed as a percentage of the population in the relevant age group for that level. The *net* enrolment ratio is the number enrolled in a level of education who belong in the relevant age group, expressed as a percentage of the population in that age group.

Expenditures
See *Government expenditures.*

Exports of goods and services The value of all goods and nonfactor services provided to the rest of the world, including merchandise, freight, insurance, travel and other nonfactor services.

Female-male gap A set of national, regional and other estimates in which all the figures for females are expressed in relation to the corresponding figures for males which are indexed to equal 100.

Fertility rate (total) The average number of children that would be born alive to a woman during her lifetime, if she were to bear children at each age in accord with prevailing age-specific fertility rates.

Food aid in cereals Cereals provided by donor countries and international organisations, including the World Food Programme and the International Wheat Council, as reported for that particular crop year. Cereals include wheat, flour, bulgur, rice, coarse grain and the cereal components of blended foods.

Food import dependency ratio The ratio of food imports to the food available for internal distribution, that is, the sum of food production plus food imports minus food exports.

Food production per capita index The average annual quantity of food produced per capita in relation to that produced in the indexed year. Food is defined as comprising nuts, pulses, fruit, cereals, vegetables, sugar cane, sugar beets, starchy roots, edible oils, livestock and livestock products.

Gini coefficient A measure that shows how close a given distribution of income is to absolute equality or inequality. Named for Corrado Gini, the Gini coefficient is a ratio of the area between the 45° line and the Lorenz curve and the area of the entire triangle. As the coefficient approaches zero, the distribution of income approaches absolute equality. Conversely, as the coefficient approaches 1, the distribution of income approaches absolute inequality.

Government expenditures Expenditures by all central government offices, departments, establishments and other bodies that are agencies or instruments of the central authority of a country. It includes both current and capital or developmental expenditures but excludes provincial, local and private expenditures.

Gross domestic product (GDP) The total for final use of output of goods and services produced by an economy, by both residents and nonresidents, regardless of the allocation to domestic and foreign claims.

Gross national product (GNP) The total domestic and foreign value added claimed by residents, calculated without making deductions for depreciation. It comprises GDP plus net factor income from abroad, which is the income residents receive from abroad for factor services (labour and capital), less similar payments made to nonresidents who contribute to the domestic economy.

GNP per capita and growth rates The gross national product divided by the population. Annual GNPs per capita are expressed in current U.S. dollars. GNP per capita growth rates are annual average growth rates that have been computed by fitting trend lines to the logarithmic values of GNP per capita at constant market prices for each year of the time period.

Gross enrolment ratio
See *Enrolment ratio.*

Health expenditures Expenditures on hospitals, health centres and clinics, health insurance schemes, and family planning.

Health services access The percentage of the population that can reach appropriate local health services on foot or by the local means of transport in no more than one hour.

Immunised The average of the vaccination coverages of children under one year of age for the four antigens used in the Universal Child Immunization Programme(UCI).

Income share The income both in cash and kind accruing to percentile groups of households ranked by total household income.

Infant mortality rate The annual number of deaths of infants under one year of age per 1,000 live births. More specifically, the probability of dying between birth and exactly one year of age.

Inflation rate The average annual rate of inflation measured by the growth of the GDP implicit deflator for each of the periods shown.

Labour force The economically active population, including the armed forces and the unemployed, but excluding homemakers and other unpaid caregivers.

Least developed countries A group of developing countries established by the United Nations General Assembly. Most of these countries suffer from one or more of the following constraints: a GNP per capita of around $300 or less, land-locked, remote insularity, desertification, and exposure to natural disasters.

Life expectancy at birth The number of years a newborn infant would live if prevailing patterns of mortality at the time of its birth were to stay the same throughout its life.

Literacy rate (adult) The percentage of persons aged 15 and over who can, with understanding, both read and write a short simple statement on everyday life.

Logarithm One of a class of arithmetical functions, invented by Lord Napier, tabulated for use in abridging calculations by substituting addition and subtraction for multiplication and division.

Logarithmic transformation A transformation of a variable x to a new variable y by some such relation as $y = a + b \log (x-c)$. There are a number of contexts in which such transformations are useful in statistics: for example, to normalise a frequency function, to stabilise a variance, and to reduce a curvilinear to a linear relationship in regression or probit analysis.

Low birth weight The percentage of babies born weighing less than 2,500 grammes.

Malnutrition
See *Underweight, Wasting and Stunting.*

Maternal mortality rate The annual number of deaths of women from pregnancy-related causes per 100,000 live births.

Military expenditures Expenditures, whether by defense or other departments, on the maintenance of military forces, including the purchase of military supplies and equipment, construction, recruiting, training and military aid programs.

Net enrolment ratio
See *Enrolment ratio.*

North
See *South-North gap.*

Official development assistance (ODA) The net disbursements of loans and grants made on concessional financial terms by official agencies of the members of the Development Assistance Committee (DAC), the Organisation for Economic Co-operation and Development (OECD), the Organisation of Petroleum Exporting Countries (OPEC), and so on, to promote economic development and welfare including technical cooperation and assistance.

Population density The total number of inhabitants divided by the surface area.

Population growth rate The annual growth rate of the population calculated from mid-year.

Poverty line That income level below which a minimum nutritionally adequate diet plus essential nonfood requirements are not affordable.

Primary education Education at the first level (International Standard Classification of Education (ISCED) level 1), the main function of which is to provide the basic elements of education — such as elementary schools and primary schools.

Purchasing power parities (PPP)
See *Real GDP per capita.*

Real GDP per capita The use of official exchange rates to convert the national currency figures to U.S. dollars does not attempt to measure the relative domestic purchasing powers of currencies. The United Nations International Comparison Project (ICP) has developed measures of real GDP on an internationally comparable scale using purchasing power parities (PPP), instead of exchange rates, as conversion factors, and expressed in international dollars.

Real per capita expenditures The expenditures for each item in each country expressed in relation to those in the United States, based on the national currency total for each item converted to international dollars by the purchasing power for that item. This comparison of quantities for each item is based on the common set of international prices applied across all the countries that are used in the International Price Comparison Project. (See *Real GDP per capita*). These expenditures are then expressed in relation to the corresponding average expenditure of the industrial countries, which are indexed to equal 100.

Rural population The percentage of the population living in rural areas as defined according to the national definition used in the most recent population census.

Rural-urban disparity A set of national, regional and other estimates in which all the rural figures are expressed in relation to the corresponding urban figures which are indexed to equal 100.

Safe water access The percentage of the population with reasonable access to safe water supply including treated surface waters, or untreated but uncontaminated water such as that from springs, sanitary wells and protected boreholes.

Sanitation access The percentage of the population with access to sanitary means of excreta and waste disposal, including outdoor latrines and composting.

Secondary education Education at the second level (ISCED levels 2 and 3), based upon at least four years previous instruction at the first level, and providing general or specialised instruction or both — such as middle schools, secondary schools, high schools, teacher-training schools at this level and schools of a vocational or technical nature.

South-North gap A set of national, regional and other estimates in which all the figures are expressed in relation to the corresponding average figures for all the industrial countries, which are indexed to equal 100.

Stunting The percentage of children, between 24 to 59 months, below minus two standard deviations from the median height-for-age of the reference population.

Tertiary education Education at the third level (ISCED levels 5, 6 and 7) — such as universities, teachers' colleges and higher professional schools — requiring as a minimum condition of admission, the successful completion of education at the second level or evidence of the attainment of an equivalent level of knowledge.

Under-five mortality rate The annual number of deaths of children under five years of age per 1,000 live births. More specifically, the probability of dying between birth and exactly five years of age.

Underweight (moderate and severe child malnutrition) The percentage of children, under the age of five, below minus two standard deviations from the median weight-for-age of the reference population.

Unemployed See *Educated unemployed.*

Urban population The percentage of the population living in urban areas as defined according to the national definition used in the most recent population census.

Wasting The percentage of children, between 12 and 23 months, below minus two standard deviations from the median weight-for-height of the reference population.

Water access See *Safe water*.

Classification of countries

Countries in the human development aggregates

Low human development (HDI below 0.500)	*Medium human development (HDI 0.500 to 0.799)*	*High human development (HDI 0.800 and above)*
Afghanistan	Albania	Argentina
Angola	Algeria	Australia
Bangladesh	Bolivia	Austria
Benin	Botswana	Belgium
Bhutan	Brazil	Bulgaria
Burkina Faso	China	Canada
Burundi	Dominican Rep.	Chile
Cameroon	Ecuador	Colombia
Central African Rep.	Egypt	Costa Rica
Chad	El Salvador	Cuba
Congo	Gabon	Czechoslovakia
Côte d'Ivoire	Guatemala	Denmark
Ethiopia	Honduras	Finland
Ghana	Indonesia	France
Guinea	Iran, Islamic Rep.	German Dem. Rep.
Haiti	Iraq	Germany, Fed. Rep.
India	Jordan	Greece
Kampuchea, Dem.	Korea, Dem. Rep.	Hong Kong
Kenya	Lao PDR	Hungary
Liberia	Lebanon	Ireland
Madagascar	Lesotho	Israel
Malawi	Libyan Arab Jamahiriya	Italy
Mali	Mauritius	Jamaica
Mauritania	Mongolia	Japan
Morocco	Myanmar	Korea, Rep.
Mozambique	Nicaragua	Kuwait
Namibia	Oman	Malaysia
Nepal	Paraguay	Mexico
Niger	Peru	Netherlands
Nigeria	Philippines	New Zealand
Pakistan	Saudi Arabia	Norway
Papua New Guinea	South Africa	Panama
Rwanda	Sri Lanka	Poland
Senegal	Syrian Arab Rep.	Portugal
Sierra Leone	Thailand	Romania
Somalia	Tunisia	Singapore
Sudan	Turkey	Spain
Tanzania, United Rep.	United Arab Emirates	Sweden
Togo	Viet Nam	Switzerland
Uganda	Zimbabwe	Trinidad and Tobago
Yemen Arab Rep.		United Kingdom
Yemen, PDR		Uruguay
Zaire		USA
Zambia		USSR
		Venezuela
		Yugoslavia

Countries in the regional aggregates

Sub-Saharan Africa

Angola
Benin
Botswana
Burkina Faso
Burundi
Cameroon
Central African Rep.
Chad
Congo
Côte d'Ivoire
Ethiopia
Gabon
Ghana
Guinea
Kenya
Lesotho
Liberia
Madagascar
Malawi
Mali
Mauritania
Mauritius
Mozambique
Namibia
Niger
Nigeria
Rwanda
Senegal
Sierra Leone
Somalia
South Africa
Tanzania, United Rep.
Togo
Uganda
Zaire
Zambia
Zimbabwe

South Asia

Afghanistan
Bangladesh
Bhutan
India
Iran, Islamic Rep.
Nepal
Pakistan
Sri Lanka

Middle East and North Africa

Algeria
Egypt
Iraq
Jordan
Kuwait
Lebanon
Libyan Arab Jamahiriya
Morocco
Oman
Saudi Arabia
Sudan
Syrian Arab Rep.
Tunisia
Turkey
United Arab Emirates
Yemen Arab Rep.
Yemen, PDR

East and Southeast Asia

China
Hong Kong
Korea, Dem. Rep.
Korea, Rep.
Mongolia
Indonesia
Kampuchea, Dem.
Lao PDR
Malaysia
Myanmar
Papua New Guinea
Philippines
Singapore
Thailand
Viet Nam

Asia and Oceania

Afghanistan
Bangladesh
Bhutan
China
Hong Kong
India
Indonesia
Iran, Islamic Rep.
Kampuchea, Dem.
Korea, Dem. Rep.
Korea, Rep.
Lao PDR
Malaysia
Mongolia
Myanmar
Nepal
Pakistan
Papua New Guinea
Philippines
Singapore
Sri Lanka
Thailand
Viet Nam

Latin America and the Caribbean

Argentina
Bolivia
Brazil
Chile
Colombia
Costa Rica
Cuba
Dominican Rep.
Ecuador
El Salvador
Guatemala
Haiti
Honduras
Jamaica
Mexico
Nicaragua
Panama
Paraguay
Peru
Trinidad and Tobago
Uruguay
Venezuela

Countries in the income aggregates

Low-income
(GNP per capita below $500)

Afghanistan
Angola
Bangladesh
Benin
Bhutan
Burkina Faso
Burundi
Central African Rep.
Chad
China
Ethiopia
Ghana
Guinea
Haiti
India
Indonesia
Kampuchea, Dem.
Kenya
Lao PDR
Lesotho
Liberia
Madagascar
Malawi
Mali
Mauritania
Mozambique
Myanmar
Nepal
Niger
Nigeria
Pakistan
Rwanda
Sierra Leone
Somalia
Sri Lanka
Sudan
Tanzania, United Rep.
Togo
Uganda
Viet Nam
Yemen, PDR
Zaire
Zambia

Middle-income
(GNP per capita $500 to $5,999)

Albania
Algeria
Argentina
Bolivia
Botswana
Brazil
Bulgaria
Cameroon
Chile
Colombia
Congo
Costa Rica
Côte d'Ivoire
Cuba
Czechoslovakia
Dominican Rep.
Ecuador
Egypt
El Salvador
Gabon
Greece
Guatemala
Honduras
Hungary
Iran, Islamic Rep.
Iraq
Jamaica
Jordan
Korea, Dem. Rep.
Korea, Rep. of
Lebanon
Libyan Arab Jamahiriya
Malaysia
Mauritius
Mexico
Mongolia
Morocco
Namibia
Nicaragua
Oman
Panama
Papua New Guinea
Paraguay
Peru
Philippines
Poland
Portugal
Romania
Senegal
South Africa
Syrian Arab Rep.
Thailand
Trinidad and Tobago
Tunisia
Turkey
Uruguay
USSR
Venezuela
Yemen Arab Rep.
Yugoslavia
Zimbabwe

High-income
(GNP per capita $6,000 and above)

Australia
Austria
Belgium
Canada
Denmark
Finland
France
German Dem. Rep.
Germany, Fed. Rep.
Hong Kong
Ireland
Israel
Italy
Japan
Kuwait
Netherlands
New Zealand
Norway
Saudi Arabia
Singapore
Spain
Sweden
Switzerland
United Arab Emirates
United Kingdom
USA

Countries in the other aggregates

Least developed countries

Afghanistan
Bangladesh
Benin
Bhutan
Botswana
Burkina Faso
Burundi
Central African Rep.
Chad
Ethiopia
Guinea
Haiti
Lao PDR
Lesotho
Malawi
Mali
Mauritania
Mozambique
Myanmar
Nepal
Niger
Rwanda
Sierra Leone
Somalia
Sudan
Tanzania, United Rep.
Togo
Uganda
Yemen Arab Rep.
Yemen, PDR

All developing countries

Afghanistan
Algeria
Angola
Argentina
Bangladesh
Benin
Bhutan
Bolivia
Botswana
Brazil
Burkina Faso
Burundi
Cameroon
Central African Rep.
Chad
Chile
China
Colombia
Congo
Costa Rica
Côte d'Ivoire
Cuba
Dominican Rep.
Ecuador
Egypt
El Salvador
Ethiopia
Gabon
Ghana
Guatemala
Guinea
Haiti
Honduras
Hong Kong
India
Indonesia
Iran, Islamic Rep.
Iraq
Jamaica
Jordan
Kampuchea, Dem.
Kenya
Korea, Dem. Rep.
Korea, Rep.
Kuwait
Lao PDR
Lebanon
Lesotho
Liberia
Libyan Arab Jamahiriya
Madagascar
Malawi
Malaysia
Mali
Mauritania
Mauritius
Mexico
Mongolia
Morocco
Mozambique
Myanmar
Namibia
Nepal
Nicaragua
Niger
Nigeria
Oman
Pakistan
Panama
Papua New Guinea
Paraguay
Peru
Philippines
Rwanda
Saudi Arabia
Senegal
Sierra Leone
Singapore
Somalia
South Africa
Sri Lanka
Sudan
Syrian Arab Rep.
Tanzania, United Rep.
Thailand
Togo
Trinidad and Tobago
Tunisia
Turkey
Uganda
United Arab Emirates
Uruguay
Venezuela
Viet Nam
Yemen Arab Rep.
Yemen, PDR
Zaire
Zambia
Zimbabwe

Industrial countries

Albania
Australia
Austria
Belgium
Bulgaria
Canada
Czechoslovakia
Denmark
Finland
France
German Dem. Rep.
Germany, Fed. Rep.
Greece
Hungary
Ireland
Israel
Italy
Japan
Netherlands
New Zealand
Norway
Poland
Portugal
Romania
Spain
Sweden
Switzerland
United Kingdom
USA
USSR
Yugoslavia

Primary sources of data

Data for the topics in italics have been taken from more than one major source.

Food and Agriculture Organization of the United Nations (FAO)
Arable land. Calorie supply. *Deforestation*. Food imports. Food production. Fuelwood. Land area. Livestock.

Habitat
Persons per habitable room

Institute for Resource Development
Breastfeeding. Child malnutrition.

International Centre for Urban Studies
City population density.

International Labour Organisation (ILO)
Labour force. Unemployment.

International Monetary Fund (IMF)
Budget surplus/deficit. Inflation. International reserves. *Military, health and education expenditures. State and local expenditures on social services.*

Interparliamentary Union
Women in parliament.

Luxembourg Income Study Database (LIS)
Real GDP per capita poverty line.

Organisation for Economic Cooperation and Development (OECD)
Debt service and interest payments. ODA, received and given.

United Nations Children's Fund (UNICEF)
Breast-feeding. Child malnutrition. Immunisation. Under-five mortality.

United Nations Development Programme (UNDP)
Human development index (HDI). Selected literacy estimates. Selected real GDP per capita estimates.

United Nations Economic Commission for Europe (ECE)
Deforestation.

United Nations Educational Scientific and Cultural Organization (UNESCO)
Literacy. Primary school dropout. Radios. School, college and university enrolment. Scientists and technicians. Teachers. Television.

United Nations Fund for Population Activities (UNFPA)
Contraceptive prevalence.

United Nations Industrial Development Organization (UNIDO)
Earnings

United Nations Population Division
Birth and death rates. Dependency ratio. Fertility. *Infant and under-five mortality.* Life expectancy. *Population: total,* urban, and rural. Population density.

United Nations Statistical Office
Exports and imports. *GDP. Infant mortality. Persons per habitable room. Real GDP per capita.* Real GDP per capita expenditures. Terms of trade. *Total population.*

United States Agency for International Development (USAID)
Real GDP per capita.

World Bank
Debt service and interest payments. Domestic savings. *GDP*, GNP and *real GDP per capita*. Household income. *Military, health and education expenditures.* Population below poverty line. *State and local expenditures on social services.*

World Fertility Survey
Breastfeeding.

World Food Programme (WFP)
Food aid.

World Health Organization (WHO)
Access to health services, safe water and sanitation. Attendance at birth. *Child malnutrition.* Doctors and nurses. *Immunisation.* Low birthweight. Maternal mortality.

World Priorities Inc.
Armed forces. *Military, health and education expenditures.*

World Resources Institute
Internal renewable water resource.